Estimating Excavation

by Deryl Burch

Craftsman Book Company
6058 Corte del Cedro / P.O. Box 6500 / Carlsbad, CA 92018

For Maudie, Morgan and Dad

In memory of Mom

A tribute to Grandma Hamn

Looking for other construction reference manuals?

Craftsman has the books to fill your needs. **Call toll-free 1-800-829-8123** or write to Craftsman Book Company, P.O. Box 6500, Carlsbad, CA 92018 for a **FREE CATALOG** of over 100 books, including how-to manuals, annual cost books, and estimating software.
Visit our Web site: http://www.craftsman-book.com

Library of Congress Cataloging-in-Publication Data

Burch, Deryl
   Estimating excavation / by Deryl Burch.
      p.   cm.
   Includes index.
   ISBN 0-934041-96-2
   1. Excavation -- Estimates.   I. Title.
TA730.B87   1997
624.1'52'0299--dc21   97-26971
                                                               CIP

©1997 Craftsman Book Company
Fourth printing 2007

Contents

1 **Get Started Right**   5
- Why calculate quantities? 6
- Reading plans and specifications 8
- Accuracy is essential 10
- Record keeping 14

2 **The Site Visit**   17
- Review the plans first 17
- Make the visit productive 18
- Site visit for a sample project 24

3 **Properties of Soils**   31
- Soil testing 31
- Soil classifications 31
- Soil characteristics 34

4 **Area Take-off by Plan and Profile**   41
- Cut and fill sections 42
- Understanding surveys 43
- End area calculations 47
- Calculating the volume 54

5 **Reading Contour Maps**   61
- Planimetric and topographic maps 61
- Understanding contour lines 62
- Locating unmarked points 66
- Monuments and bench marks 67

6 **Area Take-off from a Topo Map**   71
- Comparing the contour lines 71
- Estimating with a grid system 72
- Doing the take-off 75
- Calculating cut and fill areas 80
- Using worksheets in a take-off 85
- Shortcuts for calculating quantities 91
- The equal depth contour method 99

7 **Irregular Regions & Odd Areas**   103
- Finding area using compensating lines 104
- Finding volume using total area and average depth 109
- Using compensating lines with a coordinate system 110
- Using the trapezoidal rule 117

8 **Using Shrink & Swell Factors**   127
- Soil states and their units of measure . .127
- Using shrink/swell factors in earthwork estimates 128
- Estimating the number of haul trips . . .130
- Using material weights to customize shrink/swell factors 131
- Using soil weights to calculate equipment load factors 131
- Pay yards 133

9 Topsoil, Slopes & Ditches — 135
- Dealing with topsoil 135
- Calculating net volumes for earthwork 139
- Slopes and slope lines 142
- Estimating trenches 147

10 Basements, Footings, Grade Beams & Piers — 155
- Estimating basement excavation quantities 155
- Finding volume - outside basement walls 157
- Calculating the total volume for basement excavation 164
- Sample basement estimate 170
- Estimating ramps 175
- Grade beams and piers 178

11 All About Spoil and Borrow — 181
- Underlying costs of spoil and borrow .181
- Spoil and borrow volume calculations .183
- Calculating the volume of a stockpile .184
- Finding the volume for a stockpile of unknown height 186
- Calculating volume for a stockpile of set area 190

12 Balance Points, Centers of Mass & Haul Distances — 195
- Balance points to an excavation estimator 195
- Balance points to an engineer 196
- Reducing haul distances 197
- Calculating haul distances 200

13 Earthmoving Equipment: Productivity Rates and Owning & Operating Costs — 209
- Machine power 210
- Machine speed 214
- Machine production 218
- Productivity calculations for a simple dirt job 221
- Owning and operating costs 230
- Calculating the overhead 234
- Adding the profit 237
- Bid price per cubic yard 237

14 A Sample Take-off — 239
- General specifications 240
- Doing the take-off 242

15 Costs and Final Bid for the Sample Estimate — 335
- The bid preparation process 336
- Overhead 340
- Machine selection 340

Blank Worksheets — 434
- Grid square area and volume 435
- Grid take-off – existing contour 436
- Grid take-off – proposed contour 437
- Grid square calculation sheet 438
- Cut and fill prism calculations 439
- Quantities take-off sheet 440

Index 441

1 Get Started Right

Construction cost estimating is demanding work, no matter what type of construction is involved. But I think estimating earthwork is the hardest of all. Why? Two reasons. First, excavation has more variables and unknowns. You don't know what's down there until you start digging. Second, you have to rely on information from many sources — some of which may not be accurate.

That's why every earthwork estimator needs special skills:

- The ability to read plans and specifications
- An understanding of surveying and engineering practice
- A facility with mathematical calculations
- The ability to anticipate environmental and legal issues
- An abundance of good common sense.

If you can bring common sense to the task, this manual will show you how to do the rest.

I'll help you develop all the skills every good earthwork estimator needs. Of course, I can't cover everything on every type of job. But I'll include the information most earthwork estimators need on most jobs. Occasionally, you'll have a job that requires special consideration. But if you understand the principles I'll explain here, you should be able to handle anything but the most bizarre situations.

In this, the first chapter, I won't do much more than touch on a few important points you should understand:

1) Why you have to estimate quantities
2) The importance of plans and specs
3) Working accurately
4) Keeping good records

After making these points in this chapter, I'll describe a step-by-step estimating system, from making the site survey to writing up the final cost summary. I'll teach you a *process* for making consistently-accurate earthwork estimates. Part of this process is calculating the cubic yards to be moved. That's the heart of every earthwork estimate. I'll cover quantity estimating in detail. Then I'll explain how to find labor and equipment costs per unit. We'll also consider soil and rock properties and how the equipment you use affects bid prices.

Why Calculate Quantities?

In the past, many smaller dirt jobs were bid as a *lump sum* rather than by the cubic yard. Dirt contractors based their bids on guesses — what equipment was needed and how long should it take? They didn't bother estimating soil quantities. Making estimates this way overcame one problem: most excavation contractors didn't know how to estimate soil and rock quantities.

I think those days are over. Fuel and labor costs are too high now. And the competition is too intense. There's too much risk in "seat-of-the-pants" guesses. A few mistakes, a couple of surprises and you're going to be looking for some other type of work. Only the best survive for long in this business. Most of the survivors know how to make accurate bids by the cubic yard. Fortunately, making good quantity estimates isn't too hard if you've mastered a few simple skills. I hope that's why you're reading this page.

I've found that good earthwork estimators are good at calculating earthwork quantities. Here's why:

First, no one's going to do it for you. You have to do it yourself or it's not going to get done. Many engineers, architects, and even some builders know how to figure soil and rock quantities. But few take the trouble to do it. Instead, they depend on you, the earthwork estimator, to do it.

Second, earthwork contractors who don't bid by the cubic yard usually end up in court. That can cripple any company. It's common for the actual amount of dirt moved to be more or less than expected. The best way to protect yourself is to bid by the cubic yard. If you have to move more dirt than the plans show, you'll get paid more. It's as simple as that.

Third, most owners, engineers and architects request excavation bids based on the cubic yards moved. That's now the accepted procedure for most projects, from single-family homes to roads and commercial jobs.

General and Special Quantities

If you agree that excavation bids should be based on quantity estimates, the next step should be obvious. We have to start every estimate by figuring the quantity of soil to be moved.

I recommend you start the estimate for any project, no matter how large or small, by dividing excavation quantities into two categories:

General quantities include any work where you can use motorized equipment such as scrapers, hoes and loaders at their designed production rate.

Special quantities include anything that requires special care or lower production rates. Examples are most rock excavation, nearly all hand excavation, backhoe work around sewer lines, underground utilities, or existing structures. Naturally, prices for special quantities are higher than prices for general quantities.

Keeping these two quantities separate protects you. Most excavation contracts have a clause that covers extra work. Unanticipated rock deposits, special soil problems and unusual trenching problems are extra work that you should be paid extra for. If you've bid a higher price for special quantities, you'll get paid at that price per cubic yard for the additional work. Otherwise you could end up chipping out rock at the price of moving sand.

Calculating Cubic Yard Cost

Formula for costs per cubic yard

Here's the basic formula for costs per cubic yard:

Labor and equipment cost per hour multiplied by the hours needed to complete the work, divided by the cubic yards of material to be moved.

Does that seem simple? It's not. You may know your hourly labor and equipment costs right down to the last penny. But estimating the time needed is never easy. And calculating volumes for sloping and irregular surfaces is demanding work.

Notice several things about the formula for computing costs per cubic yard.

First, it's based on labor and equipment costs for *your* business. That's important and I'll have more to say about it later.

Second, it assumes you know the quantity of soil or rock to be moved. That's going to take some figuring.

Third, even after you've calculated the cost per hour and quantity of soil, you're not finished. You need to estimate the time needed. Usually that's the hardest part. To do it, you have to decide on the equipment (*method*) to use.

Of course, the quantity of material (*yardage*) is a very important part of our cost formula. But the excavation method (*type of equipment*) also has a major influence on cost. The most expensive equipment (cost per hour) will usually be the most productive (move soil at the lowest cost). But the machine with the largest capacity isn't always the best choice for every outhaul. I'll explain why later. For now, just understand that making good equipment selections will help reduce costs.

Reading Plans and Specifications

Nearly every significant excavation project that's let out for bid will be based on a set of plans. Plans are scale drawings that show the finished project. Plans are supplemented with written descriptions called specifications (*specs* for short). Specs explain in words what the plans can't or don't show. Ideally, the plans and specs, read together, should answer every question about the job. They shouldn't leave anything unclear or subject to interpretation. The better the job done by the engineer or designer, the more likely the plans will be clear and complete.

Plan reading is an important skill for every earthwork estimator. But this isn't a book on plan reading. If you need help with reading plans, if you don't understand the plans and drawings in this manual, pay a visit to your local library. They'll probably have several basic plan-reading texts to choose from.

As an excavation estimator, you're expected to understand every detail in the plans and specs for the jobs you bid. That's why they're worth careful study. Read these documents completely. Note everything that affects your excavation work. Some engineers and architects aren't very well organized. They may put instructions and notes almost anywhere on the plans. Read every page carefully, regardless of what you think it's about.

Pay particular attention to notes that spell out the contractor's responsibility. For example, you may find a note somewhere on plans that relieves the engineer or architect of responsibility for damage to utility lines. The note probably says:

NOTE: While every precaution has been taken to show existing utilities in their proper location, it is the contractor's responsibility to determine their actual location. No assumption should be made that no other utility lines fall within the limits of construction.

If you suspect utility lines may be a problem, ask the utility companies to locate their lines for you. Most will be happy to do that at no cost. But they may want ample advance notice.

Also pay attention to notes on natural obstacles (such as rock) or anything that's buried on the site. Is there an abandoned underground storage tank or old basement in the area to be excavated? The plans may also mention drainage problems and unsuitable soil deposits, probably in the cross-section drawings or special provisions of the specs.

Search the plans and specs for everything that may affect cost. That's always your starting place. But it's not the end of your search. Many cost items won't show up in either the plans or specs. For example, you'll have to find out from the city or county building department what permits will be required. Also, city, county or federal law may set minimums for wages, employee benefits and insurance coverage.

Here's another pitfall to watch for: Who pays to have the project staked out by a surveyor or engineer? In most cases, the designer will pay for surveying — the first time. If you knock over any survey stakes during actual work, you'll probably have to replace them at your own expense. Work as carefully around the stakes as possible. But if job layout makes it impossible to avoid moving stakes, allow enough in your bid to pay for another survey.

Make sure you understand how you'll be paid. On larger projects, you're usually paid per cubic yard, based on the difference between the original soil cross section and the cross section when work is finished. We'll talk more about cross sections later in this book.

On many smaller projects, your payment may be based on the engineer's estimate of yardage. If that's the case, look for a provision in the specs that gives you an option to have final cross sections made at your own expense. Experience will help you decide if a final set of cross sections is to your advantage. But I recommend that you always take off quantities yourself. Don't assume the plans are right. Anyone can make a mistake, but you could end up paying the price.

Undercutting

Undercutting is removing additional dirt from an area below the finished grade line. There are several situations where this is necessary. The most common is where a rock ledge is close to, but not above, the finished grade line. Figure 1-1 shows a typical situation. Most structures can't be built directly on rock. If the rock weren't there, you would excavate just to the finished grade line and be done. Because the rock is just below finished grade, you have to cut deeper. That's the undercut. Then you have to backfill with suitable material such as compacted dirt. The dirt provides a buffer between the rock and the foundation.

There's probably nothing in the specifications that gives you the right to collect for undercutting and backfill. But it's expensive work and the cost shouldn't come out of your pocket. Where undercutting may be necessary, include it in your bid item *per cubic yard cut*.

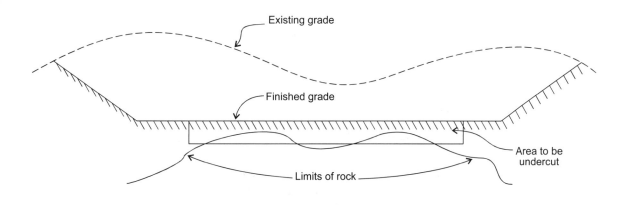

Figure 1-1
Undercutting for rock

Undercutting is also needed for underground utilities such as storm drains and sanitary sewer lines. Most plans will show only a designated flow line elevation. Based on the plans and judgment, you'll have to decide how much and what type of bedding to install below the pipe. Each cubic yard of bedding requires a cubic yard of undercutting. Figure 1-2 shows an example. Undercutting may also be required on roads, parking lots and sidewalks — anywhere there's a load on the soil.

Overfilling is the opposite of undercutting. When backfilling a large area, you can usually bring the backfill right to grade without doing any cutting away of excess backfill. But in a small area, it's usually easier to bring the area above the final grade line by 2 to 4 inches, then cut off the excess. This is still called undercutting. Of course, you can't expect to get paid for removing the 2- to 4-inch excess. But it's still a cost of the job.

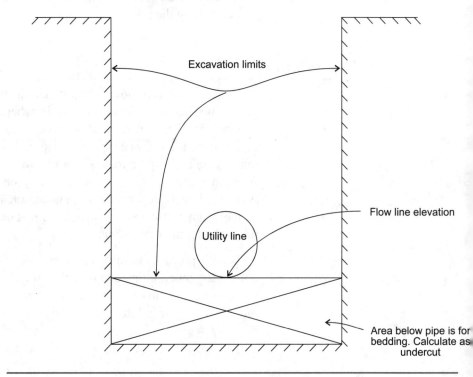

Figure 1-2
Undercutting for pipe bedding

Accuracy Is Essential

Accuracy is the essence of estimating. If you can't work accurately, you're in the wrong business. But don't get me wrong. I don't mean that we're going to account for every spadeful of soil on every estimate. There are times when you can ignore small differences in elevation. On most jobs these small plus and minus areas will average out to almost nothing. But a 1-inch mistake in elevation over the whole job can cost you thousands of dollars. Even $1/16$-inch error over a few acres can hurt you.

Here's an example. Assume you're bringing in fill on a city lot that measures 125 feet by 150 feet. Because of a mistake in grade, your estimate of imported soil is wrong. It leaves the entire site 1 inch below the specified finished grade. How much more soil is needed to correct the 1-inch mistake?

Here's the formula for volume:

Formula for volume

Volume (in cubic feet) = length (in feet) x width (in feet) x depth (in feet)

In this example, you know the length and width in feet but the depth is 1 inch. To use the formula, convert 1 inch to a decimal part of a foot. You can either refer to the conversion chart (see Figure 1-3) or divide 1 by 12, since 1" = $1/12$'. Either way, 1 inch equals 0.0833 feet.

Now you're ready to use the formula for volume:

Volume (CF) = 125 x 150 x 0.083

= 1,556.25

How many cubic yards is that? Since there are 27 cubic feet in a cubic yard, divide the cubic feet by 27:

Volume (CY) = 1,556.25 ÷ 27

= 57.6 *CY*

Trucking in almost 58 cubic yards of soil won't be cheap. If imported soil costs you $25 a cubic yard, your 1-inch mistake is a $1,450 error. That could make the difference between profit and loss on this job.

Inches	Decimal feet	Inches	Decimal feet
1/16	0.0052	7/8	0.0729
1/8	0.0104	15/16	0.0781
3/16	0.0156	1	0.0833
1/4	0.0208	2	0.1667
5/16	0.0260	3	0.2500
3/8	0.0313	4	0.3333
7/16	0.0365	5	0.4167
1/2	0.0417	6	0.5000
9/16	0.0469	7	0.5833
5/8	0.0521	8	0.6667
11/16	0.0573	9	0.7500
3/4	0.0625	10	0.8333
13/16	0.0677	11	0.9167

Figure 1-3
Inches to decimal feet conversion chart

Your Estimating Procedure

The more organized and logical your estimating procedure, the more accurate your estimates will be. If you have the tools, papers and information you need close at hand, you're off to a good start. Then you can focus your attention and concentration on producing an accurate estimate. If you're cramped for space, uncomfortable, and trying to work without all the equipment and information you need, errors are almost inevitable.

Start by organizing an efficient work area. It should be large enough so you can lay out all the plans on a table and still have room to write and calculate. Provide enough light to make reading comfortable, and keep the work area free of shadows. This is especially important when working with transparent overlays or other light-duty paper where you might mistake shadows for lines.

A good calculator is a must. I recommend buying a calculator with both a digital *and* a paper printout. You need the printout to check your figures. Make sure you have an engineer's scale and drafting triangles for checking and drawing lines, a small magnifying glass, tape for holding overlays, and the normal collection of pencils, erasers, and paper.

Although it's not essential, I like using a light table. You can place a drawing on it, overlay it with another paper, and see through both of them. It's great for working with plan and profile sheets, or overlays on grid or take-off sheets.

Later in the book we'll talk about using a planimeter to take off quantities. Although it's relatively expensive, a good planimeter will soon pay for itself. Take care to select one that's sturdy and has all the needed instructions and attachments.

A computer is even more expensive, but more and more estimators are using one. There are programs on the market today that can handle anything from simple calculations to a complete estimating program, with cross sections, quantities and printouts. But no program can take the place of an estimator who understands estimating procedures and practices. That's the purpose of this book.

There are two advantages to using a computer. The first is *time*. That's an estimator's most valuable asset, and a computer can help make your time more productive. Second, a computer makes it easier to keep cost figures for equipment and labor. Records from past projects and estimates can make current estimates more accurate.

If you don't currently have a computer, don't jump in without doing some research first. There are many computers on the market, tons of software, and hundreds of dealers. Take the time to make yourself familiar with the options. Talk to dealers. More important, talk to other estimators who use computers to do their estimating. Read trade magazines, especially the ads for estimating software. *And don't go out and buy a computer and then look for estimating programs to run on it*. First, choose the estimating program you like, and *then* buy the computer that will run that program. Otherwise, you may find the computer you bought won't run the program you like.

When you've got your work area and equipment set up to work efficiently, you're on the path to accurate estimates. To stay on the path, it's important to approach the work with a logical and organized procedure. That speeds up the work and reduces mistakes. Let me describe the method that works for me. I think it'll work for you, too.

When starting a project, first read all documents describing the job. Take notes on any situation that's not a normal work requirement. Are there utilities that must not be disturbed? Do the documents indicate specialized material types from log borings? Do they stipulate any arrangement for rock on the site? Look for any special provisions set out by the designer. Then head out for a field visit. That's the subject of the next chapter.

After returning from the field, review the documents again, looking for unusual situations that the site visit brought to your attention. Then make a complete written outline of all work that needs to be done in the order in which it will be performed. Set up files for each separate section. Make a list of additional data such as quad sheets, local conditions, and any other information you need to gather.

Here's the order I usually use.

1) Consider any drainage, traffic or work zone protection work that needs to be done. Are there any on-site streams that must remain open, or roadways to maintain? These would probably be lump sum items, not items you'd take off quantities for. Just make sure you don't miss any of these special items.

2) After studying the plans and the site, you should have a good idea if there's enough fill on the site, or if you'll need a borrow pit. Or will you need a place to put excess material off-site? Begin now to make arrangements for needed sites, sampling of material for approval by the engineer, and purchasing any material that's needed.

3) Now consider the topsoil requirements. Review the material sample, the requirements for replacement, and availability of storage area on site. Calculate the amount of usable material and the amount of waste that must be disposed of.

4) Will there be any special excavation, like rock work or the removal of existing structures or facilities? Make sure you include all work and any special equipment you'll need. Will you need to rent equipment? What about rock drills, blasting material, or cranes?

5) Begin calculating the general quantities with the cut or fill work over the entire project. Start in the same place and proceed throughout the project the same way for every estimate. One way to make sure you cover all of the project is to set up a grid system with a corresponding file system. As you finish work in each grid, mark it off, file it, and move on to the next grid.

6) Next, calculate all the utility lines, keeping the figures for each area separate. Be especially careful in estimating the tie-in between new and existing lines. Allow a little extra time for lines that aren't exactly where the plans show them to be.

7) Then consider the roads, parking lots, and paved or special drainage ditches. Again, keep the quantities for each separate. One note of caution: Remember to consider the base and sub-base when figuring final elevations.

8) Buildings, basements, sidewalks and other similar structures are next. After you've calculated each structure separately, add them all together to get a structure total.

9) Finally, calculate the topsoil. And don't forget that if you've used a borrow pit, you may have to place topsoil there also.

10) Now you're ready to start putting together all that information to come up with a realistic quantity total for the complete project. Fill out the final quantities sheet. Remember to attach all worksheets, scratch paper and calculator printouts so you can recheck your totals.

Now review your final sheet, looking for potential problem areas. If possible, have someone else check all your calculations and extensions. If that's not possible, set the estimate aside and go through it again a few days later. You'll have a fresh approach that may spot errors or omissions.

The last step is to go through all the documents and make sure they're in order. Then file them. Don't throw anything away — not even the scrap paper. Why are those records valuable? Keep reading.

Record Keeping

Once you've learned to read plans carefully and work accurately, there's still one more important step: record keeping.

Think of your estimates as accumulated wisdom. Treasure them. Keep them handy. Make sure they're easy to understand. They should show how each figure was developed. Why? There are at least four reasons:

First, planning the work is a big part of every estimator's job. You can't estimate any type of earthwork without making decisions about equipment. Once you've selected equipment for estimating purposes, document your choice on the estimate worksheets.

If your bid is accepted, you'll probably want to do the work with the same equipment assumed in the estimate. What if months have gone by and you can't remember how the figures were developed? You have to start selecting equipment and estimating costs all over again. If the equipment assumed in your estimate isn't the same as the equipment actually used, comparison of estimated and actual costs may be meaningless.

Second, you're going refer to most estimates many times over months or even years. You shouldn't have to guess about how each figure was developed. That wastes time and can exhaust your patience. I've seen estimators who should know better use the back of an envelope to figure special quantities. After entering the final cost, they usually discard the envelope. Later, if there's a question about the estimate, how can you verify the figures? They're gone!

Third, old estimates are invaluable when compiling new estimates. Every estimate (especially if you actually did the work) provides a frame of reference for future jobs — even if labor and equipment costs have changed.

Fourth, every estimator makes mistakes. That's no embarrassment. But repeating mistakes is both foolish and expensive. The best way to avoid repeating mistakes is to preserve every scrap of estimating evidence — in a neat, tidy, well-organized file. Make notes on what worked and what didn't. Review those estimates and notes when estimating similar jobs. Save everything. Someday you may want to write a book. I saved my notes and estimates and wrote a book. You're reading it.

Using Public Records

To the professional estimator, there's no such thing as too much cost information. Collect all the estimating data you can. It helps if you know where to look for it. I canvass city and county engineering departments, public works departments and maintenance departments for whatever information they can provide. They know about bid prices, soil conditions, abandoned streets, utility lines, sewer and water problems. Use the resources available from your city and county government.

Aerial maps at the county tax office and contour maps from the United States Geological Survey offer clues to possible water and soil problems. There are USGS offices in most states. They're often located in the capitol, or in cities with universities. Check your local phone book or local engineering groups for the address of the nearest office. City, state and county highway departments will have information on soil problems they've found under highways in the area.

What If You Don't Have Plans?

Up to this point, we've assumed that you're bidding the job from plans and specs provided by an architect or engineer. But you may be asked to bid on a small job that wasn't designed by an engineer or architect. Then you'll have to create your own plan. It may also be up to you to determine quantities and prepare a contract.

In any case, always figure soil quantities and get a written contract on every job, large or small. The responsibilities and liabilities are all yours, so plan and execute your bid with care. Use the procedures and guidelines in this book — *even if there are no plans*.

If the owner doesn't have a plan prepared by an architect or engineer, collect as much information as possible from the owner. Does he or she know of any soil problems at the site? Is it your responsibility to request the survey and staking? Are any permits needed? When should the job be completed? Where are the utility lines? What conditions might delay the work?

Whether the job is big or small, whether you've got no plan or a very complete plan prepared by the best engineering firm in the state, make a visit to the site part of your estimating procedure. That's important — important enough to be the subject of an entire chapter. And that's the next chapter in this book.

2 The Site Visit

A site visit is an important part of every earthwork estimate. If you skip this important step, your estimate is just a guess. In this chapter we'll cover how to prepare for that important visit and what to look for when you get there.

Review the Plans First

Before you go to the site, take time to review the plans completely. Make an itemized list of any special problems or unusual requirements you pick up from the plans and specifications. Take that list with you, and check each item while you're in the field.

The amount of information provided on the plans will determine how much work you have to do to prepare for the site visit. If you have a complete set of plans and specifications, it's easy to list the questions that need answers. But if it's a small project with not much earthwork, the plans may not tell you all you need to know. Then it's up to you to work up the quantities and requirements for your part of the job.

Most engineers and architects are very good at what they do. But unless the project has a lot of excavation or is specialized, like highway construction, they often don't furnish complete data in the earthwork area. It's up to you to make sure that the plans accurately reflect conditions at the site itself. If you have any construction experience, you know that the way things look on paper and the way they are in the field are sometimes different.

When I go into the field on a site visit I take along two lists. The first is a list of specific questions based on the current plans and specifications. The second is my standard checklist for site visits. You'll find it at the end of the chapter. Of course, my checklist may not be exactly what you need. But every estimator needs a checklist to work from. If you don't have one that works for you, start with mine, then add any items you feel need to be there. Maybe you overlooked something once and don't want to do it again. Put it on the list.

Make the Visit Productive

Your visit to the site can make a significant difference in the amount of the bid — and the size of your profit. That's why professional estimators often earn their annual salary from just one job. They can analyze the job site to anticipate problems that might interrupt work scheduling, situations that require specialized equipment, or shortcuts to speed the work along. Then they work up bids that guarantee the contractor healthy profits.

Several years ago, a friend of mine was estimating a large shopping center project. It involved moving about half a million yards of material, including more than 300,000 yards to be hauled from the site. The designated disposal site for the material was 2½ miles away by the major road. There was a much shorter route — less than a quarter mile — but it crossed a bridge with only a 5-ton rating. The other contractors all bid the job using the 2½ mile haul route. Except my friend. He got in touch with the county that owned the bridge and made them this proposition. He would remove the existing bridge and replace it with an arch culvert if they would just pay for the pipe. He'd cover the labor and equipment. Of course, they were happy to oblige. Using the much shorter haul distance, his company won the bid. They made enough profit to pay for the bridge installation and more. The estimator earned his salary on this one project alone.

On another project, the estimator earned his keep by steering his company clear of a bad situation. The project was a large subdivision in a rural area. A general provision said that even though the plans didn't show any utility lines in the area, the contractor was responsible for any lines and for keeping uninterrupted service if any were encountered. Even though it was a dry period in late summer, the estimator noticed that one area had lush grass growing on it. Suspecting either a spring or sewer system leak, he took a sample and had it analyzed. Sure enough, it was raw sewage. Several older homes in the area had septic tanks installed in a line, with the discharge in this open area.

He included the cost of correcting this problem in his bid. None of the other estimators did, so naturally, their bids were lower. The contractor who "won" this bid paid for it dearly.

Unfortunately, few of us with many years of estimating under our belts can gloat. We've all had at least one instance where we won a bid by forgetting or not noticing something.

It takes knowledge and experience to make the site visit productive. Most of the know-how comes from experience on past projects. Even an inexperienced estimator, however, can use common sense to come up with a more cost-effective way to do the job.

Use the site visit to plan the construction scheduling and to anticipate equipment and labor requirements. The actual conditions of the site will dictate the type of equipment needed and the way the work is done. Let's look at some of the things you'll consider during the site visit, beginning with the accessibility of the site.

Accessibility

First, consider the physical location of the site. How remote is it? What roads or streets lead to the site? Are there any one-way streets leading to the site? All these will have a direct bearing on the work. If the site is isolated or undeveloped, with poor or nonexistent streets, it will take longer — and cost more — to move equipment and material in and out of the job site.

If you'll have to bring dirt in or take it out, consider the distance to the borrow or dump site. And I don't mean to make a guess. I mean to measure it with your odometer. In fact, I recommend driving the route several times, using different roads to find the shortest and best route.

If the surrounding streets carry heavy traffic, it will slow down the movement of equipment to and from the work site. Will traffic problems require the use of one or more flagmen? Look for any other safety-related problems that might require additional manpower. Check with local authorities to find out how you're required to handle the traffic.

Is the site near any homes or businesses? That will affect any blasting that might need to be done. Is there a noise ordinance and is it enforced? What about bridges? Are there any low-weight-limit bridges or narrow bridges that you can't use to bring equipment or material to the job? Take complete notes during the site visit on any variable that will affect your bid.

Degree of Job Difficulty

When you've surveyed the accessibility, turn your attention to the site itself. Are there any steep slopes that would require unusual equipment? Is the area open, or are there obstructions like buildings, trees, sidewalks, or utility lines in the way? Any of these will slow down production. If specialized equipment is needed, will it be available in the area, or will you have to bring it in from a distance?

This is a good time to decide what size and type of earthmoving equipment to use. Consider whether there's enough room for the equipment to turn and move economically. While the size of the job might warrant a 20-yard-capacity scraper, is it too large to operate around the obstructions? Steep or unstable slopes usually mean you're going to have to use tracked machines instead of wheeled. As a rule of thumb, you'll have to use track machines on any slope that's greater than 3 in 1. When making your decision, consider the ground conditions, traction, and the distances and directions you'll have to move. And remember that track machines have a slower working speed. We'll talk more about working on slopes later in the book.

Surface Conditions

Drainage problems, steep slopes, dense vegetation, and sharp or large rocks scattered on the surface will all hamper production. Drainage is one of the biggest problems. What's going to happen to the water that now drains across the project area? You may have to provide drainage channels to reroute water during construction. But you can't divert water onto streets or roads. Will you need a special permit for temporary channel relocation during construction from the city, county or state?

On some jobs you'll need to estimate the volume of trees and brush to be removed. Most plans mark the trees that need to be removed, but they seldom give the volume. There are so many variables that your best estimate will just be an educated guess. But here's a formula that should give you reasonably accurate estimates. It assumes that large trees will be cut into truck-size lengths.

The total volume of material has two parts; the volume of the tree trunks, which is called the *base volume*, and the volume of the foliage. To find the foliage volume you first need to know the foliage area. You can find this from aerial photographs, or by measuring it in the field.

There are two numbers, called *constants*, which you also use to make the answer you get for total volume more correct. In the calculations below, they're called constant A (0.1) and constant B (0.04).

Look at Figure 2-1. We're going to use this for a sample project later in the chapter. For now, we'll just use it to calculate the total volume of the brush. The brush area is 1,800 feet long and 60 feet wide. Measurements taken at the site establish an average height of 35 feet.

The first step is to find the foliage area in square feet:

Using constants to find total volume of brush

$$\begin{aligned}\textit{Foliage area} &= \textit{width of foliage x length of foliage}\\ &= 1{,}800 \times 60\\ &= 108{,}000 \text{ SF}\end{aligned}$$

Next, you need the volume of the base:

$$\begin{aligned}\textit{Base volume} &= \textit{foliage area x constant A}\\ &= 108{,}000 \times .1\\ &= 10{,}800 \text{ CF}\end{aligned}$$

Now, you need the volume of the foliage:

$$\begin{aligned}\textit{Foliage volume} &= \textit{foliage area x average height x constant B}\\ &= 108{,}000 \times 35 \times .04\\ &= 151{,}200 \text{ CF}\end{aligned}$$

Finally, you're ready to find the total volume:

$$\begin{aligned}\textit{Total volume} &= \textit{base volume} + \textit{foliage volume}\\ &= 151{,}200 + 10{,}800\\ &= 162{,}000 \text{ CF}\end{aligned}$$

To convert this to cubic yards, divide by 27. There will be about 6,000 cubic yards of loosely packed material to haul off.

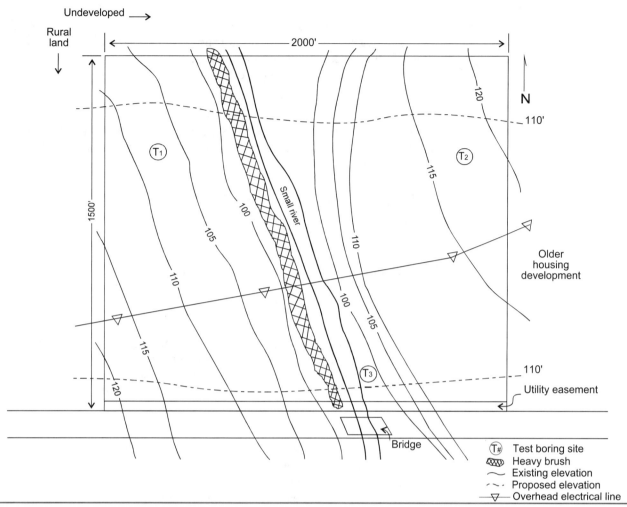

Figure 2-1
Sample earthmoving job

Subsurface Conditions

Even if you have the results of soil tests on the site, the actual conditions of the soil below the surface are really anybody's guess. Because soil testing is very expensive, most jobs don't do a lot of it.

Water running on the surface indicates underground water seeps. If the work limits are below the local water table, you'll have to pump water from trenches and excavation portions of the job. Also look for unsuitable materials (soil that's unstable under load) if there are any stream beds on the site. Many developers like to build housing projects along these stream beds. But in the past, these were often local dumping areas. The governing bodies may have approved dumping old rock, dirt or other material in these areas to fill them up to the grade of the surrounding areas. If you suspect this is true, you may want to request additional soil boring in the area. At least add a clause in your bid covering changes in soil stability.

The Site Visit **21**

Utilities

Try to determine if utility lines are shown in the correct location on the plans. Utility lines sometimes aren't where the plans show them. A variation of just a few feet can make a big difference in time when working in a confined area.

If there are existing storm or sewer lines, check the manholes for condition, material and depth to flow line. Also check for size, direction and number of inlets and outlets in the manhole. Compare this with the plans. Check for overhead wires that would be in the way of working equipment. Will temporary electric or phone connections be needed during the construction period? If any utility lines have to be relocated, find out how much advance notice the company needs to move them. What costs or permits are the responsibility of the contractor?

When you've located the utility lines, it's a good idea to mark the location permanently. The flags used by the utility companies are likely to be destroyed or misplaced during construction. I recommend using survey-type ties to mark them. Surveyors use them to "tie down" points so they can be reestablished later on.

Look at the electrical line in Figure 2-2. It starts at point A on the left at the bottom of the easement, then goes up to point B at the top of the easement, and on out in a straight line to point C. After the electrical company places flags along the line, we'll tie only the points where it changes direction: points A, B, and C.

All that's needed is a tape measure about 100 feet long, a hammer and some markers. You can make a marker by folding a 12-inch piece of colored survey flagging over several times, until it's about 2 inches wide. Then push a concrete nail through the middle. These markers are called *heads*. A red marker is called a red head, green is called a green head, and so on.

To make the first tie, stand on the road shoulder line facing point A on the electrical line. Move to the left a few feet and drive a red head in the pavement. Repeat this process by moving to the right a few feet.

Draw a circle like the one in Figure 2-2. Then measure from point A to each red head and record the distance. In our sample, the point on the left is 29.2 feet, and on the right 19.6 feet. Record these distances as well as the mark they're measured from. In this case they're both measured from a RHIP (red head in pavement). You could also place the heads on buildings or trees.

To find point A again, extend a tape measure from the left point 29.2 feet. At the same time extend a tape measure from the right point 19.6 feet. Where the two come together is point A. Using the same process, we've tied down the phone line with green heads and the water line with yellow heads.

Project Size

Is the site large enough to allow for all the storage room needed? Is a site office required on the plans? If so, what are the requirements? Consider whether building materials and equipment can be stored on the job site without interfering with the work. Finally, is there room on the site to store topsoil or unsuitable excavated material that has to be removed?

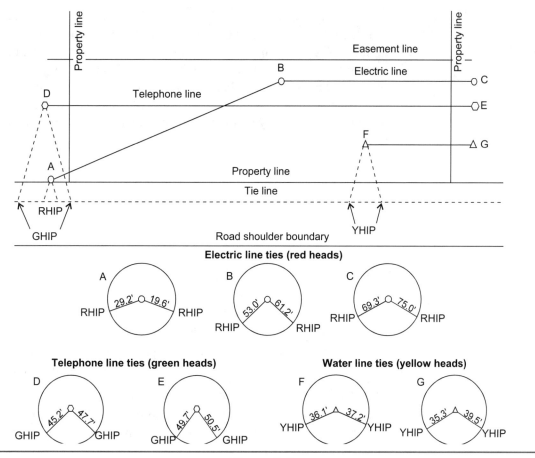

Figure 2-2
Tying down utility lines

Local Needs

There are several questions to ask if the job's in a location you're not familiar with. First, locate local suppliers of fuel, repairs, parts and any other operational needs. Find out their policy on credit or payments. Get an agreement in writing if possible. If you plan to use local workers, are qualified people readily available? What wages are expected? Are there other projects in the area that will be competing for labor?

Traffic Control

If the project will need traffic control, check with the local authorities to see what they require. Most of them spell out traffic control requirements very clearly. There are exact standards for barricades, delineators, flashing lights and other safety precautions. Some area authorities require a barricade log. That means additional labor costs to patrol and repair broken traffic control devices every day, including Sundays and holidays.

The Site Visit

Security

If the job site is isolated or in an area with a high crime rate, you may want to hire a security company. Vandalism to equipment or material, theft, and destruction of completed work can be a major financial loss. Most of it won't be covered by insurance. That makes it a cost of doing business. Be sure that cost is included in your estimate. In high-risk situations, the cost of the security company may be small compared to the cost of repairing equipment or replacing material.

Also consider public safety. Your job will probably draw sidewalk superintendents. Everyone loves watching heavy equipment at work. Will you need protective fencing around the area? Or is there a better way to keep people out of danger?

Existing and Imported Soil

When you've evaluated all of these variables, it's time to look at the soil itself, both the existing soil and any soil that must be trucked in. Wet and heavy soil costs more to move than dry and light soil.

Check the compaction requirements. The more compaction needed, the more time required for rollers, the more rollers needed and the bigger the rollers have to be.

If fill isn't available on-site, locate a source of suitable material close to the job. If unsuitable material has to be trucked away and dumped, find a disposal site and get it approved.

Site Visit for a Sample Project

Figure 2-1 shows a drawing for a small project. The owner wants to install 8-foot diameter metal culvert along the existing stream bed, using the excess material to cover the pipe and bring the area to a level grade at elevation 110. Test borings were taken at points T_1, T_2, and T_3 (Figure 2-3). There are no engineering plans or specifications except the drawing, which was prepared by a surveyor to show existing conditions. The owner added the proposed 110 elevation grade lines. If I were estimating this job, here's how I'd handle the site visit.

Before going to the site, I'd prepare a short list of specific questions and get a copy of the site visit checklist. Here are some of the questions I'd include in my specific list.

1) The property borders a four-lane highway.
   A) Will access be permitted onto the highway? Where? By whom?
   B) Is the highway divided? If so, how far in each direction is a turnaround point or street?
   C) What's the speed limit? Will trucks entering the highway be a safety problem?

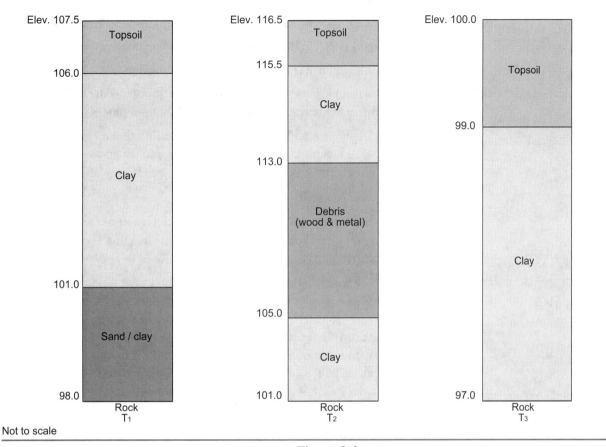

Figure 2-3
Test borings

D) Heavy trucks entering the site will probably damage the shoulders. What are they made of and how will we repair when finished?

E) Will drainage pipe be needed during construction?

F) Will work be close enough to the road to require barricades?

G) Will a permit be needed to get on road right-of-way?

2) The heavy brush and tree line along the western side of the creek will present several problem areas.

A) How large is the vegetation? What equipment will be needed for grubbing?

B) What type of trees? Can they be sold for their material? Are there any firewood types?

C) It would appear that there is little or no room on the site to stockpile debris and trees. Is there a place close by? Is permit burning allowed in this area? If not, where is the closest landfill that accepts trees?

D) What is the approximate volume of trees and brush? (We did this calculation earlier in the chapter.)

3) The stream or small river itself will need to be addressed.
    A) Is it a protected stream? Check with the Corp of Engineers.
    B) Could flooding during construction cause problems?
    C) Can equipment cross it during the first work phase?
    D) The test borings show rock possible at about the stream flow line. Inspect the entire stream bed for rock outcropping.
4) Utility easements and lines. Contact each utility that may have lines in the area. Ask them to locate these lines prior to the site visit. Request a copy of their construction drawings in the area if they're available.

    During the site visit:

    A) Make survey ties to all utility lines that have been located. This way they can be relocated later when the location flags have been destroyed.
    B) Make note of all surrounding utilities and their types. See if they look like they're in the easement. Never take anything for granted. In this project, the owner said the overhead electrical line shown on the drawing is abandoned. But what about the easement? Get a written abandonment notice from all utilities involved. Also check at the local Recorder of Deeds office for any other easement or restrictions that might be tied to the property.
    C) During the visit, I find that there's another line on the poles that's not electrical. Investigation reveals that it's a television cable company's line. Because the electrical company said their lines were abandoned, we probably wouldn't be legally responsible. But why risk it? It's better to be a good neighbor, and save yourself time. Try to foresee and prevent these problems.
5) The test borings shown in Figure 2-3 will tell a lot. I'd examine them closely in the office, and then in the field watch for evidence of past dirt work that could be a problem.
    A) Two things are evident from the test borings. First, there's about the same amount of topsoil in T_3 (bottom land along the river) and T_1 (high ground). That indicates there's been very little flooding. Any significant flooding would have left larger deposits of topsoil near the river when the water receded.
    B) Second, T_2 shows a section about 8 feet deep with particles of wood, metal and other deleterious material. This would indicate a dump site. The contour lines on the east side of the creek increase more rapidly than those on the west. The lack of any major vegetation and the presence of an older housing area just east of the area are good clues that the area was once a dump. When the homes were built, trees, building debris and other items were pushed into the valley. Then the area was covered with soil. The only way to be sure is to order additional test borings.

6) It looks like there's enough clay on the site to reach compaction requirements. A check with the company that did the borings might get some additional soil information.

7) Check boundary line agreement.

   A) View each boundary line. Check the field location of fences, structures, trees, streets adjacent to the project site. If a survey has been done, check the way the corner points line up with the surrounding property.

   B) If there's a large discrepancy between the survey points and the existing evidence, check with property owners to work out this problem.

In general, look at each and every item on the plans and in the field. View each with the movement and construction work in mind so you can anticipate any problems. Don't overlook anything, don't assume anything and *get everything in writing* that concerns any other individual or company. And use your checklist. I promised you a copy of my version. It's on the following pages.

Because the condition of the soil is so important to the estimating process, I'll devote the next chapter to an in-depth study of soil problems and their effects on the final quantities.

Site Visit Checklist

Job No _____ Location _____ Date _____
Weather _____

Plans:
_____ Do the plans and drawings match what the site looks like?
_____ Are they accurate in reference to direction?
_____ Do they show the surrounding properties in degree needed?
_____ Do they show everything needed? If not list what you need.
1. _____
2. _____
3. _____
4. _____
Comments _____
_____
_____
_____

General Specifications:

Individual Item	Agree/Not	Needs
1. _____	_____	_____
2. _____	_____	_____
3. _____	_____	_____
4. _____	_____	_____

Traffic:
_____ Is there traffic movement in or around the site?
_____ Will traffic barriers be needed? _____ Will a flagman be needed? _____ How long?
_____ Will one way or dead end streets affect construction?
_____ Will schools or other special zones affect construction time?
_____ Will rush hour traffic be a problem?
_____ Are there traffic counts available for busy streets?
_____ Are there restrictions such as bridges, culverts, etc?
Comments _____
_____
_____
_____

Clearing and Grubbing:
_____ Are limits of clearing and grubbing defined? _____ Are they shown on plans?
_____ Is there room for permit burning? _____ Is there salvageable wood?
_____ Location of disposal area _____
Comments _____
_____
_____
_____

28 *Estimating Excavation*

Utilities:
_____ Are there utilities on site?_____ Do they agree with plans?
_____ Do they need to be located in field?_____ Are all normal utilities accounted for?
_____ Will connections be necessary? If so, which ones?
1. _____
2. _____
3. _____
4. _____
_____ Will relocation be necessary? If so, which ones?
1. _____
2. _____
3. _____
4. _____
List name, phone for each utility company
1. _____
2. _____
3. _____
4. _____
_____ Will temporary service be needed? If so, which ones?
1. _____
2. _____
3. _____
4. _____
_____ Are utilities near site that will be needed during construction?

Comments _____
_____

Drainage:
_____ Is there drainage across property now?_____ Is it taken care of in plans?
_____ Will drainage increase or decrease when project is completed?
_____ Will flow need to be continued during construction?
_____ Will temporary structures be needed?_____ Are private easements involved?

Comments _____
_____
_____
_____
_____
_____
_____
_____
_____
_____
_____
_____
_____

Sanitary or Storm Sewer:
_____ Are there any sanitary or storm sewer lines on property?
_____ Are they to be saved?_____ Are they to be removed?
_____ Will continuous flow need to be maintained?
_____ Will connections need to be made? If so, to which ones, how?
1. _____
2. _____
3. _____
4. _____
_____ Inspect all manholes, drop inlets or other structures. Note size, structure type, materials, depth, number of inlets and outlets, their locations, and approximate flow.
_____ Will additional right-of-way or easement be needed to make connections or ties?

Comments _____
_____
_____
_____

General Appearance:
_____ Does general layout fit plans and surrounding area?
_____ If dry period, is area dry? _____ Are wet spots apparent?
_____ Will noise be a problem to surrounding neighborhoods?
_____ What about pedestrian safety?_____ Parking area for workers?
_____ Does the type of topsoil, and or vegetation match that shown on the plans?

Miscellaneous comments _____
_____
_____
_____
_____
_____
_____
_____
_____
_____
_____
_____
_____
_____
_____
_____
_____
_____
_____
_____
_____
_____
_____

3 Properties of Soils

Geology, the study of the Earth's subsurface, is important to every earthwork estimator. Of course, there's no way that I can cover all the technical details in one chapter. And you don't have to be a soils engineer to estimate earthwork. But you do need to understand some basic principles about soil and rock.

In this chapter, we'll discuss how these traits — stability, compaction, moisture content, drainage and soil movement — affect the final quantities on earthwork projects. They determine what type of equipment you'll use, how long the job will take, and the working rules that will apply.

Soil Testing

Testing by soil engineers is expensive. That's why there will be plenty of information available on large jobs, but little or no data for small jobs. Later we'll talk about where you can find whatever information exists. But first, let's look at how to understand and use the information.

Until the actual excavation starts, there's no way to know for sure what's under the surface. To make educated guesses, soil engineers drill boring holes at specified locations throughout the site. They auger a hollow pipe into the ground and remove samples of the soil or rock they encounter. After recording the depth of each layer of material in a boring log, they send it off to a lab to be classified.

Figure 3-1 shows a boring log. There's a project layout on the top of the page showing the locations of the boring holes. The rest of the page shows the actual elevations and depths of the soil and rock specimens removed from the

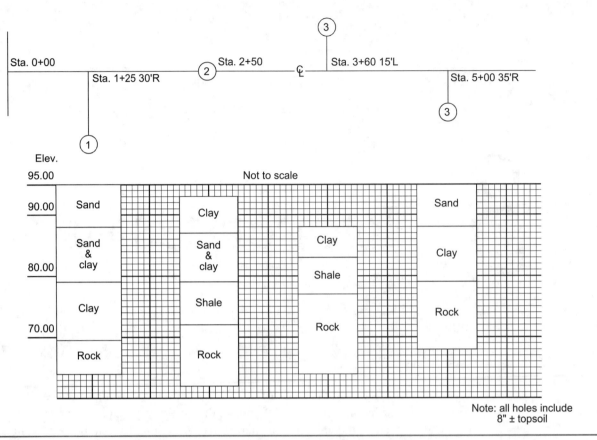

Figure 3-1
Boring log

boring hole. While this method doesn't always identify all of the soils on the site, it's the best information available on the typical job. There could be small deposits of foreign material on the site between the test holes. But most soil layers cover a relatively large area, so engineers can estimate the approximate locations and size of the various soil layers.

Soil Classifications

Classifying soil is a difficult and highly technical problem. All soil is a combination of one or more classifications. While all soils run in layers, the actual makeup of each layer can vary widely. For instance, one location may have a mixture of 60 percent clay and 40 percent sand. In a similar location, the soil might be 20 percent clay and 80 percent sand. That's why it's important to get all the information you can about soil conditions in your project area.

Your best source of information about soil conditions in a given area is probably the local American Soil Conservation Service (ASCS) office. They're located throughout the country, usually several in each state. Look for the Soil Conservation Service in the Federal Government pages in your local phone book. Your local office should have a book showing the limits and makeup of the individual layers, and some information about water runoff and slope stability.

For example, there's a soil called Knox in northwest Missouri. We don't know the origin of the name, although it was probably named after the person who first identified it. This soil has very distinctive characteristics, including the ability to stand almost vertical without eroding. There may be similar soil in other areas with a different name. And there may be soil called Knox in another state with totally different characteristics. So don't rely on what you think you know about soil types. Always get and use local soil information. Use the descriptive list of soil types below as an introduction to the subject.

1) *Bedrock* is sound hard rock in the undisturbed state. It's in its native location and is usually massive in size.

2) *Weathered rock* is rock that has weathered to the stage between bedrock and soil. It will have seams, and is often broken up into small deposits with soil in the seams.

3) *Boulders* are fragments of rock that have broken off of the bedrock. Anything over 10 inches in diameter is called a boulder.

4) *Cobbles* are smaller rock in the 2-inch to 10-inch size range.

5) *Pebbles* are even smaller, ranging from $1/4$ inch to 2 inches in size.

6) *Gravel* is a mixture of small rock particles ranging from $1/4$ inch up to 6 inches in size.

7) *Pea gravel* is a mixture of particles $1/4$ inch or less in size.

8) *Bank run gravel* is a mixture of sand and gravel that's excavated directly from the Earth's surface.

9) *Sand* is small rounded particles of weathered rock. It's usually graded into fine, medium or coarse sizes.

10) *Silt* is made up of very fine particles of rock, often having the texture of baking flour.

11) *Clay* is made up of very fine particles of inorganic material.

12) *Hardpan* describes many different mixtures of gravel, sand, and clay that have a hard texture.

13) *Till* is a mixture of sand, gravel, stones, silt, and some clay.

14) *Caliche* is a mixture similar to till, only it's held together by desert salts such as calcium carbonate.

15) *Shale* is a soft grey stonelike substance.

16) *Loam* is a mixture of sand, silt, or clay and organic matter. Another name for this is topsoil — because if there is any topsoil on an undisturbed site, it will be on the surface.

17) *Adobe* is a heavy clay.

18) *Gumbo* is a fine claylike mixture.

19) *Mud* is a mixture of various earth materials and water.

20) *Peat* is partly-decayed organic material.

21) *Muck* is a mixture of organic and inorganic material.

22) *Loess* is a siltlike material that occurs in small deposits where it was carried by blowing winds.

Soil Characteristics

In most instances, boring logs will reveal that the soil types occur in layers, one on top of another. Each of these soil types will behave differently when wet, or when handled in a specific way. Figure 3-2 is a chart that shows how to identify the various soils in the field, under both wet and dry conditions. The column headed *Cast* indicates the tendency of the soil to retain its shape after it is squeezed in the hand. *Ribbon*, in the next column, shows the ability of a soil to be rolled out into a ribbon or "worm" using the palm of the hand on a hard surface.

Soil type	General appearance	Cast		Ribbon
		Dry	Wet	
Sand	Granular appearance, free-flowing when dry	N	Y	N
Sandy loam	Granular soil; mostly sand mixed with some silt and clay, free-flowing when dry	Y	Y	N
Loam	Uniform mixture of sand, silt and clay; gritty to the touch, somewhat plastic	Y	Y	N
Silt / loam	Mostly silt mixed with some sand and clay; may have clods, but clods are easily crumbled to a powder	Y	Y	N
Silt	Contains at least 80% silt particles; has clods that grind to a very fine, flour-like powder	Y	Y	Y
Clay / loam	Fine textured soil, more clay than in silt loam (see above), may be lumpy; when dry resembles clay (see below)	Y	Y	Y
Clay	Fine textured soil, large masses may be broken into smaller very hard lumps, but does not pulverize well or easily	Y	Y	Y
Organic soils	Soil lacks any discernible structure, consists of plant fiber and decomposed organic matter, muck and peat included	N	N	N

Figure 3-2
Fill classification of soils

Stability

Engineers study soil makeup to learn about the stability of each type of soil. Will it distribute the building load evenly? Will it stand or slide when formed into a slope? This is just as important to the estimator as it is to the engineer. How well a particular soil will stand on steep slopes determines the type of equipment you'll use and how you'll move and place fill material. This is especially important where deep trenching is required. If the soil is unstable, you'll have to plan for shoring or lining the trench walls. Get as much information as possible before you begin your estimate.

Visit the site with an eye open for anything that suggests unstable soil conditions. If there are creeks on the site, how has the water affected the creek banks? Banks that are straight up and down indicate good stability. If they're sloped, is the degree of slope uniform throughout the site? If not, there may be a layer of unstable material. Are there visible seams that show different soils? Look for crumbly material or shale, which is less stable than clay.

When you do the site visit, be prepared to take some samples. Take along a small shovel, a large spoon, water, and a piece of thick glass about 6 inches square. Find a spot along a creek where the topsoil level is easily accessible, or dig a hole down through the topsoil. The topsoil is usually a dark, fine-grained material. When it's moist you can roll it into a ball between your palms. But if you keep rolling it, it will soon dry out and crumble.

Take a sample of each separate layer you encounter and try to roll a ball with each layer. Add a little water if the sample is too dry. The most stable materials will stay compacted in a ball even with continuous rolling.

Soil is more stable and compacts better when it has the ability to cling together. This characteristic is called *plasticity*. There's a simple test you can do to get an idea of how plastic the soil is. Take a small ball of the material and wet it until it's almost saturated. Place the ball of soil on the glass. Start rolling it back and forth, making a worm out of the material. Move it back and forth till you have a worm about 5 inches long. Cut it into two or three pieces, then roll the pieces back into a ball and repeat the process. If you can make the worm, cut it, and reroll it several times, you have a soil with good adhesive abilities.

There are, of course, scientific tests to determine the liquid and plastic limits of soil. The liquid limit is the point where it goes from a stable adhesive soil to a liquid. The plastic limit is the opposite — the point where it goes from a stable adhesive soil to a semiadhesive and crumbly soil. The soil has to be between the two limits to compact well.

Compaction

Probably the single most important soil trait is its density, or compaction. Soil is made up of many particles of different sizes. The closer together these particles are, the more stable the soil. After the engineers have classified the type of soil and the loadbearing needs of specific areas within a project, they can calculate the required density of the soil for each area. The *required density* is the degree to which a soil needs to be tamped down to make it as solid as it was in its original state. This requirement is usually expressed as a percentage, with 100 percent representing the maximum possible compaction.

The compaction test, called the *Standard Proctor Test*, determines how much a soil can be compacted. That figure will be used as a guideline throughout the project. The in-place material must be compacted to a certain percent of the "Proctor."

The Standard Proctor Test is known as either American Association of State Highway and Transportation Officials (AASHTO) test designation T99-70, or American Society of Testing Materials (ASTM) test designation D-698. Testers fill a cylindrical steel mold, 6 inches in diameter and 7 inches deep, with material in three separate layers, or *lifts*. Each lift is compacted by 25 blows from a 5.5 pound, 2-inch-diameter hammer falling from a distance of 12 inches. This Proctor Test is usually specified for fill material placed under buildings, sidewalks, utility trenches, and landscape areas.

They use a Modified Proctor Test (AASHTO 180-70 or ASTM D-1557) on fill material in areas that will carry heavy loads, like highways, airport runways, and so on. The modified test uses the same cylindrical mold, but the material is placed in the mold in five lifts instead of three, and each lift is compacted 25 times with a hammer weighing 10 pounds, falling from a height of 18 inches.

The Proctor is first run on soil that's relatively dry. They keep adding water and running repeated tests until the compaction reaches close to 100 percent and then drops off. Most tests peak at about 90 percent with a moisture content of about 8 percent.

The percent of compaction in the Standard Proctor is compared to the existing soil in its undisturbed state. Because the test may compact the soil until it's more dense than in its original state, it's not uncommon to see Proctor numbers that exceed 100 percent.

Moisture

The amount of moisture in soil plays an important part in the compaction process. That's why you need to understand the reaction of soil and water when they're mixed together.

To reach the required compaction, you must control the three variables: density, moisture content, and compaction effort. Look at Figure 3-3. The four curves on the graph show moisture and density information for the same soil sample under different compaction efforts. The unit dry weight on the left side represents the density. It's expressed in pounds per cubic foot. Moisture is shown as a percentage of dry weight.

All Proctor curves will show a well-defined peak. That peak indicates the maximum density for a given compaction effort at a certain moisture content. This condition is known as the *point of optimum moisture*. Soil that is too wet or too dry must be brought into this range by adding water to dry material or drying out wet material. For most soil, that means adding water or drying it until the moisture content is about 8 percent. But that's easier said than done.

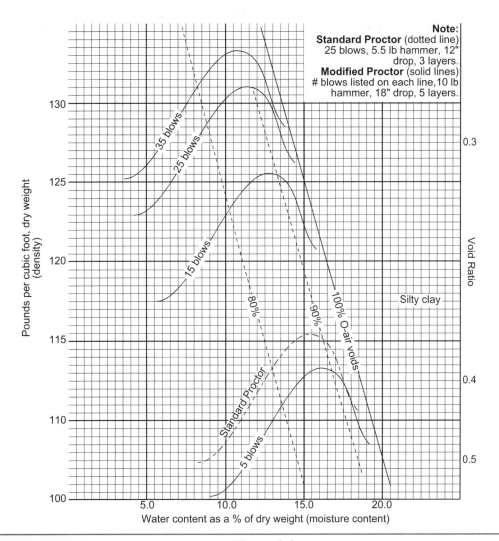

Figure 3-3
Compaction test diagrams

The moisture content of the soil may vary a great deal, even within one project. Different soils have different water-holding capabilities. The elevation, shading, weather, and many other factors make it possible to have moisture contents ranging from 2 or 3 percent to 20 percent or higher. When these soils are mixed together during construction, it's difficult to estimate the resulting moisture content. It takes an experienced superintendent and machine operators to make the job run smoothly. They can tell by the feel of the soil as they roll it between their palms and the way the machines respond just what kind of material they're working with.

Most fill is laid in 6-inch lifts. If the fill is dry, water trucks have to spray water on the material before compaction work can begin. If it's wet and muddy, the material can be dried by disking the material and blading it back and forth several times to let the sun and air dry it. If time is a problem, it's possible to add drier material to reduce the moisture content.

Properties of Soils

The moisture content is a critical point for the estimator. Dirt that's too dry or too wet requires additional handling with expensive equipment and labor. That raises the cost per cubic yard. On large jobs, you probably have the results of soil tests. On smaller jobs, there may be little if any data available. Here you're playing with fire. Surprises are inevitable, and they won't be pleasant, or cheap. Where you don't have data, take the time to investigate all available sources for information on the area. Here are some places you can get information:

1) Check with local residents. Do they have any information on water tables? If they've farmed the area, how did the soil react?

2) Contact the local county or city engineer. See if they have soil tests from areas near your project.

3) Contact local utility companies. From burying their utility lines, they may know if there are problem areas.

4) Check with local engineering firms, especially any that specialize in soil engineering.

5) Check the project site for clues to the amount of water in the area. Swampy areas, lakes or streams indicate high moisture. Lush growth of trees, grass or other vegetation also shows moisture. Lack of vegetation, barren ground, or sand indicates a lack of moisture in the area.

We've been focusing on the Proctor test to measure soil compaction, but there's another way to express density: the void ratio. The fewer the number of voids, the more dense the material. If you're curious, try this experiment. Fill a water glass to a certain mark with marbles. Then fill it up to the same level with water. Remove the marbles, and measure the water. Then fill the same glass to the same point with sand, adding water to bring it up to the same level. Remove the sand, and measure the water. There will be a lot less water in that glass than there was when it held marbles. Why? The sand particles are smaller and a lot closer together, so there's less space between them (voids) than the marbles.

For the same reason, a mixture of several different types of soil will usually compact better than each of the separate soils would compact separately.

The Easy Percolation Test

Engineers run a percolation test to find out if soil can handle sewage discharge from private septic lines or effluent from large treatment plants. You can perform a simple version of this test that yields a good indication of soil characteristics. Figure 3-4 shows the setup.

Dig a hole about 6 to 8 inches in diameter and 3 to 4 feet deep. Record the depth and type of soils you encounter. Perform the ball-rolling test I described earlier on a small sample of material from the bottom of the hole. Then place a stake on each side of the hole and nail on a cross piece, as shown in the illustration. Fill the hole with water to just below the start of the topsoil layer. Measure and record the distance from the cross member down to the water

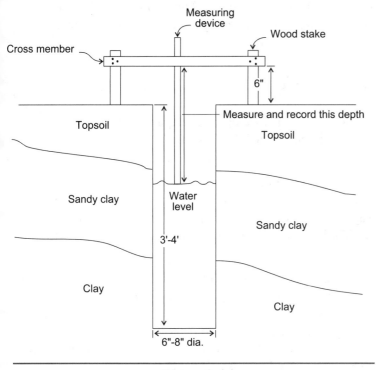

Figure 3-4
"Perc" test setup

line. After one hour, again measure and record the distance from the cross member to the water line. Repeat the process at two, four, and 24 hours.

This isn't an exact test, but it stands to reason that if the water disappears from the hole in the first hour or two, the soil is very porous. If it disappears in four hours, it is probably porous but with some stability. If there's still water in the hole after 24 hours, the soil is probably clay or some other material that compacts well. Notice I said *probably*. Maybe there's a high water table, or a layer of rock that prevents the water from draining off quickly.

Of course, the more holes you dig, the more likely you'll get meaningful results. But you can't depend on the results of this test alone. If you combine all the bits and pieces of information, you should be able to draw a fairly accurate picture of the existing ground conditions.

Drainage and Soil Movement

The drainage on the project — before, during, and after construction — is affected by the soil's water-holding capabilities. Rains or upstream drainage can cause problems in porous soils that absorb water easily. It takes longer for the soil to dry out, delaying the project completion.

During construction, each day's work must be left in a condition that allows the best possible drainage. Ponding water takes longer to dry out. In areas with steep slopes, high water runoff and other drainage problems, you need to allow time for building temporary drainage ditches to carry the water around or away from the work area. While this isn't a pay item in the contract, it pays for itself because there's less delay after a rain.

The soil's makeup and moisture content also affect the way it moves. Soil with a high sand content is more easily moved by wheeled equipment. Dense clay and other high-moisture soils require tracked equipment in most instances.

Wet material can be pushed and loaded by machines in larger amounts than dry material, which tends to spill over. But the wet material doesn't push as easily or as smoothly. It usually takes a pusher tractor to help load the scrapers. The wet material also won't dump smoothly from the scrapers. You may need an additional blade or dozer to level the material down into the lifts.

When material is wet, allow extra time for extracting stuck machines, and for track cleaning. In extremely dry or sandy soil, the equipment needs to be serviced more often than usual. That raises the hourly operating cost. In a later chapter, we'll talk about calculating operating cost, including working in various conditions.

Compaction Testing

Every good estimator knows that compaction requirements determine the type and amount of equipment needed. But not all of them take into consideration the amount of testing and when it will be done.

Engineers and architects require compaction tests on many jobs. After all, the tests assure them that the material is being placed and compacted correctly. That's the only way they can be sure that the soil will support the structure under construction. These tests are done under the supervision of a certified soils engineer.

On large jobs, the plans and specs will spell out how many compaction tests are required, and whose responsibility they are. A common requirement is one test for each 5,000 square feet of fill on each 6-inch lift. On a government project, the controlling agency will usually do the test or hire a private firm. Either way, they'll absorb the cost. On large private jobs, the owner will generally pay a soils engineer to do the test. If the contractor wants additional tests, he'll probably have to pay for them.

On smaller projects, there may not be any tests required. But the contractor will be held responsible for improper compaction if the structure settles later. That's why many contractors pay a soils engineer to do the tests, or else perform some of the simple tests themselves. If there's any doubt at all, the investment in soil testing is well worth it.

Each test takes about an hour to run. But stopping work while it's being done is expensive. Most contractors schedule around the tests, so workers aren't idle. They may bring in one 6-inch lift, compact it, then move over and work in another area while the test is being done. It the test fails, they'll have to remove the material, replace it and recompact. Since they only work one lift at a time, they only have to remove one lift if a test fails.

There are two other crucial soil properties that affect every earthwork estimate — swell and shrinkage. We'll cover those in a later chapter. For now, let's begin learning how to calculate earthwork quantities. In the next chapter we'll cover area take-off by plan and profile.

4 Area Take-off by Plan and Profile

There are two common ways to calculate excavation quantities: the plan and profile method, and the contour method. In this chapter we'll take a look at the plan and profile method. It's the method I prefer for estimating a large project, or a project where the final grade line is fairly uniform throughout. Most road construction projects, as well as many large housing tracts and shopping centers, should be estimated using the plan and profile method.

The plan and profile method has several advantages. It's the most accurate and the easiest method to use. It also simplifies quality control during the actual earthwork. The contractor can calculate quantities any time during construction by restaking the project and shooting new elevations. Then he can figure the quantities removed and quantities remaining. Most contractors leave the survey stakes in place as long as possible to facilitate this restaking.

In this chapter I'll begin to introduce you to take-off methods estimators use to calculate soil and rock quantities. The procedure is called *take-off* because you take quantities off a plan and profile or cross section sheet and transfer them to an estimating worksheet.

Before we begin, let me clear up one area of possible confusion. You may come across several names for the finished ground line. For some reason, engineers, architects and contractors all use different terms for this. Engineers usually call it the design elevation, design plan, or finished profile. Architects refer to the proposed elevation, future elevation, or final elevation. Contractors talk about grade line, or final grade line. But fortunately, they all agree that the grade of the ground before starting work is the existing grade line or elevation line.

Cut and Fill Sections

The designer of any earthwork job has two objectives: first, to create a relatively flat finished surface that allows good drainage; second, to move as little material as possible, import no soil to the job and haul none away. That's called balancing the cut and fill quantities. We'll talk about that later in the book. For now, we'll focus on the difference between cut and fill, and how it affects the take-off.

Every earthwork job begins with a projected profile — an imaginary line that you'll create by moving the dirt around. You'll have to cut away material that's higher than the profile and add dirt (fill) in lower areas. It's important to keep the cut and fill quantities separate while you're doing the take-off.

Figure 4-1 shows the mechanics of a cut and fill operation. Imagine a fish aquarium that's been filled with sand poured in at random. The object is to get it level, as shown by the finished grade lines.

In Figure 4-1, the baseline represents the project as it's laid out. The plane B-B' represents the level, finished surface. The lines A-A and B-B are two edge-on views of plane B-B'. To create that level surface, you'll have to cut off some hills and fill in some holes. Lines A-A and B-B show where the cuts and fills will go.

Figure 4-2 is a cut and fill cross section representing the end view (A-A) of Figure 4-1. Note that although the existing grades appear rounded in Figure 4-1, as they are in the field, in Figure 4-2, we've used straight lines to plot these points. That's the accepted procedure. The quantity difference between the straight and rounded lines will balance out over the total project.

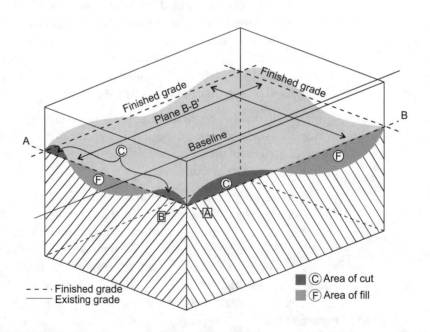

Figure 4-1
A simple cut and fill problem

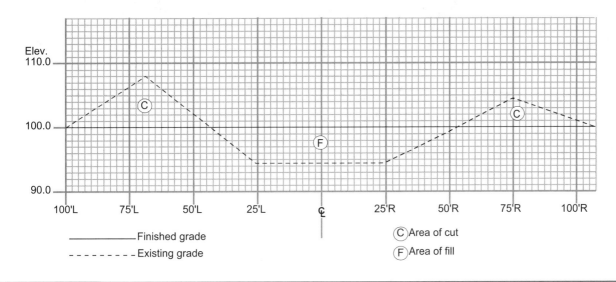

Figure 4-2
Cut and fill cross section

Understanding Surveys

As you might guess by now, an excavation estimator's job requires reading survey maps created by land surveyors. You should understand what surveyors do and the meaning of the maps they create. That's why I'll devote part of this chapter to surveying. Of course, I'm not going to make you a professional surveyor. I'll just provide the essentials: how the surveyor arrived at the elevation points on the worksheets, and how they affect your take-off quantity.

Plan and Profile and Cross Section Sheets

Surveyors and engineers work with two types of paper when doing earthwork design: plan and profile sheets and cross section sheets. Both are created on lightweight paper that's easy to reproduce, usually 22 by 36 inches in size. Plan and profile sheets are blank on the top half to allow room for the layout or any design needs. The bottom half is for plotting the points. It's divided into 1-inch squares drawn with heavy lines. Each 1-inch square is divided into 100 smaller squares drawn with lighter lines. The cross section paper is composed entirely of the plotting squares. Both sheets have a place for project name, dates, changes, and name of the person who did the work.

When plotting a cross section, be careful about the scale. Select a scale appropriate for your project. Consider these variables:

- Overall width and length of the job site
- Difference in elevation between the highest and lowest points
- Frequency of the cross section layout stations
- Degree of accuracy needed

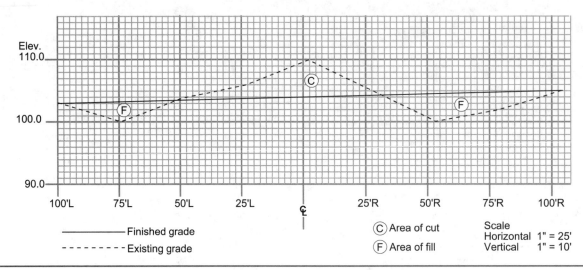

Figure 4-3
Cross section drawn at appropriate scale

First establish the baseline. Try to put the baseline in the center of the page if you can. After it's located, leave room on either side of the sheet for calculations and for recording cut and fill notes. Now choose a scale that will allow you to plot the entire cross section. For the average project (200 to 400 feet wide), using a scale of 1 inch = 25 feet will work for the *horizontal scale* (along the bottom of the section from left to right). A 400-foot section would be 16 inches long. For smaller projects, 1 inch could represent 10 feet, while in a larger project you might choose 1 inch = 100 feet.

The *vertical scale* is normally a lot smaller. Using a smaller vertical scale makes the picture clearer and makes plotting and take-off more accurate. You'll most often see 1 inch = 5 feet (or 10 feet). Here again, the difference between the highest and lowest points will determine the scale to use. A 2-foot rise in a vertical scale of 1 inch = 100 feet would be almost imperceptible. In fact, it would be less than the width of a pencil mark. On the other hand, a 2-foot rise on a 1 inch = 5 feet scale is almost half an inch.

Ideally, the scale you choose should make the drawing fill the space available — both vertically and horizontally. The larger the scale, the more accurate the section and the easier the calculations. Figure 4-3 shows a cross section drawn at about the right scale. Of course we've reduced it here.

Field and Office Procedure

A surveying or engineering crew will stake out a baseline on the project. It's usually in the center of the project for roadways, and along one or two sides in areas where sites are smaller or more cluttered. Then they put in stakes at 100-foot intervals along this baseline. The stakes are called stations and are written *Sta. 0+00*, for example. Sta. 0+00 is the beginning station. Sta. 1+00 is read as "station one plus balls." It's 100 feet (plus double zero) from the start. Sta. 4+00 is 400 feet. Sta. 192+00 is 19,200 feet from the beginning of the project.

Along this baseline, the staking party will also put in a stake every place the ground either rises or falls significantly. They measure from the previous 100-foot stake and give the point a location based on that stake plus the distance to the elevation change. In Figure 4-4, a small creek crosses the baseline three times, the first time between stations 0+00 and 1+00. The first stake on the edge of the creek is 50 feet from the 0+00 station, so it's Sta. 0+50.

When the stakes are set, the staking party runs a set of *levels* on the stakes. They can read the actual elevations, using a bench mark, or simply assign the beginning stake (0+00) an arbitrary value, such as 100.00. Then they shoot the elevation of the rest of the stakes and assign them an elevation that's above, below or the same as the first stake.

After the field work is done, they take the distances and elevations back to the office and plot them on graph paper. This sheet is called a plan and profile sheet. Figure 4-5 shows a typical example.

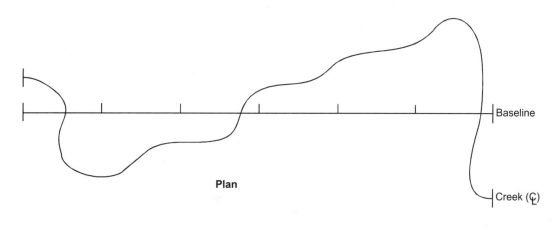

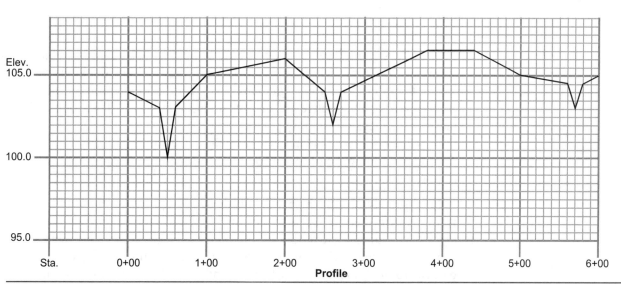

Figure 4-4
Surveyors stake at significant elevation changes

Area Take-off by Plan and Profile **45**

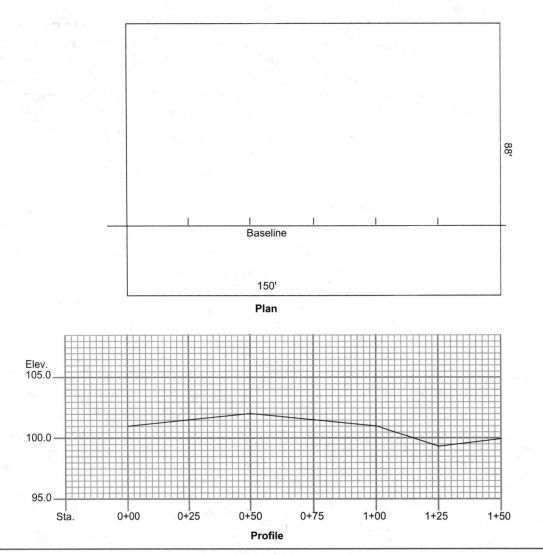

Figure 4-5
Plan and profile sheet

Then the staking crew returns to the field and does what they call cross section work. From each stake on the baseline, they measure out to the right or left (or to both sides) far enough to get past the limits of the project. They may go even farther where there are possible drainage problems. Then they measure the elevation at those points. In most cases they don't place stakes there. They just record the distance from the center stake, and the elevation. Then they return to the office and plot the information on a cross section sheet (Figure 4-6).

Finally, the designers lay out the finished, or proposed elevations. Then they plot the finished elevations onto the cross section sheets that already show existing elevations. The result typically looks like Figure 4-7. This figure shows cross-section views from two stations with existing and finished grades plotted. To make the difference between them very clear, two kinds of lines appear in Figure 4-7. The broken line plots existing elevation, and a solid line shows the proposed elevation.

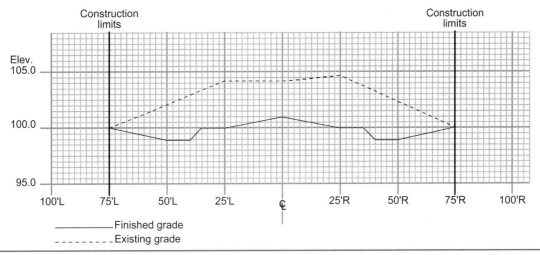

Figure 4-6
Cross section sheet, highway project

When the cross section sheets are finished, you're ready to begin calculating the amount of dirt you'll have to move. As you can see, each cross section shows an enclosed area, bounded on one side by the existing ground line, and on the other by the proposed ground line. The first step in figuring the quantities is to find the area of each cross section. Once you've found the cross section area, multiply by the length to find the volume. Remember that volume equals the area times the length.

End Area Calculations

A cross section drawing like Figure 4-6 shows what the road would look like if it were sliced open across its width. To find the volume to be cut, we'll start by finding the end area for a section. There are several ways to calculate this area. I'll show you three methods: first using a planimeter, then a measuring strip, and last the arc method. The most accurate one uses the polar planimeter — also known as a *buggy*.

Using a Planimeter

The planimeter is an instrument commonly used by engineers and estimators to measure area. But it measures only area. You have to multiply by the depth to convert to volume figures.

A polar planimeter is an instrument that measures area by tracing the boundaries. A planimeter consists of two arms and a movable carriage that links the arms. The pole arm ends in a sharp, weighted point, that's called the anchor point. The second, or tracing, arm has a stylus, or point, at the end, used to trace the area's outline. The carriage also contains a roller mounted on a drum. The drum's circumference is a scale, also called a read disc, dividing the drum into 100 parts. As you trace around the boundaries of the area, the roller follows and the number of revolutions made by the roller registers on the read disc. The standard ratio of roller revolutions to read disc is 10:1.

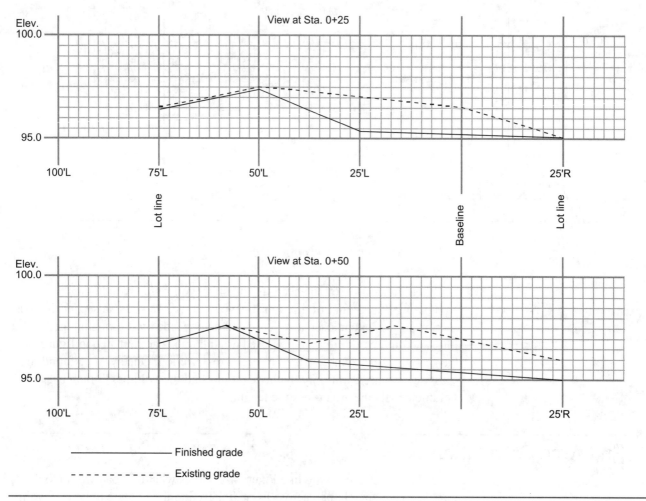

Figure 4-7
Existing and proposed profiles

Let's look at how to properly use a planimeter, starting with your work surface. I recommend covering it, whether it's wood, steel or a plastic laminate, with a sheet of cardboard. The anchor point tip is quite sharp, and easily leaves holes in wood. On a hard surface, like steel or plastic, the cardboard protects the point. Trying to use a planimeter on these surfaces may bend, or even break, the anchor point. A bent anchor point produces incorrect readings, and a broken one results in an unusable instrument.

Spread out the plan or map so it lies absolutely flat. Make certain the sheet has no wrinkles or buckles anywhere. Once the sheet is perfectly flat, keep it that way by taping it to the cardboard.

Use the following guideline to position the planimeter anchor point:

1) Place the anchor point outside of the area to be measured

2) Place the anchor point to allow tracing of the entire area perimeter

If the area is too large to cover in one sweep of the planimeter, divide it into several smaller areas that you can cover in one pass. Then add the results together.

Choose a starting point that's easy to remember. Set the roller vernier to zero, or record its current reading. Begin tracing the outline of the area in question. Work your way around the area perimeter, moving clockwise, until you're back at your starting point. Follow the boundary lines carefully and closely. A deviation adds an error to your result.

The next step is reading the results from the read disc. Read the result direct from the disc, if you set the vernier to zero before you started the run. If you didn't zero the disc, the difference between the start reading and the current reading equals the area.

Many professionals like to go over the area two or three times, then average the results to get a more accurate figure.

The Planimeter Constant

Most instruments have a planimeter constant of 10.00 square inches. This means that when the main disc reading is 1.00, the area is 10 square inches. Some instruments give readings in metric measure, and others have different scales, but the procedure is the same. This constant value is usually printed on the instrument itself, and in the instruction book that comes with the planimeter.

If you don't know the constant for the instrument you're using, here's how you can find it. Lay out an area of known value. Since most planimeters have a constant of 10 square inches, use a 2-inch by 5-inch area. Run the planimeter around this 10-square inch area and you should read 1.0 on the disc.

If you don't, you can recalibrate the instrument by finding the new constant. If you run around the area and come up with a figure of 0.910 instead of 1.0, use this formula to find the new constant:

Formula for planimeter constant

$$C = A/N$$
$$= \frac{10 \text{ SI}}{0.910}$$
$$= 10.989 \text{ square inches}$$

Where:

C = *planimeter constant*

A = *area*

N = *final planimeter reading*

Finding the Area

To find the area from the planimeter reading, use this formula:

Area = C (the planimeter constant) x N (final planimeter reading)

Let's look at a brief example. Assume the roller of a fixed-arm planimeter with a constant of 10.00 is set at zero. After following the perimeter clockwise, the reading is 2.55. The formula for area is:

$$Area = C \times N$$
$$= 10.00 \times (2.55 - 0)$$
$$= 25.5 \text{ square inches}$$

The scaled off area is 25.5 square inches.

Using a Measuring Strip

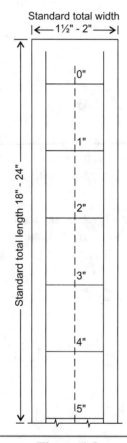

Figure 4-8
A measuring strip

There's a faster way to find the area when the degree of accuracy isn't as important. Simply use a measuring strip that you can make yourself. Start with a piece of clear plastic $1^1/_2$ to 2 inches wide and 18 to 24 inches long. Choose a transparent plastic that you can write on with ink and that won't smudge if you try to rub the ink off. Draw two lines about 1 inch apart down the length of the strip. Then mark off 1-inch increments, starting with zero at the top. You'll end up with a line of 1-inch squares, as in Figure 4-8.

To begin measuring, place the strip on the plan and profile sheet as shown in Figure 4-9 (step 1), with the zero line on the bottom of the area line and the left line even with the first whole 1-inch line. Then place the point of a sharp instrument where the centerline intersects the zero line. A pin will work but the point on the end of drafting dividers is better. Then move the strip up the page until the pointer is on the top line of the area you're measuring. Hold the strip steady and move the pointer back to the bottom line (step 2). Then move the strip over 1 inch to the right (step 3) and repeat step 2. Keep repeating steps 2 and 3 until you reach the end of the area line (step 4). If you're interrupted before you finish the entire cross section, stop and mark your spot, then start again.

When you're ready to read your measurement, lay the strip beside a drafting ruler and read the last point you marked with your pointer. Figure 4-9 (step 5) shows a reading of 4.0 square inches.

With this method, you're simply building a running total of the square inches of the end area. With a little practice, you can read right off the cross section sheets, without having to use the measuring strip. When using a measuring strip, as in all take-offs, be sure to keep cut and fill measurements separate.

Using the Arc Section

A third method, the arc section, is similar to the measuring strip because you measure each 1 inch of horizontal area and build a cumulative total. Figure 4-10 shows a cross section worked up this way. You divide each whole or partial 1-inch section in half with a vertical line that goes through both the

50   *Estimating Excavation*

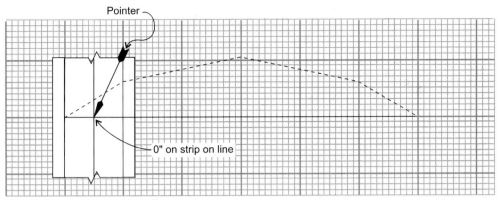
Step 1: Place pointer on zero line.

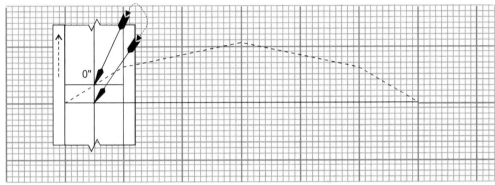
Step 2: Slide strip up to area line, then move pointer back to bottom line.

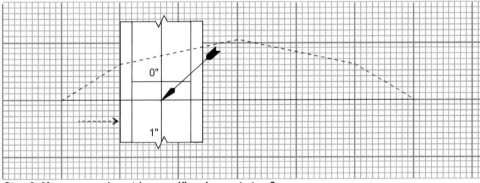
Step 3: Move measuring strip over 1" and repeat step 2.

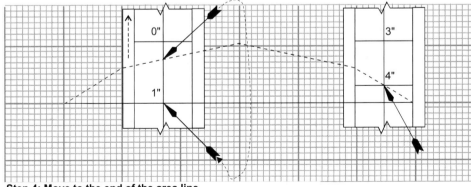
Step 4: Move to the end of the area line.

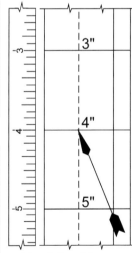
Step 5: Read the last pointer point from a ruler held next to the strip. The reading here is 4.0 inches.

Figure 4-9
Using the measuring strip

Area Take-off by Plan and Profile

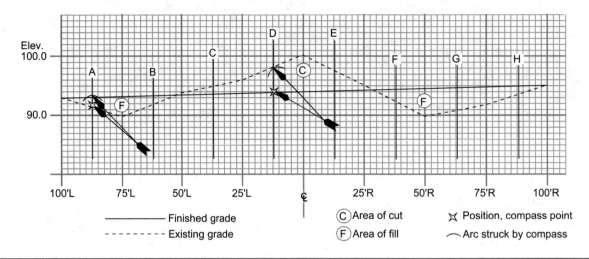

Figure 4-10
Using the arc section

existing and proposed profile. Then label each of these division lines with a letter (A, B, C and so on) as in Figure 4-10. Put the point of a drawing compass on the bottom line where it intersects the division line. Set the compass to strike an arc that runs through the point where the division line and the top line intersect. Then strike this arc. Let's suppose that we just struck arc A in Figure 4-10. Before we strike arc B we need to make a copy of arc A on a worksheet. My worksheet appears in Figure 4-11. For the time being, set the compass aside. Be careful not to change its setting. Let's take a look at this worksheet first. Then we'll cover the nuts and bolts of copying arcs from the cross section sheet onto this worksheet.

There are four main points to note when you look at the worksheet in Figure 4-11.

1) "Sta. 1+00" appears twice on the worksheet to separate fill areas, top half, from cut areas, bottom half.

2) Just to the right of "Sta. 1+00" is a line with several arcs marked on it. We'll call it the measuring line.

3) All arcs on the worksheet intersect the measuring line.

4) The name of an arc stays the same when it's copied from a cross-section sheet onto a worksheet.

Using the worksheet is easy. Start by filling in the station name. Since the first arc we're copying is A, the station name is "Sta. 1+00." Next retrieve the compass, place its point on the far left end of the measuring line (marked by a filled circle in Figure 4-11), and strike a copy of arc A that crosses the measuring line.

Now we'll go back to the cross section sheet and strike arc B. Adding a copy of arc B to the worksheet we'll use a slightly different process. Here's

52 *Estimating Excavation*

Station name		Inches	Constant	Cut	Fill
Sta. 1+00	●A■B)) F) G) H)	1.45	1		1.45
Sta.					
Sta.					
Sta. 1+00	) C) D) E	0.69	1	0.69	
Sta.					
Sta.					

Figure 4-11
Arc section take-off worksheet

why. Arc B, like arc A, is from a fill area, as well as being from the same station. That means you add arc B to the same measuring line as arc A. To add arc B, place the compass point on the intersection of arc A and the measuring line. (In Figure 4-11 this point's marked with a filled square.) Then go ahead and strike the arc and label it B.

To copy the cut arcs (C, D, and E) follow the same steps. However, this time you'll use the "Sta. 1+00" measuring line that's in the bottom half of Figure 4-11.

When you've finished marking arcs, use a drafting scale or ruler to measure from the beginning of the line to the last arc, then record this length on the worksheet under Inches and again under either Cut or Fill. Each measuring line represents the total square feet of either the cut or fill on the job. In our example, the scale is 1 vertical inch equals 10 feet and 1 horizontal inch equals 25 feet. So each square inch equals 250 square feet (10 x 25 = 250). There are 1.45 square inches of fill, or 362.5 square feet (1.45 x 250 = 362.5).

Notice the column headed Constant in Figure 4-11. This is the width of the area that each arc measures. In this example, the constant is 1 inch. In relatively flat areas, you could use a wider constant.

You may need to use several of the lines on the worksheet to finish all the areas. And make sure you keep the cut and fill sections separate. Notice that the cut and fill sections are labeled on Figure 4-10.

Both the arc section and measuring strip method are just approximations, not accurate take-offs. They're based on the assumption that the slope of a particular section will be roughly equal on both sides of the centerline of that section. Look at Figure 4-12. The area of triangle A is equal to the area of triangle B. Measuring along the centerline is good enough. But if the slope in triangle B is steep and the slope of triangle A is shallow, measuring along the centerline isn't going to be very accurate. Fortunately, slopes are usually more or less uniform and small errors tend to cancel out.

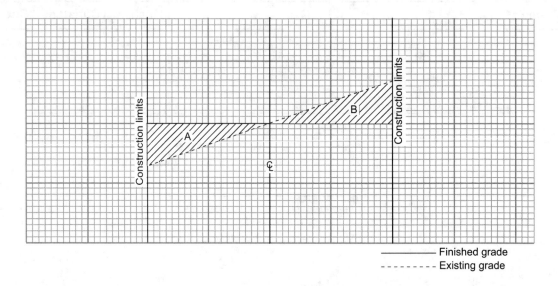

Figure 4-12
Dividing an even slope with a centerline to form two triangles with equal areas

Calculating the Volume

Formula for scale factor

You now know three methods for calculating the end section area. The next step is to convert area on paper into actual area and then convert that into cubic yards of earthwork. Remember, the planimeter and the measuring strip or arc methods measure only the paper area. To convert them, you need to multiply the square inches of end area by the scale factor. Here's the formula:

Scale factor = V scale x H scale

Where:

V scale = the vertical height scale (plan inches to actual feet)

H scale = the horizontal distance scale (plan inches to actual feet)

You apply this scale factor to each square inch of end area. For example, assume you've got a cut area with a planimeter reading of 2.95 square inches taken from a plan that has a vertical scale of 1 inch = 5 feet and a horizontal scale of 1 inch = 25 feet. To calculate the scale factor for this cut section:

Scale factor = V scale x H scale

Scale factor = 5 x 25

Scale factor = 125

To find the area of cut or fill, multiply the scale factor by the number of square inches. If the scale factor is 125 and we measured an end area of 2.95 square inches:

125 x 2.95 SI = 368.75 SF

There are 368.75 square feet in that particular cut section.

54 *Estimating Excavation*

Converting to Volume

We're finally ready to use all of this information to work up the actual cubic yards of earthwork. We'll always use cubic yards to calculate volumes and excavation costs for earthwork.

Cubic measure is length times the width times the height. In our take-off so far, we've been working in two dimensions, width and height, to find the square feet of area. To make the conversion to cubic measure, we need to know the length.

We'll calculate volume the same way we figured area: break the task into many small measurements and then add or subtract them to get the final figure.

Earlier in this chapter, we talked about how the surveyors choose the stations for cross sections: They take them at each significant change in the ground slope. That helps make our calculations more accurate. For volume measurements, we'll select a point between each two stations as the third measuring point. We'll calculate dimensions at each point separately, total them, and then divide by 2 to find the average. The formula for this is:

Formula for average area

Volume = EA1 + EA2 x Sta. L ÷ 2

Where:

EA1 = end area of one station

EA2 = end area of the next station

Sta. L = distance between the two stations

We're adding the area of one station to the area of the second station, then dividing by 2 to average them. Finally, we'll multiply the average area by the length.

In Figure 4-13, let's suppose that Sta. 1+00 has a fill area of 206.0 square feet. Sta. 1+65 has a fill area of 400.0 square feet. (Of course this isn't drawn to scale. The cut and fill data are for this example only.) Let's plug those numbers into our formula:

Volume = (206.0 + 400.0) x 65 ÷ 2

= 606.0 x 65 ÷ 2

= 19,695 cubic feet

Notice that this answer is in cubic feet. To convert to cubic yards, divide by 27.

19,695 ÷ 27 = 729.44 cubic yards

You can either convert to cubic yards as you compute each station, or wait and do it at the end after you've averaged all the stations. I think it's easier and less confusing to convert at each station.

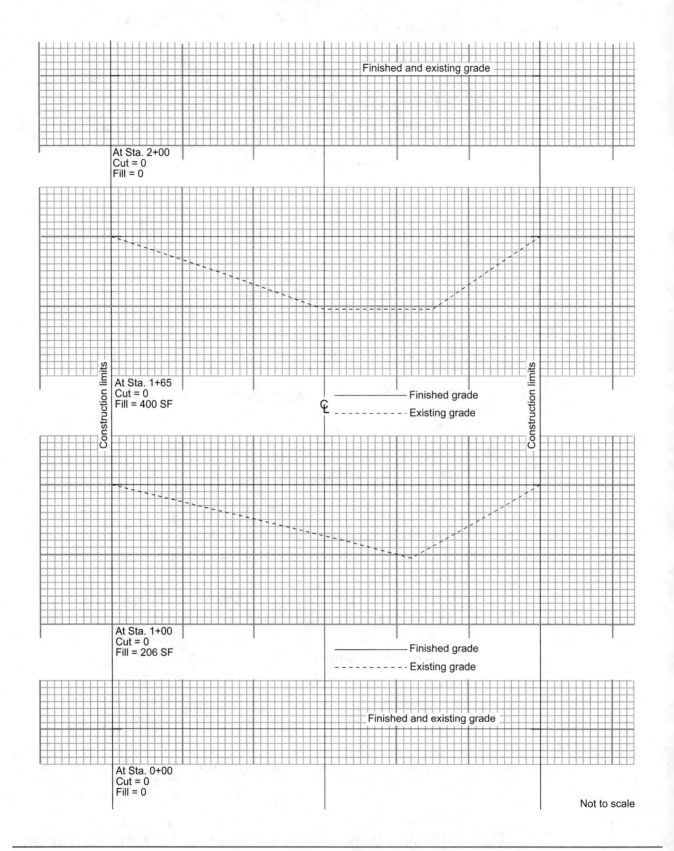

Figure 4-13
Calculating an average volume between two stations

56 *Estimating Excavation*

Beginning and Ending Stations

Designers usually want to make a smooth transition from the surrounding ground elevation to the finished project. That means they try not to have any earthwork at the beginning and ending stations. You'll seldom see any area calculations at these stations. But there's a trick here. In your volume calculations for these two stations, you'll still average them with the adjoining station. If Sta. 0+00 has an area of 0 and Sta. 1+00 has an area of 400 square feet:

Volume = (0 + 400) ÷ 2 x 100
= 20,000 cubic feet

20,000 ÷ 27 = 740.74 cubic yards

A Practical Example

Figure 4-14 shows a series of cross sections from a total of six stations. Each cross section shows areas of both cut and fill. The project is a parking lot located on hilly terrain. The end areas of the cross sections have been calculated with a planimeter. We'll use these end areas, and the scale factor (derived in Figure 4-14B) to calculate total earthwork quantities for the project. The quantity sheet, Figure 4-15, records and summarizes my calculations.

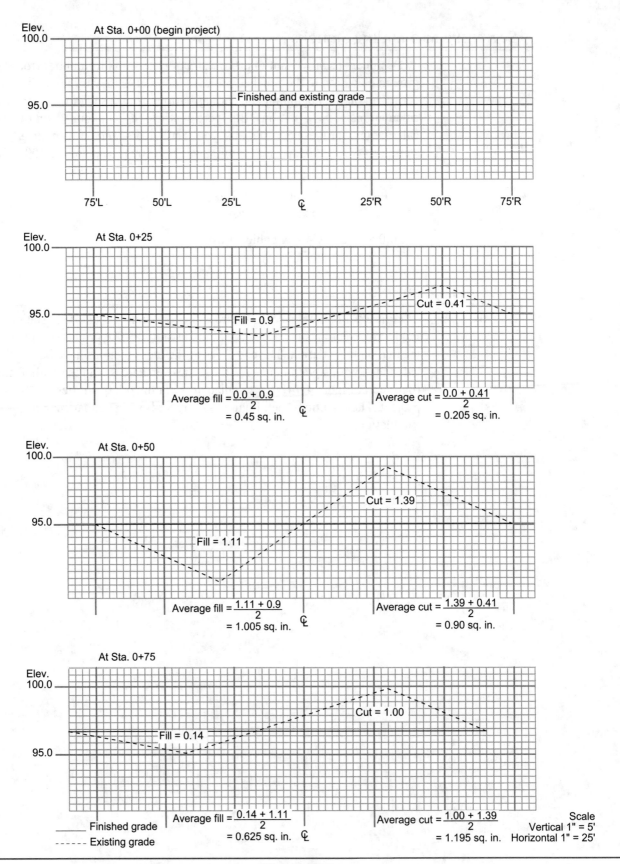

Figure 4-14
Cross section worksheet 1

58 *Estimating Excavation*

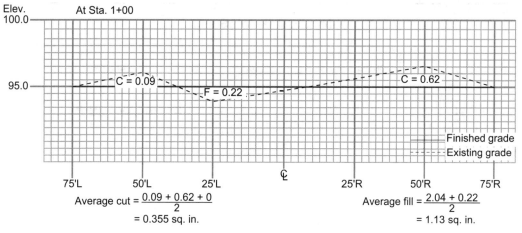

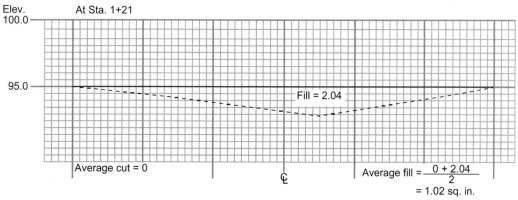

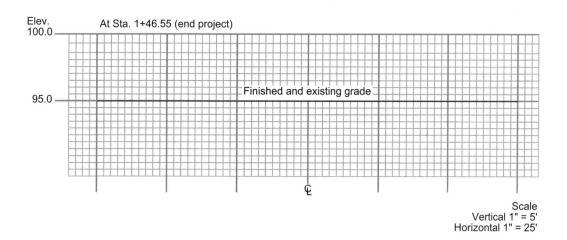

Average fill and cut calculations between Sta. 0+75 (previous page) and Sta. 1+00

Average fill = $\frac{0.22 + 0.14}{2}$ = 0.18 sq. in.

Average cut = $\frac{0.62 + 1.00}{2}$ = 0.81 sq. in.

Scale factor calculation

V scale : 1" = 5'
H scale : 1" = 25'

Scale factor = V scale x H scale
Scale factor = 5 x 25
= 125

Figure 4-14 (continued)
Cross section worksheet 2

Quantities Take-off Sheet

Project: City lot #000-A **Location:** 123 A St. Anytown **Owner:** James Smith

Estimate prepared by: D. Burch **Date:** 10/10/99 **Checked by:** Charles A. Rogers

Take-off method: Polar planimeter **Sheet:** 1 of 1

1st sta.	2nd sta.	Distance 1st sta. to 2nd sta. (feet)	Cut AEA* (sq. in.)	Fill AEA* (sq. in.)	Scale factor**	Total area cut (SF)	Total area fill (SF)	Volume cut (CY)	Volume fill (CY)
0+00	0+00	0			125				
0+00	0+25	25	0.205	0.45	125	25.63	56.25	23.73	52.08
0+25	0+50	25	0.90	1.005	125	112.5	125.63	104.17	116.32
0+50	0+75	25	1.195	0.625	125	149.38	78.13	138.32	72.34
0+75	1+00	25	0.81	0.18	125	101.25	22.5	93.75	20.83
1+00	1+21	21	0.355	1.13	125	44.38	141.25	34.52	109.86
1+21	1+46.55	25.55	0	1.02	125		127.5		120.65
1+46.55	1+46.55	0			125				
Totals		146.55	3.465	4.41		433.14	551.26	394.49	492.06

Notes: * AEA = average end area, for calculations see Figure 4-14.
 ** For scale factor calculations, see Figure 4-14.

Figure 4-15
Quantities take-off sheet

5 Reading Contour Maps

Every good earthwork estimator has to be good at reading contour maps. In this chapter I'll explain the essentials of contour map reading: how they're prepared, what the symbols mean and how to find the information needed to calculate earthwork quantities.

Planimetric and Topographic Maps

You'll use two types of maps when preparing estimates. *Planimetric* maps show the position of natural and man-made features of the Earth's surface. A road map is a planimetric map. It shows the Earth's surface in two dimensions and doesn't give us much information about the third, the elevation of the ground.

A contour or topographic map (or *topo* map) shows most features of the planimetric map plus the contours of the Earth's surface. Contour lines on a topo map show the third dimension (elevation of the Earth's surface) that's missing on a planimetric map. This added dimension is referred to as *relief*. I'll say more about relief on topo maps later in this chapter.

Datum is a term used to define what we know about specific points on a map. On both planimetric and topo maps, there are two main types of datum. *Horizontal datum* is information on the location of specified points on a horizontal plane. For example, a point at the beginning or ending of a street is defined by its horizontal datum. The earthwork estimator uses horizontal datum, of course. But vertical datum tends to be much more important. *Vertical datum* is the distance up or down from a given reference point, most often sea level.

A government agency provides the *National Geodetic Vertical Datum*, a calculation based on the average sea tide at a specific time at 26 tide-monitoring stations throughout the United States and Canada. The average of these points is considered to be sea level and is assigned the elevation of zero. Every point on the Earth's surface can be assigned an elevation above, below, or at this level. Map elevations in the U.S. and Canada are based on this National Geodetic Vertical Datum.

You'll probably find an example of the National Geodetic Vertical Datum at your local general aviation airport. At the airport near my home, the control tower has a sign stating that the runway is 1023 feet above sea level. On charts for pilots, this point is shown as "elevation 1023." This means that a particular point on the field is 1023 feet above sea level. Unless specified otherwise, consider any elevation you see on a topo map as being feet above or below sea level. An area that's below sea level will be labeled on the map as a minus number (-250.0) or 250.0 *below* sea level.

Relief Marking on Topo Maps

Topo maps use *relief markings* (symbols, contour lines, color changes, and shading) to show natural earth features and man-made changes like buildings, railroads, highways, and dams. But only contour lines actually show points of equal elevation.

The Department of the Interior U. S. Geological Survey publishes topo maps of the United States on quadrangle sheets (called *quad sheets*). The scale for these maps is either 1:24,000, or 1:100,000. Figure 5-1 shows part of a USGS 1:24,000 quad sheet. A full 1:24,000 quad sheet covers an area of about 65 square miles.

The USGS also publishes an illustrated, color pamphlet: "Topographic Map Symbols." In it you'll find all the symbols used on USGS topo maps both illustrated and described. For a free copy write to:

U.S. Geological Survey
P. O. Box 25286
Denver, CO 80225

Topo maps have many uses in construction. Engineers use them to design drainage structures, plan streets, curbs, gutters and so on. You'll use them to take off elevation points to find the amount of earth to be moved.

Understanding Contour Lines

A contour line on a topo map connects points of equal elevation. These contour lines are your best source of information on the shape of the earth at the building site. On small jobs, you may figure earthwork quantities from a topo map that has only the project boundaries laid out. The builder and engineer probably haven't given much thought to how the dirt work should be done or how much earth has to be moved. They leave that up to you.

62   *Estimating Excavation*

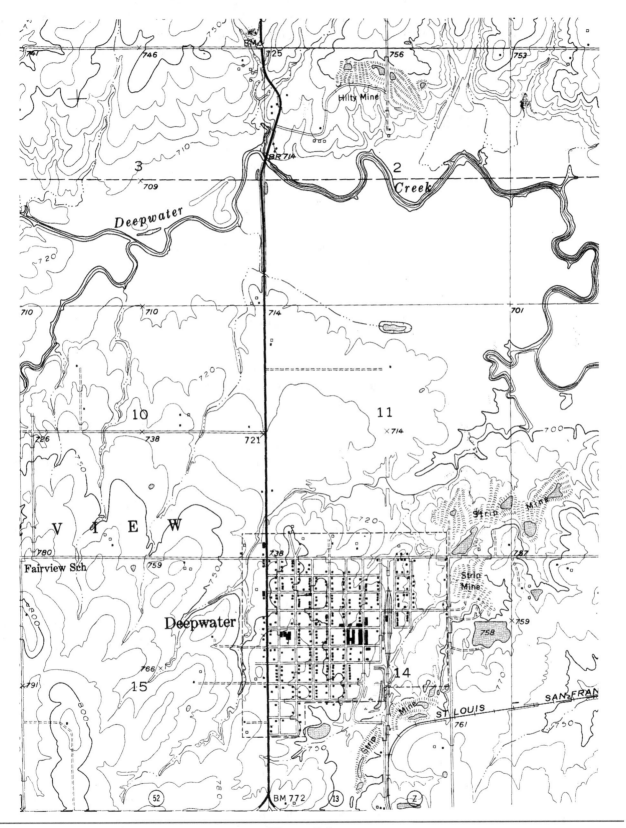

Figure 5-1
Portion of a USGS 1:24,000 quad sheeet

Reading Contour Maps 63

You'll begin earthwork calculations by laying out a project centerline on a topo map of the site. Then you'll write finish elevations on the same map right beside the existing elevations. The difference between the two is the amount of soil that needs to be moved. Making those calculations is commonly called the *pull-off*.

Characteristics of Contour Lines

A contour line is an imaginary line following a specific elevation throughout the area of the map. Figure 5-2 shows a simple example. The elevation is the same at all points around the edge of a lake. A contour line drawn at the elevation of the lake would follow the shoreline exactly. You can think of all contour lines the same way.

Contour lines make the map a little harder to read. But without them, you wouldn't be able to estimate excavation quantities. Reading topo maps takes a little practice. Learning will be easier if you remember these properties of all contour lines.

1) Contour lines are almost always drawn freehand.

2) Contour lines connect points of the same elevation.

3) Contour lines never touch another contour line unless the earth's surface is nearly vertical, and they cross only where there's an overhanging cliff.

4) Every contour line closes (returns to where it began) eventually. Of course, you may need several adjacent map sheets to follow a particular contour line all the way around to where it began. Some contour lines may continue for miles before closing on themselves.

5) Contour lines never break or split into more than one line.

6) The closer the lines are together, the steeper the slope. The farther the lines are apart, the flatter the slope. Look at Figure 5-3.

7) When a contour line crosses a valley or gully, it forms a V, with the V pointing uphill. Figure 5-4 shows contour lines crossing a stream bed.

8) When a contour line crosses the top of a ridge, it also forms a V. The V points downhill. Try picturing Figure 5-4 without the broken line and arrow. Now you know what contour lines that cross a ridge look like on the downhill side.

9) Contour lines which close on themselves on the same map represent a hill or a depression.

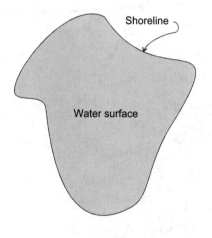

Figure 5-2
Contour line defining the shoreline of a lake

Contour Interval

Contour lines can be used to represent any difference in elevation, such as 1 or 10 or 100 or 500 feet. This is known as the *contour interval*, the difference in elevation between one contour line and another. A relatively flat area might use a contour interval of 1 foot. Each line shows a 1 foot difference in elevation. In the mountains, a topo map may have an interval of 500 feet.

64 *Estimating Excavation*

*In this figure, regardless of width, the space between adjacent vertical lines equals a 1-foot change in elevation. The spaces' different widths indicate changes in relative slope.

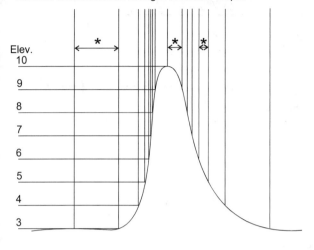

Figure 5-3
The closer together the contour lines, the sharper the rise or fall of the terrain

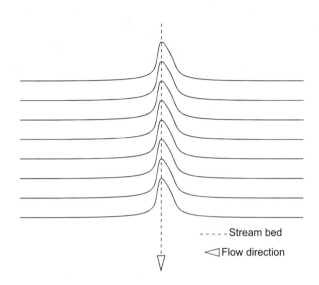

Figure 5-4
Contour lines form Vs where they cross a stream bed, valley, or ridge

Figure 5-5 shows how a small mound of dirt might look in both a profile and top view. Notice that the contour lines are closer together where the mound is steeper. Figures 5-6A and 5-6B show two mounds with the same shape, but different heights.

The two profiles seem identical, but one look at the contour interval tells a different story. Figure 5-6A does show a mound. It's about 65 feet high. The other mound (Figure 5-6B) is really a mountain. It's nearly 12,500 feet high.

Make a point of noting the contour interval anytime you use a topographic map. This data is easy to find on most maps. The USGS quad sheets, for example, list contour interval right in the center of the bottom margin.

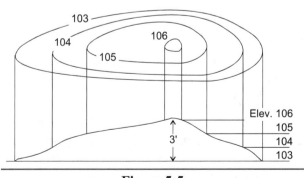

Figure 5-5
A small mound shown in a top view with contour lines, and in profile with elevations

Many topographic maps also include intermediate contour lines. These secondary contour lines give you a more detailed picture of the terrain. In Figure 5-7, the four lighter lines between the dark lines (labeled 50 and 60) are intermediate contours. Typically, intermediate contours have no elevation tags, and there's no listing of their interval. Fortunately, both of these are easy to figure out. For example, the four light lines in Figure 5-7 divide the area between 50 and 60 into five smaller areas. So the interval used here is 2 feet. Reading from 50 to 60, the intermediate elevations are: 52, 54, 56, and 58. Other types of lines, broken or dashed, also denote intermediate contours on maps.

Reading Contour Maps **65**

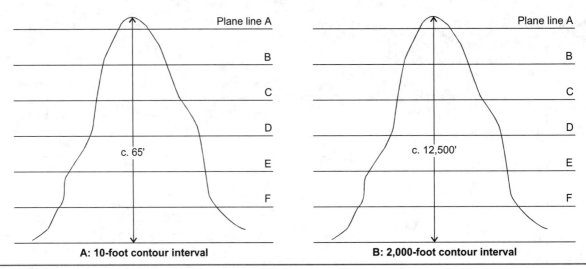

Figure 5-6
Contour intervals demonstration

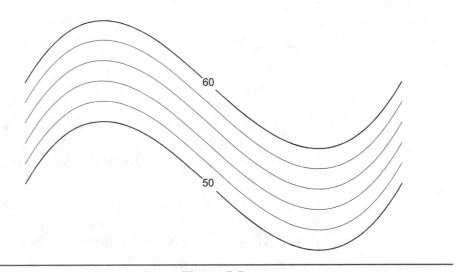

Figure 5-7
Intermediate (light) and major (dark) contour lines

Locating Unmarked Points

You'll often need to know the elevation of a point that doesn't fall exactly on a contour line. Instead, it's part way from one contour line to the next. How do you estimate the elevation at that point?

There are two ways to estimate the elevation of a point that isn't on a contour line. If you have to find only a few points, it's easiest to measure and calculate. First, find the interval between the two contour lines on either side of the point in question. Then measure the distance between the lines using an engineer's scale, and assign a value to each mark on the scale between the two lines.

66 *Estimating Excavation*

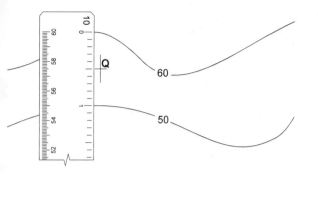

Figure 5-8
Using an engineer's scale to estimate the elevation of a point

Let's try this out using Figure 5-8. We want to find the elevation of point Q. The contour interval in Figure 5-8 is 10 feet. On the engineer's scale that's equal to 10 marks. If 10 marks equal 10 feet then 1 mark equals 1 foot. The scale's 0 mark lines up with the 60 foot contour line. Point Q lines up with the fifth mark *down* from the 0. That's a 5-foot loss from the 60-foot zero point. Point Q's elevation is equal to 60 – 5, or 55 feet. Now picture turning the scale end for end, so the 0 mark lines up with the 50 foot contour line. Point Q now lines up with the fifth mark *up* from the 0. That's a 5-foot gain from the 50-foot zero point and the elevation at point Q still equals 55 feet.

Of course, this is only an estimate. It assumes the ground slopes uniformly from one contour line to the next. That's not always true. But it will usually be about right. Any errors will tend to cancel out if you're figuring elevations at several points.

When you have to figure elevations at several intermediate points, it's easier to make a simple tool you can use over and over. I like to use a plain rubber band, at least ¼ inch wide and 4 inches long. With the band relaxed, mark a beginning point with black ink and another point 1 inch away on the band. Then use an engineer's scale to mark nine fine ink lines on the band between the beginning point and the 1-inch mark. Now you've got a stretchable tool for estimating the elevation of any point where the distance between contour lines is 1 inch or more.

To use it, set the first mark at the contour line on one side of the point of unknown elevation. Stretch the band past that point until the top mark is on the next contour line. Then count the marks between one contour line and your point. Multiply the number of the mark (in tenths) by the contour interval. That's the difference in elevation from the contour line to the point. For example, if your point is at mark 3 and the interval is 10 feet, multiply 0.3 by 10 feet to get 3 feet.

Bench Marks and Monuments

By now we know that contour lines connect points of equal elevation. But how can we be sure what elevation each contour line represents? Fortunately, all that's been figured out for us. Since early in the history of this country, the federal government has made surveys and set survey monuments that all surveyors now use. Every populated part of the country (and a lot of unpopulated areas) have been surveyed and marked with monuments. The engineering department in your county or city can identify the location of monuments in your area.

Figure 5-9 shows a U.S. Geodetic Survey bench mark monument. These bronze markers are embedded in either concrete or rock. The cross at the center marks the exact location of a reference point with a known elevation. The

Figure 5-9
A U.S. Coast and Geodetic Survey bench mark monument

bench mark in Figure 5-9 doesn't show the actual elevation. Other bench marks may show the elevation. But a topo map that includes this bench mark shows the marker's location and that point's exact elevation above mean sea level.

Placing Survey Markers

Surveyors usually place survey reference points on a solid surface that won't be affected by earth movement during normal freeze-thaw cycles. Good locations for survey markers include concrete footings, heavy spikes driven into power poles or large trees, or exposed natural rock outcroppings or ledges. Because survey markers have to be used regularly during design and construction, they should be as close to the project as possible without being in the way of construction.

You should be familiar with two types of bench marks that surveyors place.

Types of bench marks

A *permanent bench mark* is as precise as possible given the conditions of the project. They're normally placed about every 300 feet on relatively flat terrain. Where the terrain or obstacles make moving from one bench mark to another time-consuming, they may be spaced as close as every 50 feet. They're always designated with the standard notation *BM*.

Temporary bench marks (designated as *TBM*) aren't as accurate or as stable as regular bench marks. They're established for a short period of time, or for a specific portion of the work within a project.

There are also two different types of elevation numbers used by the estimator: real elevations and project elevations.

Types of elevation numbers

Real elevations are the actual elevation of the points above sea level set from existing known elevations.

68 *Estimating Excavation*

Project elevations are commonly used for engineering, estimating and construction because the actual elevation above sea level is usually of no practical interest. The engineer will pick a point (such as the top of a curb), identify it as the beginning BM, and assign some elevation to this point.

To make the math easier, the beginning project elevation is usually assigned the number 100 or 1000. Most engineers use a number high enough so that every elevation used when designing and building the project will be a positive (rather than a negative) number. It's easier to add and subtract positive numbers. It also makes a mistake less likely.

Contour Profile

In the next chapter we'll begin working with *contour profiles*. A contour profile shows what the surface of the earth would look like if half of it were neatly cut away at some point. The original contour profile is the earth's shape before any work is done. The final contour profile is how it should look when excavation work is completed. That's the topic of the next chapter: taking off quantities from topo maps.

6　Area Take-off from a Topo Map

If I were asked to identify the most important chapter in this book, this is the one I'd choose. Taking off quantities from topographical maps is the heart of the earthwork estimator's job.

In this chapter I'll add more information to the already-complicated topo maps we're using. I'll add a second set of lines showing the proposed final grade. These additional lines tend to make the map even more confusing. But they're an essential part of understanding the work to be done.

I'll also show you how to estimate soil quantities by comparing contour lines, the best way to do your calculations, and I'll suggest some problem areas to watch for.

Comparing the Contour Lines

A contour line is a simple two-dimensional representation of a three-dimensional land form. A *template* is another name for the finished contour line, often used when the finished contours are very flat or gradually sloping. Figure 6-1 shows the difference between the two, and how they might look used together.

Line A-A in Figure 6-1 is known as a *pick-up* line. Assume that a road will be built along this line, with the middle of the road along line A-A. At each point where A-A crosses a contour line, there's a projection from the line down to the *graphic of contour*. In this case, the graphic of contour connects elevations from 104 through 101. There's also the *template*, which plots the finished grade line.

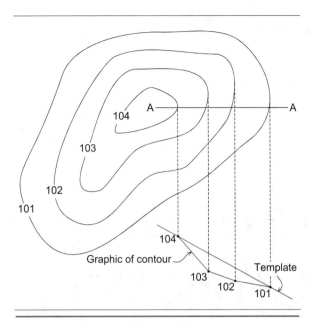

Figure 6-1
Contour lines and a finished grade template

You should understand that Figure 6-1 is just an illustration to show how the lines work together. On an actual job, of course, both lines are plotted on graph paper.

On topographic maps, don't expect a particular type of line, such as solid or dashed, to always mean the same thing. There aren't any hard and fast rules that apply when it comes to topo map symbols. One map may use solid lines to plot existing grades and dashed lines to show the finished grade. However, the next map you use is as likely to reverse the meanings as it is to repeat them. All mapmakers are free to choose the types of lines and symbols they prefer. But it's also the mapmaker's responsibility to assign a meaning to each element used, and to provide users of the map with a key. This key is called a *legend*, and one appears on every map. The legend lists each type of line and symbol found on the map and its assigned meaning. Get in the habit of checking the legend on every map you use. Make sure you know the map symbols, grid square information and location designation system before you start work.

This book mirrors the variety found in real topo maps and site plans by being inconsistent. You'll find that solid lines don't always show finished grades, and broken lines don't necessarily mark existing grades. I've also used a variety of ways to designate locations, grid corners and so on.

No matter what types of lines appear on a topo map, one set of contours always shows the existing grades. This is your "before" picture of the site. Another set of contours marks the finished grades on the same map — these contours are the "after" picture of a job site. The difference between these sets of contour lines represents the quantity of material moved as cut or as fill.

Estimating with a Grid System

It's easier to estimate work if you divide the job into many small sections and estimate the earthwork for each section. Then total the excavation for all sections to find quantities for the whole job.

There are three advantages to using a grid system:

1) It's easier to see the work area.

2) Calculations are easier when figuring small areas.

3) Your work is more accurate when you can average quantities from several small work areas.

The grid helps you focus on a smaller part of the topo map, simplifying the task. Begin by laying out a grid of small squares on a piece of lightweight tracing paper or film. Figure 6-2 shows a simple grid.

72 *Estimating Excavation*

When your grid is done, choose the part of the topo map you'll be working on and lay the grid over that part of the map. Note the grid position on the topo map. Record the map page or section number and the grid square designation when doing the calculations for each square.

Figure 6-3 shows a section of a contour map with a 50-foot grid overlay. At first glance this may look like a hopeless tangle. Let's take it step by step and then you'll see for yourself that it's not all that complicated.

Start with the grid lines. The grid pattern should extend out to the limits of the project, including all areas where earth will be moved. In this case the grid sheet is square. But you can make a set of grids large enough to cover any shape you want. If a small portion of the work area extends beyond the grid sheet, cover it with a single grid square, or several in a row.

There are two rules that apply when estimating with a grid system:

1) To make the calculations easier, all grid squares should be connected to adjoining squares by at least one common line.
2) Each individual grid square must have its own identification.

Choosing Your Identification System and Scale

The method you use for identifying grids is entirely up to you. Use any method that's convenient to label each square in your grid system. But avoid cluttering the drawing with confusing information. Most estimators use the letters of the alphabet instead of numbers. That's what I've done in the following examples. Letters won't be mistaken for, or confused with, contour

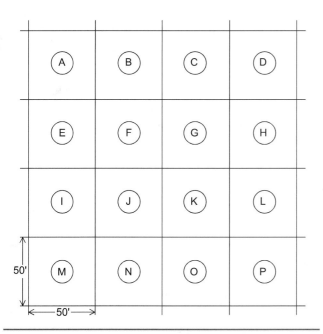

Figure 6-2
A simple 50' x 50' grid

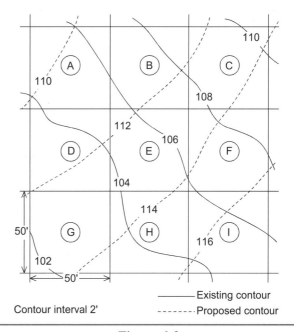

Figure 6-3
A partial site plan overlaid with a grid

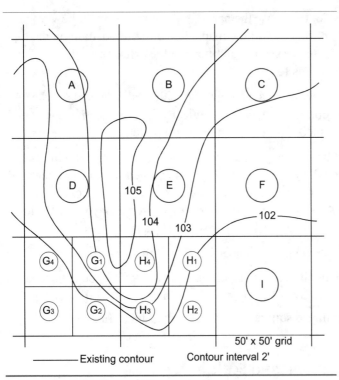

Figure 6-4
Dividing the grid for steep areas

line elevations. And the letter system expands easily to fit any size grid. You can use double letters (AA, AB) or add a number to the letter (A_1 to Z_1; A_2 to Z_2; A_3 to Z_3). Find a system that makes sense to you and is easy to follow. Then stick with it. Being consistent is the best and easiest way to minimize errors.

The scale you use will depend on the size of the plan sheets, the difference in elevation between contours, and your skill. If there's not a great difference between the highest and lowest contour lines, it's safe to use a large grid. However, if the difference is large, you're better off using a small grid. Figure 6-2 uses a 50-foot by 50-foot grid that's scaled so 0.6 inch equals 50 feet. Figure 6-3 uses a different scale, 0.8 inch = 50 feet.

Where the map shows sharp changes in ground contour, divide a grid square into four smaller squares. Look at Figure 6-4. I've divided grid squares G and H into four equal sections and used numbers from 1 to 4 to identify each.

Reading the Contour Lines

Now consider the contour lines themselves. In Figure 6-3 the existing ground contour is shown with solid lines. Ignore, for a moment, everything but the solid contour lines in Figure 6-3. Notice how the lines show a gradual increase in elevation from 102 in the lower left corner to elevation 110 at the upper right.

Next, concentrate on the finished contours shown by the dashed lines. See how the contours increase from elevation 110 in the upper left corner to elevation 116 in the lower right corner. Of course, this map is relatively simple. Most of your jobs will be more complex. But the map shows the points I want to emphasize.

Subcontour Lines

When an area is relatively flat, the contour lines will be far apart. That makes it hard to precisely establish zero lines and other reference points. To make your job easier, establish points midway between the contour lines that you'll connect into *subcontour lines*. They have the same characteristics as contour lines but show midpoints that are useful when making quantity calculations.

74 *Estimating Excavation*

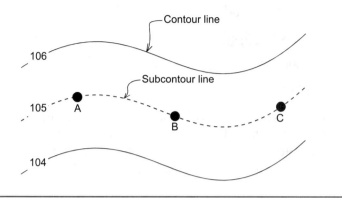

Figure 6-5
Plotting a subcontour line

In Figure 6-5, notice the wide distance between contour lines 104 and 106. To lay out the subcontour line, use a scale (or your rubber band device) to find and mark several points exactly half-way between the 104 and 106 contour lines. You'll see in Figure 6-5 that I marked three midpoints, labeled them *A, B* and *C*, and connected them with a dashed line. That's the 105 foot contour line. Although I guessed at the path of the 105 foot contour, chances are good that it's close enough for most excavation estimating work.

Doing the Take-off

To find the excavation quantity, you need to know the elevation of the existing and finished contours for each square in the grid. To find these elevations, start by finding both elevations at each corner and then calculate their average.

If all the contour lines were level, comparing the two elevations would be simple. Figure 6-6 shows exactly this situation. Grid square A measures 50 feet on each side and has a single existing elevation (a flat base) and a single proposed elevation (a flat top). The difference between existing and proposed elevations in Figure 6-6 is identical at all four corners, 20 feet. Calculate the volume in cubic yards using this formula:

Formula for volume in CY

Volume (CY) = (Length x width x height) ÷ 27

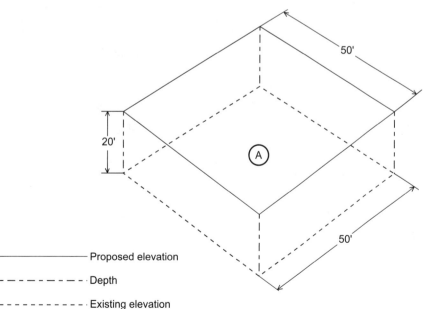

Figure 6-6
Calculating the volume of a grid square

Substitute the numbers from Figure 6-6, and we find:

Volume (CY) = (50 x 50 x 20) ÷ 27

= 50,000 ÷ 27

= 1,851.85 CY

In the real world it's never that easy. Grid squares rarely have just one contour line for either elevation, and as for contour lines lining up with the grid corners, forget it. You'll have to do some calculating to find the corner elevations. Another name for that process is *interpolation*.

Interpolating Corner Elevations

Figure 6-7 shows a single grid square laid over part of a topo map. Points A, B, C and D mark the corners of the grid square. In an ideal situation, each corner would fall precisely on some contour line. Finding those elevations is as easy as copying a number. Figure 6-7, however, comes closer to reality. All four points fall between contours. To find the corner elevations, we interpolate. That's an educated guess, based on measuring how far away the corner is from the two closest contour lines. We assume that the closer the corner is to a contour line, the more similar its elevation is to the contour line's elevation.

Let's start out with a simple interpolation problem. Say that we want to know the elevation of point Z in Figure 6-8. To find this elevation we interpolate, because Z falls between two contour lines. In this example I used an engineer's scale to measure distances, but a ruler or the rubber band device described in Chapter 5 would also work. Whatever tool you use, remember to keep it perpendicular to the contour lines.

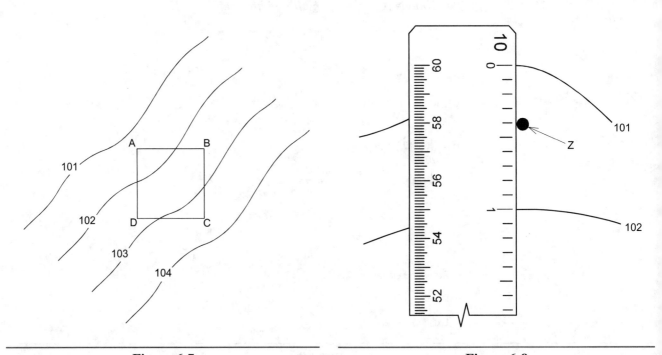

Figure 6-7
Single grid square overlaying existing contours

Figure 6-8
Using a scale to interpolate an elevation

76 *Estimating Excavation*

The contour interval in Figure 6-8 is 1 foot, and on my scale that measures as 10 units. The next step is to measure the distance from Z to each of the contour lines. Measuring from the 101 foot contour line to Z, I get a distance of 4 units. Measuring from Z to the 102 foot contour line, I get a result of 6 units. That's all the data we need to interpolate Z's elevation. Let's take it step by step:

Step 1

Find the value of one scale mark in feet. We already know (from Figure 6-8) that 1 foot is equal to 10 marks on my engineer's scale. So one scale mark is equal to $^1/_{10}$ foot, or as a decimal 0.1 feet.

Another name for a number like the one we just found is *constant*. A constant makes it easier to change a measurement made or calculated in one unit system to its equivalent in another unit system. In this example, I measured in units of scale marks. However, we need an elevation measurement in units of feet. Because we've found the constant, all we do to change scale marks to feet is multiply two numbers:

Scale marks x constant = feet

Step 2

Assume a smooth, even slope exists from the 101 foot contour line to the 102 foot contour line.

Step 3

We already know how many scale marks separate point Z from each contour line: Z to 101 = 4, and 102 to Z = 6. We want to change those measurements to feet at this point. Here's our chance to use the constant we found earlier:

4 x 0.1 = 0.4'

6 x 0.1 = 0.6'

In this example the results run to just one decimal place. Results often run to three or more decimal places but I recommend always rounding them off to just two places. That's accurate enough for our purposes, as I'll explain later.

Step 4

There are two ways to go in this last step. Whichever method you choose, the math is simple and the answer comes out the same. Each method uses a different set of data — one method uses subtraction, and the other uses addition. You're the one doing the estimate so it's up to you to choose. But before we look at the options, I have a few words of warning for you. Each option uses a different set of data. When you choose an option, you're also choosing the set of data you'll use. The link between data and option makes it very important that you don't mix data, or methods. (Details follow the example.) And now for the options:

Option 1 Z's elevation is equal to the sum of the lower contour line elevation (101') plus the separation distance in feet (0.4'). Or:

101 + 0.4 = 101.4

Option 2 Z's elevation is also equal to the upper contour line elevation (102') minus the separation distance in feet (0.6'). Or:

102 - 0.6 = 101.4

Point Z's elevation is clearly 101.4 feet.

Now we'll move on and tie up the loose ends. First, here's why rounding doesn't make the take-off inaccurate. There are two reasons. The first is relative size. Just compare the error's size to the size of the other elements. Here's an example. Suppose that in Step 3 of the interpolation we multiplied 8 by 0.0385. The answer is 0.308 which I'll round to 0.31

By rounding 0.308 to 0.31 I added 0.002 feet to the interpolated elevation. That's too small to make a significant difference in the total amount of earth you have to move.

The second reason rounding doesn't hurt take-off accuracy is that rounding lowers a value as often as it raises one. In rounding that's done consistently to a group of numbers, about half will go up in value and the other half will go down. In the end, all the tiny errors (+ and -) cancel out. Their net effect is zero.

But don't get the idea that interpolation's foolproof. It isn't. Remember that warning in the interpolation example? Here it is again, just in case:

Each option uses a different set of data. Don't mix data or methods.

The unwary estimator can get careless or confused and add instead of subtracting (or vice versa) or do the right calculation using the wrong elevation or measurement. Any of those mistakes results in big errors. At best, you'll look and feel foolish, and careless. At worst, you stand to lose the job, or win the job and lose your shirt. Your best defense against these kinds of errors is to work carefully and systematically. Always follow the same sequence of steps.

Another way you'll avoid errors is by taking the time to check your work. Here are two quick and easy rules to make sure your interpolation results are accurate:

- *Add* distance (measured contour to point) to the *low* elevation.
- *Subtract* distance (measured point to contour) from the *high* elevation.

As an earthwork estimator, your biggest use for interpolation is in finding grid square corner elevations. Test your interpolation skills now, using the single 50-foot by 50-foot grid square shown in Figure 6-9. Interpolate both existing and proposed elevations for each corner. When you're done, check your results against those in Figure 6-10. If they match within ± 0.1, good work! You're ready for the next step — calculating excavation volumes. If you're answers didn't match those in Figure 6-10, review this section before going on. Then try your hand at the interpolations again.

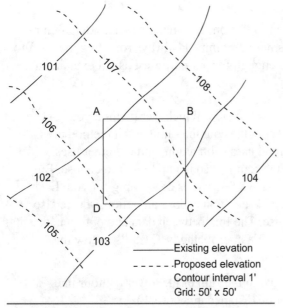

Figure 6-9
Single grid square and partial site plan

78   *Estimating Excavation*

Corner	Existing elevation	Proposed elevation
A	101.86	106.50
B	102.63	107.43
C	103.46	106.71
D	102.72	105.93

Figure 6-10
Interpolated corner elevations for Figure 6-9

Calculating Volume Using the Cross Section Method

Figure 6-11 is a three-dimensional projection showing the same grid square as Figure 6-9. This odd-looking shape is a *truncated prism*. Notice that it's made up of two planes. One plane, based on the existing elevations, has corners labeled A, B, C, and D. The second plane, based on the proposed elevations, has corners labeled A_1, B_1, C_1, and D_1. The area between these two planes represents the excavation volume. As an earthwork estimator it's your job to calculate that volume. It's the difference between the existing and proposed elevations.

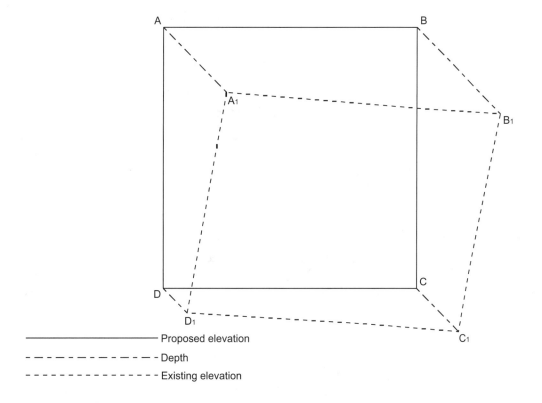

Figure 6-11
This truncated prism is a three-dimensional view of the grid square in Figure 6-9

We've already found both elevations for each corners. The rest is simple subtraction:

A) 106.50 - 101.86 = 4.64

B) 107.43 - 102.63 = 4.80

C) 106.71 - 103.46 = 3.25

D) 105.93 - 102.72 = 3.21

The sum of the four depths divided by 4 gives us the average depth. Here's the math:

Average depth = (4.64 + 4.80 + 3.25 + 3.21) ÷ 4

= 15.9 ÷ 4

= 3.975 feet

We'll round that off to two decimal places and call it 3.98 feet.

That's all the data we'll need to calculate excavation volume using the cross section method. Here's the formula we'll use:

Volume (CY) = (Grid length x grid width x average depth) ÷ 27

Figure 6-9 supplies the following:

- Grid length = 50'
- Grid width = 50'

We found that average depth equals 3.98 feet. Now simply plug the numbers into the formula:

Volume (CY) = (50 x 50 x 3.98) ÷ 27

= 9,950 ÷ 27

= 368.52 CY

The proposed elevations in Figure 6-9 are higher than the existing elevations. That mean that means that we're talking about 368.52 cubic yards of fill.

Calculating Cut and Fill Areas

The system I've just described will work when the grid square is either all cut or all fill. But in some cases you'll have to calculate both cut and fill in the same grid square and separate the totals into different areas. You may come out with minus numbers. Adding minus numbers is the same as subtracting them.

Figure 6-12 shows a grid square that covers both fill and cut areas. Picture a hillside that you'll cut down on one side and fill in on the other to end up with a flat area. The grid square in Figure 6-12 also consists of two planes. Points A_1, B_1, C_1 and D_1 define the existing elevation plane. Points A, B, C and D define the proposed elevation plane. The points where the planes intersect are *zero points*. At a zero point the existing and proposed planes have exactly the

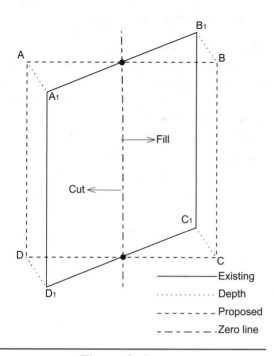

Figure 6-12
Grid square divided into cut and fill areas with a zero line

same elevation. A line connecting two or more zero points is a *zero line*. A zero line divides a project, such as that in Figure 6-12, into two areas: an area of cut and an area of fill. In Figure 6-12 the cut area is to the left of the zero line and the fill area is to the right of the zero line.

It's easy to divide a project into cut and fill areas if you follow these steps:

Step 1

Locate and mark the zero points (intersection of existing and proposed contour lines with the same elevation).

Step 2

Connect the zero points from one edge to the other of the grid square and you have a zero line that divides cut from fill.

The exact path of a zero line depends on the paths of the existing and proposed contour lines. However, as Figure 6-13 shows, if the existing and proposed contour lines form a square, a rectangle, or a parallelogram, then the zero line is a diagonal.

Here are some other rules that will help you understand how a zero line works.

- The zero line runs through all locations where existing and proposed contour lines of the same elevation intersect. There's also a zero line where earthwork is stopped by the presence of manmade or natural structures, such as curbs or walls.

- At any point where a zero line intersects a contour line, there will be another contour line of the same elevation at the point of intersection.

- Like any contour line, a zero line also eventually closes on itself as shown in Figure 6-14.

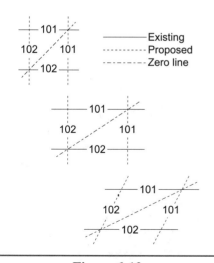

Figure 6-13
Examples of zero line paths

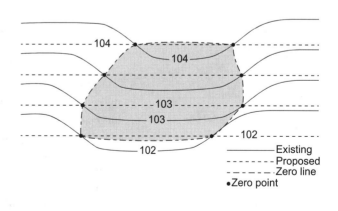

Figure 6-14
A complete or closed zero line

Area Take-off from a Topo Map **81**

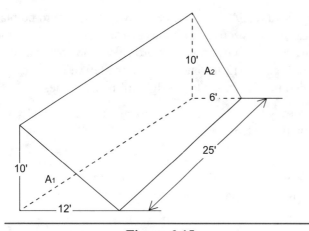

Figure 6-15
Calculating volume using the average end area method

• • • • • • • • • • • • • •
Formula for area of triangle
• • • • • • • • • • • • • •

The Average End Area Method

We'll use several methods to calculate the volume in these cut and fill areas. For Figure 6-11 we used the cross section method to calculate volume. We subtracted the existing corner heights from the desired corner heights, divided by 4 to find the average depth and then multiplied the answer by the grid dimensions.

Another common way to figure volumes of soil is the *average end area* method. We find the average area of the two ends and multiply by the distance between them. The formula is:

Volume (CF) = [(Area 1 + Area 2) ÷ 2] x grid length

Look at Figure 6-15. The end areas are the triangles labeled A_1 and A_2. Here's the formula used to calculate the area of a triangle:

Area = ½ base x height

First we'll calculate triangle A_1.

Area (SF) = (12 x 10) ÷ 2

= 120 ÷ 2

= 60 SF

So end area for A_1 equals 60 square feet. Repeat the calculation for triangle A_2.

Area (SF) = (10 x 6) ÷ 2

= 60 ÷ 2

= 30 SF

Now you're ready to calculate the volume of Figure 6-15 using the average end area method. Here's the math:

Volume = [(60 + 30) ÷ 2] x 25

= [90 ÷ 2] x 25

= 45 x 25

= 1,125 CF

I always convert volumes to cubic yards. Why? First, I know I'll have to make this conversion sooner or later. That's because the quantities are so large that cubic yards are the only practical units to use. Second, by consistently using cubic yards for any volume right from the start, I eliminate a very large category of potential errors. If I don't convert the volume for Figure 6-15 into cubic yards now, it's too easy to overlook that fact later on. That's no minor slipup. It's a major disaster. Throughout this chapter you'll see volume calculations set up so that the result's in cubic yards. I strongly recommend that you do the same. Converting cubic feet to cubic yards is easy. You just divide by 27. For practice let's convert the volume we just found for Figure 6-15.

Estimating Excavation

$$1{,}125 \div 27$$

$$41.66 \text{ CY}$$

- Let's say that no one noticed the discrepancy in the units. Then you would use 1,125 cubic yards, instead of 41.66, in your estimate. I think you'll agree that error makes a huge difference.

Being able to calculate end areas accurately is an important skill for an earthwork estimator. Most of the time you'll find the area of regular geometric shapes: rectangles, circles, and triangles. Occasionally, however, you'll need to find the area of a less familiar shape, such as a rhombus. If you need to refresh your memory of geometry (what's a polygon and how do you find its area?), most dictionaries and encyclopedias have the area and volume formulas.

Calculating the Volume of a Trapezoidal-Shaped Prism

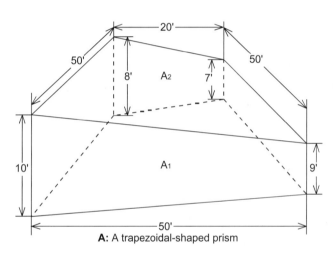

A: A trapezoidal-shaped prism

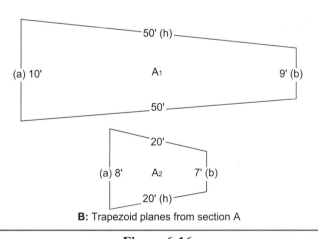

B: Trapezoid planes from section A

Figure 6-16
Calculating the volume of a trapezoidal-shaped prism

As we've seen, there are two ways to calculate *most* volumes. Notice I emphasize *most*, and here's why. The shape shown in section A of Figure 6-16 is a "trapezoidal-shaped prism." The two planes, labeled A_1 and A_2, are trapezoids. A trapezoid is a four-sided geometric shape with one pair of parallel sides and one pair of nonparallel sides. It so happens the only way to accurately calculate the volume of this shape is by using the average end area method. Do not use the cross section method. The result it gives for volume is dangerously low. Let's calculate the volume of the same trapezoidal-shaped prism by both methods and then compare the results.

Using the Average End Area Method

This time we'll work from section B in Figure 6-16. It shows only the two trapezoid-shaped planes A_1 and A_2. Notice the letters in parenthesis that appear next to the dimensions. The letters come from this formula used to find the area of a trapezoid. Here's the formula:

Area (SF) = [(a + b) ÷ 2] x h

Take another look at Figure 6-16 B and note that *a* and *b* are the parallel sides and *h* is one of the nonparallel sides in each trapezoid. Here's the math for the A_1 area calculations:

$$\text{Area A1 (SF)} = [(10 + 9) \div 2] \times 50$$
$$= [19 \div 2] \times 50$$
$$= 9.5 \times 50$$
$$= 475 \text{ SF}$$

Here's the area for A₂.

$$\text{Area } A_2 \text{ (SF)} = [(8 + 7) \div 2] \times 20$$
$$= [15 \div 2] \times 20$$
$$= 7.5 \times 20$$
$$= 150 \text{ SF}$$

Here's the formula for volume, in cubic yards:

Formula for volume by average end area

$$\textit{Volume (CY)} = \{[(\textit{Area 1} + \textit{Area 2}) \div 2] \times \textit{length}\} \div 27$$

Plug in the numbers, and you get:

$$\textit{Volume (CY)} = \{[(475 + 150) \div 2] \times 50\} \div 27$$
$$= \{[625 \div 2] \times 50\} \div 27$$
$$= \{312.5 \times 50\} \div 27$$
$$= 15{,}625 \div 27$$
$$= 578.7 \text{ CY}$$

Round off the result to 579 cubic yards.

Using the Cross Section Method

We'll work from Figure 6-16 and begin by finding these dimensions: length, width, and depth. Length is consistent, and equals 50 feet. The other two dimensions vary, so we'll find averages for both depth and width.

$$\textit{Average width (feet)} = (50 + 20) \div 2$$
$$= 70 \div 2$$
$$= 35 \text{ feet}$$
$$\textit{Average depth (feet)} = (10 + 9 + 8 + 7) \div 4$$
$$= 34 \div 4$$
$$= 8.5 \text{ feet}$$

Now we'll calculate the volume in cubic yards using this formula:

$$\textit{Volume (CY)} = (\textit{length} \times \textit{average width} \times \textit{average depth}) \div 27$$
$$\textit{Volume (CY)} = (50 \times 35 \times 8.5) \div 27$$
$$= 14{,}875 \div 27$$
$$= 550.93 \text{ CY}$$

We'll round the result off to 551 cubic yards. Now let's compare the results of the two methods of calculating volume.

- Cross section method: Total volume = 551 CY
- Average end area method: Total volume = 579 CY

The difference is 28 cubic yards. That's how much you'll underestimate the job if you use the cross section method. In this business, big mistakes come with big price tags. Always use the average end area method for this kind of calculation.

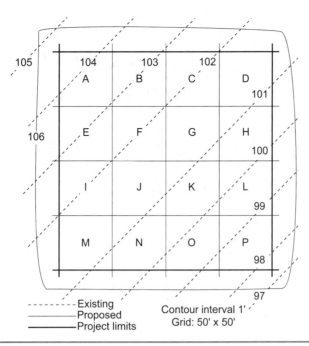

Figure 6-17
Site plan for sample project

Using Worksheets in a Take-off

We've covered the basics of doing take-offs from topo maps. It's time now to see how you can simplify, organize and streamline the process. Worksheets help you organize and simplify the whole take-off process. Constants help by making the math faster and easier. In this partial take-off, using the project layout shown in Figure 6-17, we'll use worksheets and constants. We'll start at grid square A. You'll find it easier to work from Figure 6-18. This is only the top left quadrant of Figure 6-17, slightly enlarged.

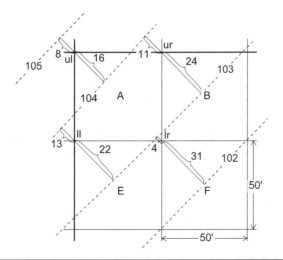

Figure 6-18
Detail, grid square A from Figure 6-17

Area Take-off from a Topo Map **85**

Individual Grid Square Area and Volume Worksheet

Figure 6-19 shows one type of worksheet I use for a quantity take-off. While it may seem extravagant to use a separate sheet for each grid square, it's a good way to get started. Paper is cheap. Mistakes are expensive. Use a little extra paper to help prevent mistakes. As your quantity take-off skill increases and you build confidence, consider combining more grid calculations on a single page. But even experienced estimators find that this type of worksheet reduces errors and makes it easier to check the work later.

The worksheet is divided into three main sections: top, middle and bottom. You use the top section to record general data such as project name/number, data and grid name. Record raw data and calculate *Existing contour* and *Proposed contour* in the middle section. Then, in the bottom section, calculate cut and fill volumes for the entire grid square.

Let's take a close look at the middle part of the worksheet now. I think you'll find following along easier if you make a copy of the worksheet. You'll find a blank copy in the back of the book. We'll be looking at the section heading by heading as well as working through some examples. We'll start with the this pair of column headings:

Existing contour / Proposed contour Copy the lines used for existing and proposed contours on your site plan. I've filled in a dashed line after *Existing contour* and a solid line after *Proposed contour*.

	Individual grid square area and volume worksheet							
	Grid square: A Area = l x w 50' x 50' = 2,500 SF							
Factors	**Existing contour (symbol: - - - - - -)**				**Proposed contour (symbol: ──────)**			
	ul	ur	ll	lr	ul	ur	ll	lr
Out	105	103	103	102				
In	104	104	104	103				
Diff	1	1	1	1				
Dist	24	35	35	35				
Out±	-8	+24	+22	+31				
In±	+16	-13	-13	-4				
Point elevation	105 - [(1/24) x 8] 105 - [0.042 x 8] 105 - 0.3 104.7	104 - [(1/35) x 11] 104 - [0.029 x 11] 104 - 0.3 103.7	104 - [(1/35) x 13] 104 - [0.029 x 13] 104 - 0.4 103.6	103 - [(1/35) x 4] 103 - [0.029 x 4] 103 - 0.1 102.9	106	106	106	106
Average elevation	(104.7 + 103.7 + 103.6 + 102.9) ÷ 4 414.9 ÷ 4 103.7				(106 + 106 + 106 + 106) ÷ 4 424 ÷ 4 106			
	Fill volume (CY)= [(average proposed elevation - average existing elevation) x grid area] ÷ 27 = [(106 - 103.7) x 2,500] ÷ 27 = [2.3 x 2,500] ÷ 27 = 5,700 ÷ 27 = 212.96 CY							

Figure 6-19
Worksheet, grid square A

Notice that below both Existing contour and Proposed contour you find the same set of four column headings. These are the names we'll use for the corner points:

ul is upper left

ur is upper right

ll is lower left

lr is lower right

In each of these columns we're going to make an educated guess about the elevation of one of the four corners.

Now let's run down the list of row headings shown in the far left column, starting with:

Factors This is a collective heading for the next six row headings.

Out and **In** Use these spaces to record the elevation of the contour lines that are outside and inside that corner. *Out* means the nearest line that's outside of the grid square itself. *In* is the nearest line that's inside the grid square. In Figure 6-18, contour line 105 is outside of grid square A at corner *ul* (upper left). Contour line 104 is inside grid square A at corner *ul*.

Most rules have exceptions and so do these. For example, take a look at corner ur in Figure 6-18. The bracketing contour lines are clearly 104 and 103. But, which one's the Out factor and which is the In factor? Notice that neither contour line is inside the grid square at this corner point. Furthermore, both contour lines do pass through grid square A elsewhere. Here's what I do:

- *In* factor = The bracketing contour line *closest* to corner *ur*. In Figure 6-18 that's the 104 foot contour line.

- *Out* factor = The bracketing contour line *furthest* away from corner *ur*. In Figure 6-18 that's the 103 foot contour line.

Or you can turn it around like this:

- *Out* factor = The bracketing contour line at corner *ur* with the *higher elevation*. In Figure 6-18 that's the 104 foot contour.

- *In* factor = The bracketing contour line at corner *ur* with the *lower elevation*. In Figure 6-18 that's the 103 foot contour.

It doesn't really matter which method you use. What does matter is consistency. Choose a way of dealing with this situation, and stick with it.

Now, let's get back to the rest of the row headings listed on the worksheet under Factors. The next one is:

Diff This is shorthand for "difference." You use this row to record the difference between *Out* and *In* factors.

Here's an example. Let's find *Diff* for corner *ul* in Figure 6-18. Remember that *Diff* = *Out* - *In*. Substitute the numbers and you get:

105 - 104 = 1

Area Take-off from a Topo Map **87**

Difference or *Diff* is always the same as the contour interval on the site plan.

Dist is short for "distance." This is the total measured horizontal distance that separates the *In* contour line from the *Out* contour line.

Figure 6-18 includes the measured distances from the corner points to each contour line. To find *Dist* for corner *ul*, for instance, all you do is add the measurements together. Here's the math:

8 + 16 = 24

Out± Use this row to record the horizontal distance you measure from the corner point to the *Out* contour line. The + and - signs show whether the change in elevation between the corner point and the contour line is positive or negative. If that seems unclear, it won't be after you follow along with these two examples taken from Figure 6-18.

Here's what we know about corner *lr*:

Out = 102

In = 103

Measured horizontal distance (102 to *lr*) = 31

At corner **lr** the **In** elevation, 103 feet, is greater than the **Out** elevation, 102 feet. That means there's a ***gain*** in elevation between corner **lr** and the 102 foot contour line. So the **Out±** factor is positive and it equals +31

The elevation change here (102 to *lr*) is positive, so:

Out± = +31

For the second example we'll use corner *ul*. Here's what we know about it:

Out = 105

In = 104

Measured horizontal distance (*ul* to 105) = 8

At corner **ul** the **Out** elevation, 105 feet, is greater than the **In** elevation, 104 feet. That means the elevation ***drops*** between corner **ul** and the 105 foot contour line. So the **Out±** factor is negative and it equals -8.

The elevation change here (105 to *ul*) is negative, so:

Out± = -8

In± Use this row to record the horizontal distance you measure between the corner point and the *In* contour line. The + and - signs serve the same purpose here as they do in the case of the *Out*± factor.

Follow along as we find the *In*± factor for corner *ul* in Figure 6-18.

Here's what we know about corner *ul*:

Out = 105

In = 104

Measured horizontal distance (104 to *ul*) = 16

At corner **ul** the **Out** elevation, 105 feet, is greater than the **In** elevation, 104 feet. That means there's a ***gain*** in elevation between the 104 foot contour line and corner **ul**. So the **In±** factor is positive and it equals +16.

These two factors are very important, so be sure you record the data correctly and use the right sign. The only way to be certain that this data's correct is to check your work. Here's how I check these factors:

1) Check the signs using the following fact. A corner point always has an *In±* factor, as well as an *Out±* factor. One of the two factors will always be negative. The other factor must be positive. If the signs match, there's an error in your work.

2) Ignoring the signs, find the sum of the two factors. The result should match the *Dist* factor for the same corner point.

Let's get back to the last two headings in the far left column on the worksheet (Figure 6-19). Both are multistep calculations.

Point elevation Calculate it for each corner point using this formula:

Point elevation (feet) = high elevation - (Diff ÷ Dist) x the negative ± factor

• • • • • • • • • • • • • •
Formula for point elevation

Earlier we saw how to find both the **In±** factor and the **Out±** factor using corner **ul** as our example. Now let's try out this formula using the data for corner **ul**.

Point elevation (feet):

= 105 - (1 ÷ 24) x 8

= 105 - (0.04 x 8)

= 105 - 0.32

= 104.7 feet

Average elevation Find the sum of the four point elevations and divide by 4. To see what this looks like, check out the *Average elevation* row on the *Existing contour* side of the worksheet for grid square A.

Perhaps you're wondering what's going on in Figure 6-19 on the *Proposed contour* side? I have to admit it looks short on data. In Figure 6-17, contour line 106 surrounds the whole project. In other words, it's flat. If it weren't, you would repeat the same calculations we just finished on the *Existing contour* side.

That leaves only the bottom section of the worksheet to cover. Let's start with a summary of the data we'll use:

■ Average existing elevation (AEE) = 103.7 feet

■ Average proposed elevation (APE) = 106 feet

■ Area grid square A, calculated in the top section = 2,500 SF

We'll begin by seeing how to tell if the excavated volume is cut or fill. Then we'll calculate the total excavation volume. To find if you're dealing with cut or fill, compare the average existing and proposed elevations. If the *existing* elevation is larger, you'll have a cut volume. If the *proposed* elevation is larger, you'll have a fill volume.

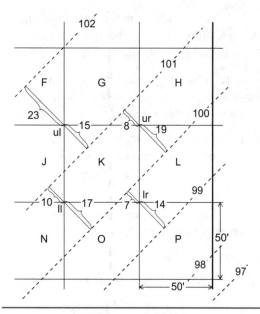

Figure 6-20
Detail, grid square K from Figure 6-17

Here's the formula you use to calculate fill volume in cubic yards:

Fill volume (CY) = [(APE - AEE) x grid square area] ÷ 27

Plug the numbers for grid square A into the formula and you get:

Fill volume (CY) = [(106.0 - 103.7) x 2,500] ÷ 27

= [2.3 x 2,500] ÷ 27

= 5,750 ÷ 27

= 212.96 CY

Now test yourself by completing a worksheet for grid square K. Make another copy of the blank form from the back of the book. You'll also find the enlarged view of K in Figure 6-20 helpful. Check your results against those shown in Figure 6-21.

	Individual grid square area and volume worksheet							
	Grid square: K Area = l x w 50' x 50' = 2,500 SF							
Factors	**Existing contour (symbol: - - - - - -)**				**Proposed contour (symbol: ─────)**			
	ul	ur	ll	lr	ul	ur	ll	lr
Out	102	100	100	99				
In	101	101	101	100				
Diff	1	1	1	1				
Dist	38	27	27	21				
Out±	-23	+19	+17	+14				
In±	+15	-8	-10	-7				
Point elevation	102 - [(1/38) x 23] 102 - [0.026 x 23] 102 - 0.6 101.4	101 - [(1/27) x 8] 101 - [0.037 x 8] 101 - 0.3 100.7	101 - [(1/27) x 10] 101 - [0.037 x 10] 101 - 0.4 100.6	100 - [(1/21) x 7] 100 - [0.048 x 7] 100 - 0.3 99.7	106	106	106	106
Average elevation	(101.4 + 100.7 + 100.6 + 99.7) ÷ 4 402.4 ÷ 4 100.6				(106 + 106 + 106 + 106) ÷ 4 424 ÷ 4 106			
	Fill volume (CY)= [(average proposed elevation - average existing elevation) x grid area] ÷ 27 = [(106 - 100.6) x 2,500] ÷ 27 = [5.4 x 2,500] ÷ 27 = 13,500 ÷ 27 = 500 CY							

Figure 6-21
Worksheet, grid square K

Some Shortcuts for Calculating Quantities

Remember the basic rule for calculating the elevation of grid squares: Total the elevation of all four corners and divide by 4. This always works, but it's not always the fastest way to get the job done. After you've gained some estimating experience, you'll learn some shortcuts. They save time, needless repetition, or are just a lot less bother. I've included a few of my best shortcuts in the next example.

The sample project is a small parking lot. Figure 6-22 is a topo map that's been made into the site plan for the project. The legend shows the contour lines and contour interval. Note the zero line, running diagonally from lower right to upper left. It connects three points where existing and proposed contour lines of the same elevation meet. You'll recall that a zero line also divides a project into an area of cut and another of fill. In Figure 6-22, left of the zero line is cut, and right of the zero line is fill. There's a grid imposed over the topo map.

For practice, I recommend that you make a photocopy of the site plan in Figure 6-22. We're going to add subcontours to the project layout. On your copy, draw existing and proposed subcontours freehand, halfway between each pair of plotted contours. The result should look like Figure 6-23. Check the added subcontours for elevations 103 and 101. These lines (existing and proposed) intersect at the zero line.

Take a look now at Figure 6-24. It shows a different system for identifying grid square corners. In this system each corner position is a number. The top right corner is 1. Move clockwise around the square, ending with number 4 at the top left corner.

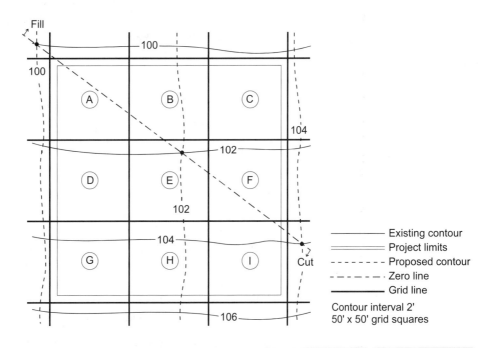

Figure 6-22
Sample project layout

Area Take-off from a Topo Map **91**

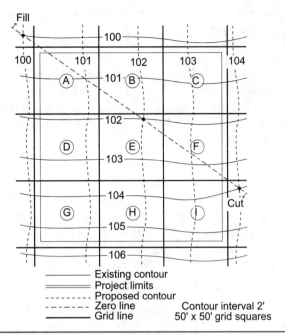

Figure 6-23
Site plan after adding the intermediate contour lines

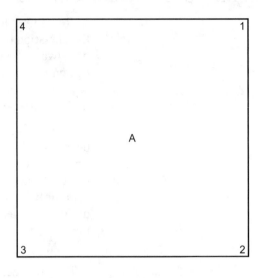

Figure 6-24
Grid square A with corner points labeled

Figure 6-25 is a completed copy of the worksheet using this method of numbering the corners. We'll use it to find the corner elevations. It's very different from Figure 6-19, so let's take a close look. At the left is a column identifying the grid squares we're using for this project. Next to the identification for each grid square are five columns and three rows to record data and calculate corner depths. Look for the headings *Element*, another name for corner point, and the corner points 1, 2, 3 and 4. Below *Element* are *Proposed*, *Existing* and *Depth*. *Proposed* and *Existing* refer, as you know, to elevations. *Depth* is their difference.

Interpolate existing and proposed corner elevations, in feet, for all nine grid squares. Compare your interpolated elevations with Figure 6-25. Then complete the worksheet by finding the difference between the two elevations and entering it in the *Depth* row.

The fill data and calculations appear in Figure 6-26. The top portion of this worksheet contains standard information: *Project, Date, By*, and "*All (cut or fill)*." This is filled in already in Figure 6-26 to read "*Fill*." The main part of the worksheet is a table with six columns. However, only three of these contain data: *Grid, Corner,* and *Total depth. Grid* refers to the grid square name. However, this list isn't complete. Three grid squares are left out: D, G, and H. Another look at the project layout in Figure 6-22 shows why. These three grid squares are entirely on the cut side of the zero line. Only grid square C is entirely on the fill side of the zero line.

Corner is the next column in Figure 6-26 that contains data. This column lists the corner points from each grid square that lie on the fill side of the zero line. Only grid square C has all four of its corner points listed in this column.

92 *Estimating Excavation*

Individual Grid Square Depth Calculations Worksheet
Job Number: 6973 **Project:** (Figure 6-22) **By:** L. Level **Date:** 06/25
Sheet 1 **of** 1   **Cut & Fill**

	Element	1	2	3	4
Grid A	Proposed	101.32	101.31	100.19	100.16
	Existing	100.11	101.54	101.86	100.22
	Depth	1.21	0.23	1.67	0.06
Grid B	Proposed	102.44	102.46	101.31	101.32
	Existing	100.2	101.75	101.54	100.11
	Depth	2.24	0.71	0.23	1.21
Grid C	Proposed	103.62	103.45	102.46	102.44
	Existing	100.17	101.73	101.75	100.2
	Depth	3.45	1.72	0.71	2.24
Grid D	Proposed	101.31	101.31	100.18	100.19
	Existing	101.54	103.48	103.5	101.86
	Depth	0.23	2.17	3.32	1.67
Grid E	Proposed	102.46	102.46	101.31	101.31
	Existing	101.75	103.5	103.48	101.54
	Depth	0.71	1.04	2.17	0.23
Grid F	Proposed	103.45	103.63	102.46	102.46
	Existing	101.73	103.47	103.53	101.75
	Depth	1.73	0.16	1.04	0.71
Grid G	Proposed	101.31	101.33	100.19	100.18
	Existing	103.48	105.40	105.53	103.5
	Depth	2.17	4.07	5.34	3.32
Grid H	Proposed	102.46	102.45	101.33	101.31
	Existing	103.55	105.40	105.43	103.48
	Depth	1.04	2.84	4.07	2.17
Grid I	Proposed	103.63	103.43	102.45	102.46
	Existing	103.47	105.31	105.29	103.5
	Depth	0.16	1.88	2.84	1.04

Figure 6-25
Depth calculations worksheet

Area Take-off from a Topo Map

		Cut and Fill Prism Calculations Worksheet			
	Project: Parking lot (Figure 6-22)		Date: 06/25 By: L. L. Level		
	All (cut or fill): Fill		Checked by: J. Jacobs		
Grid	Corner	No.	Depth	Total depth (feet)	Notes
A	1			1.21	
B	1			2.24	
	2			0.71	
	4			1.21	
C	1			3.45	
	2			1.72	
	3			0.71	
	4			2.24	
E	1			0.71	
F	1			1.72	
	2			0.16	
	4			0.71	
I	1			0.16	
Totals	13 corners			17.0*	*rounded

Figure 6-26
Fill calculations

Now look at the *Totals* line for the *Corner* column. This is where you record the number of corner points on the fill side of the zero line. In this example it's 13.

The last data column is *Total depth*, and these numbers should look familiar. They're transferred here from the *Depth* rows in Figure 6-25. Just as you'd expect, it's not a complete list. Opposite the *Totals* heading for this column, enter the sum of numbers in the *Depth* row. For our project, that's 16.95. We'll round all the depths to one decimal place, so I'll use 17.0.

The next worksheet, Figure 6-27, looks almost the same as Figure 6-26. You use it the same way, but there are important differences. All the data on this worksheet comes from the cut side of the zero line. Use the *Grid, Corner,* and *Total depth* columns just like you did for the fill calculations. This time, however, remember you're working on the cut side of the zero line.

After you've totaled the cut corners and depth, adjust them to find the average depth. Add the cut and fill corners to find the total corners. Then subtract the fill depth from the cut:

• • • • • • • • • • • •
Formula for total depth

Total depth (feet) = cut depth total - fill depth total

Total depth (feet) = 44.8 - 17.0

= 27.8 feet

Cut and Fill Prism Calculations Worksheet

Project: Parking lot (Figure 6-22) **Date:** 06/25 **By:** L. L. Level
All (cut or fill): Cut **Checked by:** J. Jacobs

Grid	Corner	No.	Depth	Total depth (feet)	Notes
A	2			0.23	
	3			1.67	
	4			0.06	
B	3			0.23	
D	1			0.23	
	2			2.17	
	3			3.32	
	4			1.67	
E	2			1.04	
	3			2.17	
	4			0.23	
F	3			1.04	
G	1			2.17	
	2			4.07	
	3			5.34	
	4			3.32	
H	1			1.04	
	2			2.84	
	3			4.07	
	4			2.17	
I	2			1.88	
	3			2.84	
	4			1.04	
Totals	23 corners + 13 corners = 36 corners			44.8* - 17.0* (fill) 27.8 feet	* rounded to one decimal place
Average depth	27.8 feet ÷ 36 corners = 0.77 feet				
Volume (CY)	(9 × 50 × 50 × 0.77) ÷ 27 = 17,325 ÷ 27 = 641.66 CY				
Round volume (full CY)	642 CY				

Figure 6-27
Cut calculations

Formula for average depth

Use the Average row to calculate average total excavation depth for the entire project:

Average depth (feet) = Total depth ÷ corner count

Average depth (feet) = 27.8 ÷ 36

= 0.77 feet

The next heading in Figure 6-27 is *Volume (CY)*. In this example the total volume of cut is greater than the total volume of fill. Their difference is the total volume of spoil to remove from the site. In the reverse situation, their difference is the total volume of fill to bring on site from elsewhere. In either case, use the following formula to calculate the volume:

Formula for total volume

Volume (CY) = # of gs x gs length x gs width x average depth ÷ 27

In this formula *gs* is short for grid square.

The key for Figure 6-22 gives the grid square dimensions. Length and width are both 50 feet. Add the other numbers and you get:

Volume (CY) = (9 x 50 x 50 x 0.77) ÷ 27

= 17,325 ÷ 27

= 641.66 CY

Round that off to full cubic yards, and you'll find the total volume of cut, less what we'll use as fill, equals 642 CY.

Use separate worksheets to calculate cut and fill until you feel comfortable using this method. Then you're ready for shortcuts — either the ones covered here or your own inventions. I use shortcuts whenever I can, and they're real time-savers. But don't jump the gun.

Let's see how you can streamline this take-off method. For starters we'll turn two worksheets (Figures 6-26 and 6-27) into one, and combine the cut and fill calculations. Figure 6-28 shows the combined worksheet, already filled in with the data from the parking lot project. You use a plus sign for fill and a minus sign for cut.

We'll also use the blank columns we didn't use in the last example to minimize the math. Here's how it works. Take a look at corner B2 in Figure 6-22. This one corner point has three other names (C3, F4, and E1). But they're all the same point, so they all have the same elevation. You don't need to list that same point four times, or calculate the same depth four times. All you do is list this point once, and then use the space in the *No.* column to indicate the multiplier and whether it's fill or cut. In Figure 6-28, find row B2. Check the data entered in the *No.* column you see "+ 4." The plus sign shows that this is fill, and 4 is the multiplier to use in the following formula:

Shortcut formula for total depth

Depth x No. = Total depth

Bring forward the *Depth* results from Figure 6-26 to the *Depth* column in Figure 6-28. Here's how the formula works with the numbers for B2:

Total depth = 0.71 x 4

2.84 feet

96 *Estimating Excavation*

Cut and Fill Prism Calculations Worksheet

Project: Parking lot (Figure 6-22)   Date: 06/25   By: L. L. Level
All (cut or fill): Shortcut   Checked by: J. Jacobs

Grid	Corner	No.	Depth	Total depth (feet)	Notes
A	1	+2	1.21	+2.42	
	2	-4	0.23	-0.92	
	3	-2	1.67	-3.34	
	4	-1	0.06	-0.06	
B	1	+2	2.24	+4.48	
	2	+4	0.71	+2.84	
C	1	+1	3.45	+3.45	
	2	+2	1.72	+3.44	
D	2	-4	2.17	-8.68	
	3	-2	3.32	-6.64	
E	2	-4	1.04	-4.16	
F	2	+2	0.16	+0.32	
G	2	-2	4.07	-8.14	
	3	-1	5.34	-5.34	
H	2	-2	2.84	-5.68	
I	2	-1	1.88	-1.88	
Totals		13 (+) + 23 (-) = 36		17.0* (fill +) - 44.8* (cut -) - 27.8 feet	* rounded to one decimal place
Average depth	-27.8 ÷ 36 = 0.77				
Volume (CY)	(9 x 50 x 50 x 0.77) ÷ 27 = 17,325 ÷ 27 = 641.66 CY				
Round volume (full CY)	642 CY spoil				

Figure 6-28
Shortcut worksheet, calculations for cut and fill

Here's another example, using corner G3. We know it's on the cut side of the zero line, and it's not a corner point for any other grid square, so under *No.* enter 1. Then carry forward *Depth* from Figure 6-26, and enter 5.34. Finish by calculating *Total depth:*

5.34 x -1 = -5.34

Repeat these calculations for each line. Then move down to the *Totals* line. In the *No.* column you'll calculate three totals:

1) Total the + items.

Area Take-off from a Topo Map **97**

2) Total the - items.

3) The sum of 1 and 2 (ignore the signs) equals the total corner count.

Your calculations for *Average* and *Volume (CY)* are the same as they were in Figure 6-27. And unless there's a math error, they'll produce the same results. This shortcut should save a lot of time. But there's a catch involved. It's easy to lose track of what's been counted and what hasn't been included in the *No.* column. Always check your work to make sure nothing has been left out or duplicated.

Finding the Volume of a Triangle

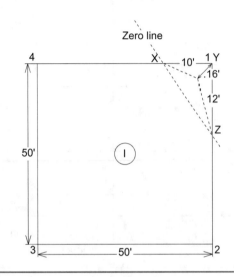

Figure 6-29
Finding the volume of a triangle

So far we've only worked with grid squares. In the real world that's not always the case. There are situations where you'll use a triangle instead of a square. A triangle, for example, is better when only a part of a grid square's area lies within the project's limits. Sometimes you need data that's more detailed or precise for a specific grid square. The best solution is to break the square into triangles (two, four or more).

Let's see how this works by calculating the volume of a triangular piece from a grid square. We'll use a portion of grid square I in Figure 6-22. Figure 6-29 is an enlarged view of grid square I. The zero line cuts through just below corner 1 in grid square I. This little triangular piece (called XYZ) is fill in a grid square that's otherwise all cut.

Here's what we know about this triangle:

- Point Y is also corner I1 (Figure 6-23), so its depth = 0.16' (Figure 6-25)

- The hypotenuse of triangle XYZ is the zero line, so depth at X and Z = 0'

- Side YZ = 12'

- Side XY = 10'

You find the triangle's volume by following these simple steps:

1) Find the average depth, using the sum of the corner depths divided by the number of corners:

   Average depth (feet) = (0.16 + 0 + 0) ÷ 3

   = 0.16 ÷ 3

   = 0.05 feet

2) Find the area of this right triangle with this formula:

   Area = base x height ÷ 2

   Area = 12 x 10 ÷ 2 = 60 SF

Formula for volume of triangle

3) Find the volume of triangle XYZ in cubic yards with this formula:

Volume (CY) = area x average depth ÷ 27

60 x 0.05 ÷ 27 = 0.11 CY

We'll round that off, and call it a cut of 0.1 CY.

Many earthwork estimators use this easy method to calculate the volume of any triangle that the zero line creates. Just be careful not to forget the rest of the grid square, after you pull the triangle out. How do you find the volume of a square that's missing a corner? The easiest way is to ignore the triangle. Just calculate the volume of the entire grid square, then subtract the triangle's volume. The result is the volume of the rest of the grid square. Let's try this out now on grid square I (Figure 6-29). Here's what we already know:

- Corner depth at 1 = 0.16'
- Corner depth at 2 = 1.88'
- Corner depth at 3 = 2.84'
- Corner depth at 4 = 1.04'

Average depth (feet) = (0.16 + 1.88 + 2.84 + 1.04) ÷ 4

= 5.92 ÷ 4

= 1.48 feet

Volume (CY) = (50 x 50 x 1.48) ÷ 27

= 3,700 ÷ 27

= 137.04 or 137.0 CY

Then subtract the volume of the triangle, 0.1 cubic yard, to find that the rest of grid square I has a volume of 136.9 cubic yards.

The Equal Depth Contour Method

There are three common ways to estimate excavation quantities from a topo map: cross sections, horizontal planes, and equal depth of equal height contour lines.

Until now, we've used the cross section method in this chapter. The horizontal plane method is worked out right on the contour map but it has three serious disadvantages. First, it's limited in its usefulness to sites where the difference between existing and proposed elevations is very large. Second, it involves even more math. Third, the results are less accurate than with the other methods. I don't use this method, and I don't recommend it for your use either.

That leaves the equal depth of equal height contour method. You use this method when conditions combine an irregularly-shaped area with a steep slope. In this situation neither of the other two methods is practical. You'll find the area of two or more segments, usually with a planimeter. Then find their average and multiply the result by the depth (normally the contour interval) between the segments.

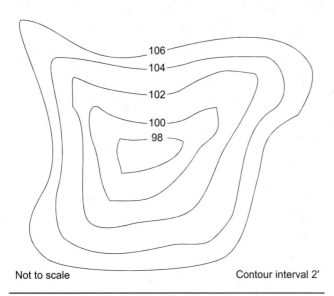

Figure 6-30
Finding the volume of a pond

Let's suppose your job site includes a very steep slope, but you don't have a planimeter. That doesn't mean you're out of luck. All you need is a little basic geometry and one property of the contour lines to make a rough estimate of the volume. Of course, the results aren't as precise as those from a planimeter. But if all you need is a rough guess, save yourself time and trouble by estimating this way.

You'll recall that all contour lines eventually close on themselves. So contour lines are circles, no matter how irregular they are. That means we can measure their length and call it a circumference. With the circumference of a circle, simple geometry produces the circle's diameter and area. The topo map gives contour interval. Combine that with a pair of consecutive areas and you have everything you need to calculate volume using the equal depth contour method. Here's an example to show how easy it really is.

Figure 6-30 shows a drained pond that's to be filled in. Your job is to make a rough estimate of how many cubic yards of material it'll take to do the job. To make it easy, we'll assume that the pond's level was lowered 2 feet at a time. This left a series of still-visible old shorelines at 2-foot intervals. In other words the shorelines are contour lines. We'll begin by measuring the length of these handy shorelines/contour lines. There are several ways to go about that. You could muck about on a muddy slope with a tape measure. A less athletic approach is using a map measuring wheel to trace the contour lines. Multiply the wheel reading by the map scale to find the actual length of the contour line. Or there's a third method. Lay a piece of string on the contour line, then measure the string.

I recommend using a worksheet, like the one in Figure 6-31. In the course of this example I'll regularly refer to the headings on this worksheet. There's nothing new or surprising in the first line. However, the headings in the second line are less familiar.

Map scale Find this data on your site plan or topo map then record it here. Figure 6-30 isn't drawn to scale. The scale I chose for Figure 6-30 is *1"= 10'*.

Contour interval You'll also get this data from your topo map or site plan. In this case it's 2 feet.

Contour line Record the elevation of the first contour line. For our example that's 106.

1) Select, and mark on your topo map, a start point on the 106 foot contour line.

2) Place one end of the string on that point, then lay the string directly on top of the 106 foot contour line.

3) Follow the contour line as closely as possible throughout its course until you return to start.

	Equal Depth Contour Volume Worksheet				
Project: Pond fill-in (Figure 6-30) Date: 6/30 By: L. Level					
Map scale: 1" = 10' Contour interval: 2'					
Contour line	Length (in.)	Circum. (ft.)	Diameter (ft.)	Area (SF)	Volume (CY)
106	18	180	57.3	2,578.5	161.8
104	15	150	47.7	1,788.8	
104	15	150	47.7	1,788.8	85.1
102	8	80	25.5	510	
102	8	80	25.5	510	23.6
100	4	40	12.7	127	
100	4	40	12.7	127	5.4
98	1.5	15	4.8	18	
				Total volume (CY)	275.9

Figure 6-31
Worksheet for Figure 6-30

4) Mark that point on the string.

5) Straighten the marked string and measure the length you've marked on it with a ruler.

Length (in.) Use this space to record the length of the string. For this example *Length* is 18.

The actual length of the contour line equals *length* times *map scale*.

Circum. (ft.) You'll use this column to record the circle's actual circumference. For the example, the map scale is 1" = 10' and the length is 18 inches, so the circumference is 180 feet.

Diameter (ft.) Use this column to record the circle's calculated diameter, after rounding it to one decimal place. Here's the formula:

• • • • • • • • • • • •
Formula for diameter of circle

Diameter of a circle = Circum. ÷ π (pi)

Assume pi (π) = 3.1416, and we find:

Diameter = 180 ÷ 3.1416

= 57.2956

For the example that's 57.3 feet

Area (SF) Record the area, rounded to one decimal place. Find the area of the circle using the following formula:

• • • • • • • • • • • •
Formula for area of circle

Area (SF) = (diameter x circumference) ÷ 4

Area (SF) = (57.3 x 180) ÷ 4

= 10,314 ÷ 4

= 2,578.5 SF

Area Take-off from a Topo Map

Finally, repeat steps 1 through 8 using the 104 foot contour line.

Volume (CY)

Here's the formula we'll use to calculate volume in cubic yards:

Formula for volume of equal depth contours

Volume (CY) = {[(area 1 + area 2) ÷ 2] x contour interval} ÷ 27

Volume (CY) = {[(2,578.5 + 1,788.8) ÷ 2] x 2} ÷ 27

= {[4,367.3 ÷ 2] x 2 } ÷ 27

= {2,183.65 x 2} ÷ 27

= 4,367.3 ÷ 27

= 161.75 CY

You'll note that Figure 6-31 is completely filled in. Think of this as an opportunity to test yourself. Do the calculations yourself for contour lines 102, 100 and 98. Then check your answers against mine.

Total volume (CY) On this line you'll simply record the sum of the Volume (CY) column. This is your rough estimate of the excavation volume for the project. For the pond job in Figure 6-31, the total volume works out like this:

161.8 + 85.1 + 23.6 + 5.4 = 275.9 CY

Round that to full cubic yards, and call it a total of 276 cubic yards of fill.

Of course, this method works just as well when you flip the pond inside out, and make it a hill. Picture Figure 6-30 with the elevations beginning at 98 and ending at the center with 106. If you think you need the practice feel free to repeat all the calculations. However, there isn't any need to do so. Here's why. The total cut to level this hill to the 98 foot contour is 276 cubic yards.

You'll always overestimate the actual volume when you use the equal depth contour method. That's because the contour line isn't a perfect circle. The more regular it is, the more accurate your results. The more irregular it is, the more inaccurate your results.

We've certainly covered much ground in this chapter. But I hope you've followed along in the examples. It's one of the ways I try to make difficult concepts easier. Before going on to the next chapter, I recommend reviewing anything that seems a bit hazy. The material we've covered so far is your foundation, so be sure it's solid before you start Chapter 7.

7
Irregular Regions & Odd Areas

Up to this point, we've only worked with areas that had simple shapes. That makes their area easy to calculate. Unfortunately, most sites you'll work with won't be nice squares or rectangles — they'll be odd-shaped.

Job sites with odd shapes usually also have other challenges, including sharp changes in grade or in contour direction. These conditions make it difficult to find area by the normal methods. But no matter how irregular the site, you'll always be able to find its area by breaking it down into simple shapes.

In this chapter you'll learn several different ways to calculate the area of an irregular shape. To find these areas, you'll need to use a few mathematical formulas which may look a little unfamiliar. But don't let them put you off. After just a little practice and a few calculations, you'll find they're not so difficult.

We'll begin by looking at Figure 7-1. If you look only at the corner elevations, it seems there's no earthwork needed here. All four corners of the grid square have the same elevation. But that's not the whole story. What about the contour lines inside the grid square? They tell you there's a 4-foot high mound inside the grid square. If the job specs include leveling this area, you need to know the volume of this mound.

Here are four ways of finding the volume of the mound:

1) Use a planimeter as discussed in Chapter 4.

2) Subdivide your grid system by breaking it into smaller squares as shown in Figure 7-2. We discussed this method in Chapter 6.

3) Use compensating lines to approximate the shape's outline and to break it down into simpler shapes made up of straight lines.

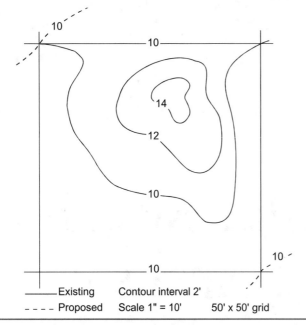

Figure 7-1
A grid square with the same elevation at all four corners

Figure 7-2
Figure 7-1 with a smaller grid

4) Use an odd-shaped grid system or the multiplane method. You can use any geometric shape for a grid system — if you also know how to calculate its area.

In this chapter we'll learn how to use both the compensating line method and the multiplane method. Let's start with the compensating line method.

Finding Area Using Compensating Lines

What's a compensating line? The compensating line of a curved line is just a straight line that's as close as possible to the curved line. To make a compensating line from any curved line, you take pieces of the curved line and replace them with straight lines. If the curved line doesn't turn very much, you can replace it with a long straight line. If it makes a sharp turn, you'll need a shorter straight line.

In Figure 7-3 we've put compensating lines around the 10 foot contour line. If a section of the contour line doesn't have many turns, or if they're very gradual, we've used fairly long compensating lines. Two good examples are lines AG and GF in Figure 7-3. Where there are many turns, or very sharp turns, in the contour line, the compensating lines are shorter, like line EF. The accuracy of your area and volume estimates depends on how closely your compensating lines follow the contour line.

After drawing the compensating lines, mark a point in the middle of the highest elevation contour. In Figure 7-3 it's point H at the center of the 14 foot contour. Connect each end of each compensating line to the center point. This

104 *Estimating Excavation*

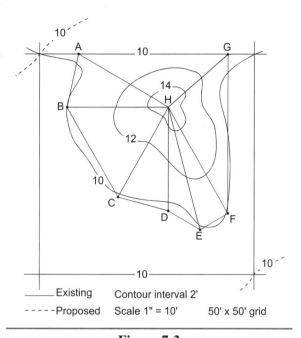

Figure 7-3
Add the compensating lines to Figure 7-1 to divide the mound into seven oblique triangles

divides the mound into seven triangles. We'll figure out the area of each of these triangles and add the areas together to find the total area in the elevation 10 contour.

All seven of the triangles in Figure 7-3 are oblique triangles; they have no angles that measure exactly 90 degrees. Triangles that include a 90-degree angle are right triangles. Here's the formula you use to find the area of an oblique triangle:

Area = (base ÷ 2) x height

Now let's define height and base. The height of an oblique triangle is the length of a perpendicular line drawn from one angle to the opposite side. The base is the side of the triangle that forms a 90-degree angle with the perpendicular. Take a look now at the oblique triangle ABC in Figure 7-4. Notice that the dashed line from angle A forms two 90-degree angles with side BC. The dashed line is the height and side BC is the base. But why isn't side AB the base? Although you can draw a line from angle C to side AB, that line won't be perpendicular to side AB. This is also true for the line you could draw from angle B to side AC. There's only one possible height and base in any oblique triangle.

Now let's calculate the area for the seven oblique triangles in Figure 7-3. Try doing the calculations for all the triangles except EHF, and check them against my worksheet in Figure 7-5.

What about EHF? We'll use a different method to find its area. That's because it's difficult to draw an accurate perpendicular in such a narrow triangle. The method we'll use to find the area of EHF works with just the measured lengths of the triangle's sides. Here are the formulas we'll use to find the area of EHF:

$$S = (EH + FH + EF) \div 2$$
$$R = \sqrt{[(S - EH) \times (S - FH) \times (S - EF)] \div S}$$
$$Area = R \times S$$

Compared to the single formula we used earlier, this looks terribly complicated. It really isn't as bad as it looks. But I'm sure you can see why I didn't calculate the areas for all seven triangles this way. You'll find all three of these formulas as well as all of the math in Figure 7-6, the area calculations worksheet for triangle EHF.

Look at the last calculation in Figure 7-5. The total area is the sum of areas of the seven triangles. That comes to 1,251 after rounding.

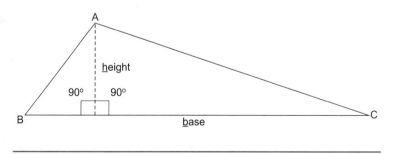

Figure 7-4
Finding the base and height dimensions in an oblique triangle

Irregular Regions & Odd Areas **105**

Triangle: AHB		
BH = b	26	**Area = ½(b x h)** (26 x 14) ÷ 2 364 ÷ 2 182
AH	25	
AB	15	
h	14	

Triangle: BHC		
BH = b	26	**Area = ½(b x h)** (26 x 17) ÷ 2 442 ÷ 2 221
BC	18	
CH	24	
h	17	

Triangle: CHD		
CD = b	17	**Area = ½(b x h)** (17 x 22) ÷ 2 374 ÷ 2 187
CH	24	
DH	22	
h	22	

Triangle: DHE		
EH = b	29	**Area = ½(b x h)** (29 x 7.5) ÷ 2 217.5 ÷ 2 108.75
DE	11	
DH	22	
h	7.5	

Figure 7-5
Worksheet — areas of oblique triangles

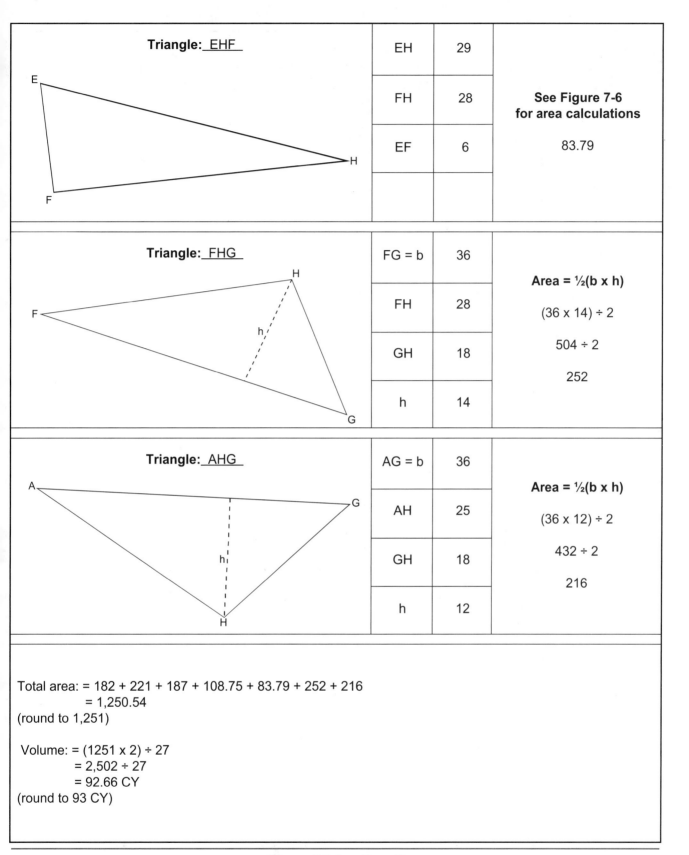

Total area: = 182 + 221 + 187 + 108.75 + 83.79 + 252 + 216
           = 1,250.54
(round to 1,251)

Volume: = (1251 x 2) ÷ 27
        = 2,502 ÷ 27
        = 92.66 CY
(round to 93 CY)

Figure 7-5 (continued)
Worksheet — areas of oblique triangles

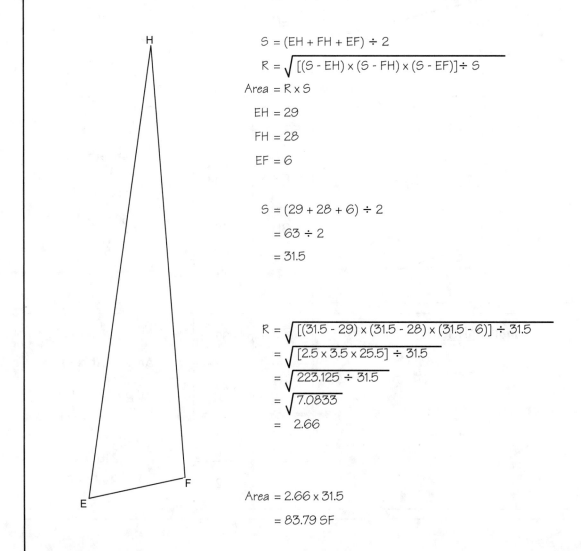

Figure 7-6
Alternate method of area calculation using oblique triangle EHF as the example

Finding Volume Using Total Area and Average Depth

To find the average depth, try thinking of the mound as a big meringue pie. Imagine you cut the pie in pieces along the lines in Figure 7-3. Figure 7-7 shows a piece cut out along the lines of triangle BHC.

If you take the piece out, it'll look like Figure 7-8. Let's use this figure to see how to find the average depth. The topo map shows that the mound slopes evenly on all sides. That means we only need to find average depth once. Points B and C have the same elevation, 10 feet. At point H the elevation is 14 feet. Here's the math:

Average depth = (14 - 10) ÷ 2

= 4 ÷ 2

= 2 feet

That's all the data we need to find the volume of the mound using this formula:

• • • • • • • • • • • • • •
Formula for volume of a mound

Volume (CY) = *(average depth x total area) ÷ 27*

Volume (CY) = (2 x 1,251) ÷ 27

= 2,502 ÷ 27

= 92.67

We'll round that off and call the volume of the mound 93 cubic yards.

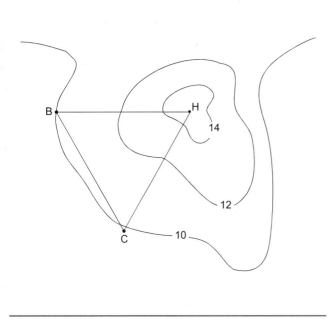

Figure 7-7
Triangle BHC

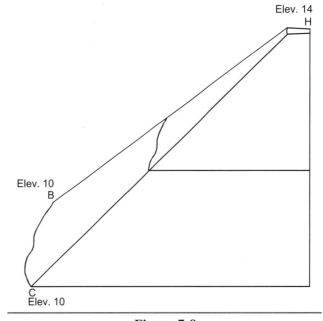

Figure 7-8
Triangle BHC "removed" to find average depth

Irregular Regions & Odd Areas **109**

Finding Volume Using Compensating Lines with a Coordinate System

Another way to find the area of an irregular area combines compensating lines with a coordinate system. Land surveyors often use this method. Again, it may seem difficult, especially the mathematical formula you use, so we'll go through it step-by-step.

Step 1

Trace the boundaries of the area in question from the original topo map or scaled site plan onto graph paper. Be careful to choose graph paper that has the same scale as the scale used on your topo map or plan. Here's an example using section A in Figure 7-9 — my tracing of the 10 foot contour line from Figure 7-1. We'll pick up the scale of 1" = 10' from Figure 7-1. Now let's suppose you have the following three sizes of graph paper:

- 1" = 5 squares
- 1" = 8 squares
- 1" = 10 squares

Which graph paper should we use? Here's a hint. There's only one wrong answer. The most obvious answer is paper with 10 squares to the inch. It makes a perfect match with the scale used in Figure 7-1. If we make the tracing onto this graph paper, one square equals 1 foot. The other right answer is to make the tracing on paper with five squares to the inch. Then each square would equal 2 feet. In Figure 7-9, I used five squares to the inch.

Step 2

Now we'll add two reference lines, one horizontal and one vertical, to the tracing. Figure 7-9 shows the usual placement for the reference lines. But notice that I said this is the "usual" placement. That means you're free to change their locations to suit the situation or yourself.

Step 3

Next we'll add unit divisions to the reference lines and label the major divisions. Always label the zero on both reference lines. Since one square equals 2 feet, five squares equal 10 feet. That's the major division used for the graphs in Figures 7-9, 7-10 and 7-11.

Step 4

Now add compensating lines to the tracing of the 10 foot contour. Mark a point each time the direction of the compensating lines changes. Then label the points using any system that makes sense to you. I prefer to use letters to avoid any chance of confusion. Here's the only rule: Name the points in order. Start with whatever point you like. Move around the contour in whichever direction you like, naming points until you're back at the start point. In Figure 7-9, I ended up with a total of seven points labeled A through G.

A: The 10 foot contour line with compensating lines and reference lines

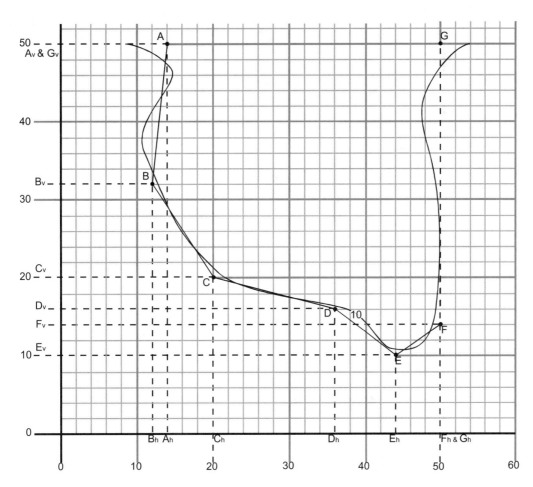

B: Worksheet and area calculations for the 10 foot contour line

Point	A	B	C	D	E	F	G	
h (scale distance)	14	12	20	36	44	50	50	
v (scale distance)	50	32	20	16	10	14	50	

Scaled distance

Formula

$$\text{Area (SF)} = [(N_v \times N+1_h) + (N+1_v \times N+2_h) + \ldots (N+N_v \times N_h) - (N_h \times N+1_v) - (N+1_h \times N+2_v) - \ldots (N+N_h \times N_v)] \div 2$$

Calculations

Area (SF) = [(50 x 12) + (32 x 20) + (20 x 36) + (16 x 44) + (10 x 50) + (14 x 50) + (50 x 14) - (14 x 32)
 - (12 x 20) - (20 x 16) - (36 x 10) - (44 x 14) - (50 x 50) - (50 x 50)] ÷ 2

= [600 + 640 + 720 + 704 + 500 + 700 + 700 - 448 - 240 - 320 - 360 - 616 - 2,500 - 2,500] ÷ 2

= -2,420 ÷ 2 (ignore the minus sign)

= -1,210 SF (ignore the minus sign)

Figure 7-9
Finding volume by the coordinate system using the 10 foot contour line

Irregular Regions & Odd Areas

A: The 12 foot contour line with compensating lines and reference lines

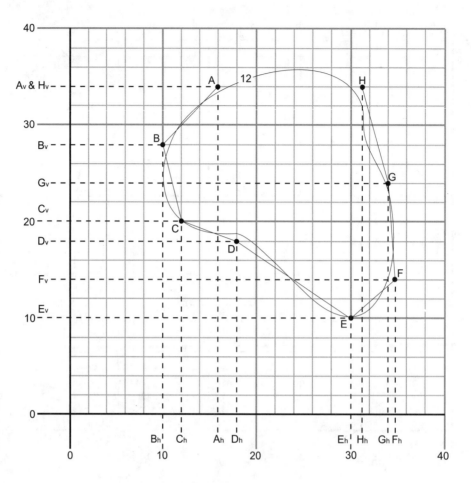

B: Worksheet and area calculations for the 12 foot contour line

Scaled distance

Point	A	B	C	D	E	F	G	H
h (scale distance)	16	10	12	18	30	35	34	31
v (scale distance)	34	28	20	18	10	14	24	34

Formula

Area (SF) = $[(N_v \times N + 1_h) + (N + 1_v \times N + 2_h) + ... (N + N_v \times N_h) - (N_h \times N + 1_v) - (N + 1_h \times N + 2_v) - ... (N + N_h \times N_v)] \div 2$

Calculations

Area (SF) = [(34 × 10) + (28 × 12) + (20 × 18) + (18 × 30) + (10 × 35) + (14 × 34) + (24 × 31) + (34 × 16)
  − (16 × 28) − (10 × 20) − (12 × 18) − (18 × 10) − (30 × 14) − (35 × 24) − (34 × 34) − (31 × 34)] ÷ 2

= [340 + 336 + 360 + 540 + 350 + 476 + 744 + 544 − 448 − 200 − 216 − 180 − 420 − 840 − 1156 − 1054] ÷ 2

= −824 ÷ 2 (ignore the minus sign)

= −412 SF (ignore the minus sign)

Figure 7-10
Using the 12 foot contour line

112 *Estimating Excavation*

A: The 14 foot contour line with compensating lines and reference lines

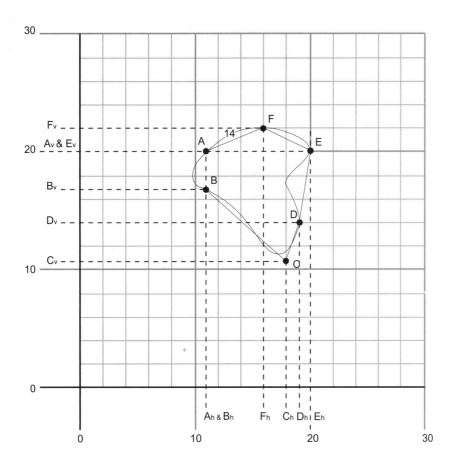

B: Worksheet and area calculations for the 14 foot contour line

Scaled distance

Point	A	B	C	D	E	F		
h (scale distance)	11	11	18	19	20	16		
v (scale distance)	20	17	11	14	20	22		

Formula

Area (SF) = [(N_v x N + 1_h) + (N + 1_v x N + 2_h) + ... (N + N_v x N_h) - (N_h x N + 1_v) - (N + 1_h x N + 2_v) - ... (N + N_h x N_v)] ÷ 2

Calculations

Area (SF) = [(20 x 11) + (17 x 18) + (11 x 19) + (14 x 20) + (20 x 16) + (22 x 11) - (11 x 17) - (11 x 11)
 - (18 x 14) - (19 x 20) - (20 x 22) - (16 x 20)] ÷ 2

= [220 + 306 + 209 + 280 + 320 + 242 - 187 - 121 - 252 - 380 - 440 - 320] ÷ 2

= -123 ÷ 2 (ignore the minus sign)

= -61.5 SF (ignore the minus sign)

Figure 7-11
Using the 14 foot contour line

Step 5

Go back to your first point (in Figure 7-9 that's point A) and draw a horizontal line from point A to the vertical reference line. Our line meets the vertical reference line right at the major division labeled 50. Let's call the intersection Av (A's vertical coordinate). So Av equals 50 feet.

I like to use solid lines for the reference lines and a dashed line to connect points and reference lines. That style's used in the graphs shown in Figures 7-9, 7-10 and 7-11. You can differentiate these lines any way you like. Drawing all the lines you add with a colored pencil is one method.

Let's return to point A now and add another line. This time we'll draw a vertical line from point A to the horizontal reference line. Label the intersection Ah (A's horizontal coordinate). Notice that Ah falls on the second division to the right of 10. We know that each division is 2 feet so Ah equals 10 + 4 or 14 feet.

Repeat this process, drawing lines from each point to both reference lines, working your way around the contour line point by point, until you return to your starting point.

Step 6

We're ready now to move from the graph section in Figure 7-9 to the worksheet. The worksheet's divided into three sections: Scaled distance, Formula, and Calculations. Use the first section to record horizontal and vertical values for each point as you read them off the scaled reference lines.

The second section gives the formula you use to find the area in square feet.

Formula for area using compensating lines

$Area\ (SF) = [(N_v \times N{+}1_h) + (N{+}1_v \times N{+}2_h) + \ldots (N + N_v \times N_h) - (N_h \times N{+}1_v) - (N{+}1_h \times N{+}2_v) - \ldots (N + N_h \times N_v)] \div 2$

At first glance this formula may look like a lot of gibberish. But don't give up. Read through the following definitions and the formula starts to make a lot more sense.

- N = the first in series of variables
- N_h = horizontal coordinate of a variable
- N_v = vertical coordinate of a variable
- $N{+}1$ = the next variable in a series of variables
- $N{+}1_h$ = the horizontal coordinate of the next variable in a series
- $N{+}1_v$ = the vertical coordinate of the next variable in a series
- $N{+}2$ = the third variable in a series
- …. = continues the sequence within a series of variables
- $N{+}N$ = the final variable in a series, of infinite length

To use this formula, break it down into smaller, more manageable parts. That's the purpose of the parentheses ( ) and brackets [ ] already included in the formula. Here's what they tell you to do:

- Do the multiplication first. Each multiplication operation is enclosed by a set of parentheses ( ).

114 *Estimating Excavation*

- Look for the brackets next. They set off a long string of addition and subtraction that you do in sequence.
- Finally, do the division.

Now take a look at the Calculations section in Figure 7-9 to see the formula in action. The first two lines are a single long equation. There just isn't enough room to string it all out on a single line. Where did this huge equation come from? It's the result when you replace the variables in the formula with the actual horizontal and vertical distances for each point. That's easy to do in just one step, after a bit of practice. But you're new at this so we'll do the replacement in two parts. Breaking this step into two parts means I can show you exactly where every number in the equation comes from. First we'll replace all of the N's in the formula with actual point references for the 10 foot contour line. Here's the result:

Area (SF) = [$(A_v \times B_h) + (B_v \times C_h) + (C_v \times D_h) + (D_v \times E_h) + (E_v \times F_h) + (F_v \times G_h) + (G_v \times A_h) - (A_h \times B_v) - (B_h \times C_v) - (C_h \times D_v) - (D_h \times E_v) - (E_h \times F_v) - (F_h \times G_v) - (G_h \times A_v)$] ÷ 2

Next we'll replace all the point references with the corresponding horizontal or vertical distance recorded in the Scaled distance section of the worksheet. Here's the result for the 10 foot contour line:

Area (SF) = [(48 x 11) + (34 x 20) + (18 x 36) + (14 x 46) + (8 x 50) + (12 x 50) + (48 x 15) - (15 x 34) - (11 x 18) - (20 x 14) - (36 x 8) - (46 x 12) - (50 x 48) - (50 x 48)] ÷ 2

Now simply work through the math. First, find the parenthesis and do all the multiplying. Second, find the brackets and do the string of addition and subtraction. If the result's a negative number, as it is in Figure 7-9, just ignore the minus sign. Third, divide the result by 2. The result is the area inside the contour line. For the 10 foot contour line in Figure 7-9 the area comes out to 1,210 square feet.

Step 7

Repeat the first six steps for each contour line. In the case of the mound in Figure 7-1, there are two more contour lines. Figures 7-10 and 7-11 show the calculations for the 12 and 14 foot contour lines. Their areas are 412 and 61.5 square feet, respectively.

We'll use the area within each of the three contour lines and the average end area method to calculate volume.

Step 8

Find the average area between adjacent contour lines. In the case of our sample mound there are two average areas to calculate. First, the average of the areas within the 10 foot and 12 foot contour lines:

Average area = (1,210 + 412) ÷ 2

= 1,622 ÷ 2

= 811 SF

Irregular Regions & Odd Areas **115**

And second the average of the areas within the 12 foot and 14 foot contour lines:

Average area = (412 + 61.5) ÷ 2
= 473.5 ÷ 2
= 236.75 SF

Step 9

Multiply the average areas by depth and divide by 27 to find volume in cubic yards. Depth equals contour interval and that's 2 feet for our sample mound. Here's the math for the volume between the 10 foot and 12 foot contour lines:

Volume (CY) = (811 x 2) ÷ 27
= 1,622 ÷ 27
= 60.07 CY

After rounding to full cubic yards that comes to 60 cubic yards.

Next find the volume between the 12 foot and 14 foot contour lines:

Volume (CY) = (236.75 x 2) ÷ 27
= 473.5 ÷ 27
= 17.54 CY

After rounding to full cubic yards that comes to 18 cubic yards.

Step 10

To find the total volume simply find the sum of the volumes. Our sample mound's total volume for the area between the 10 foot and the 14 foot contour lines is 78 cubic yards (60 + 18 = 78).

This volume assumes that above the 14 foot contour line the mound is more or less flat. If this is true it's safe to leave it out of your estimate. The volume of material is too small to have an effect on your estimate. But suppose the slope of the mound continues upward from the 14 foot contour line (without reaching the 16 foot contour line)? In that case it's wise to include this area in your volume estimate.

Let's use the sample mound and see how you find a volume for an area like this using the average end area method. Above the 14 foot contour line, the next contour line, if there were one, would be at 16 feet. Since there is no 16 foot contour line, the area inside is zero. We already know the area inside the 14 foot contour line is 61.5 SF. Assuming a depth of 1 foot, find the volume of the area above the 14 foot contour line:

Average area = (61.5 + 0) ÷ 2
= 61.5 ÷ 2
= 30.75 SF

Volume (CY) = (30.75 x 1) ÷ 27
= 30.75 ÷ 27
= 1.1 CY

We'll round that off to 1 cubic yard and add it to our earlier total. The adjusted total volume of the mound is 79 cubic yards.

You'll notice that there's a fairly large difference in the volume of the mound using the two methods. The first thing to remember is that any method is only as accurate as the placement of the points. The more points you use at even small changes of direction, the more accurate the results. Keep this in mind when you're choosing which method to use. First, you have to decide how accurate you need the results to be. Sometimes you're just looking for a ballpark figure. That would be close enough if you're just looking for a borrow pit along the project. You don't need a high degree of accuracy to find if a certain area contains the amount of material you need. But if you're working with a small area where drainage or site size restrictions are involved, you need to be more accurate.

Either of the two methods might be the best in certain situations. I prefer the compensating line method when the direction changes aren't close together and there's room to draw the compensating lines and interior triangles. When the work area is smaller, or the contour lines make drastic and frequent direction changes, the coordinate system work best.

Finding Volume Using the Trapezoidal Rule

Imagine trying to find the volume of an area with lots of twists and turns using compensating lines. If you drew a new line and a triangle for each small curve, you'd soon have too many to deal with. For a very irregular shape, that method is just too cumbersome to be practical. Instead, you can use the Trapezoidal Rule to find the area of this sort of irregular shape.

To use this method, you begin by dividing the area into strips of equal width. Then measure the length of each strip. The strips' lengths vary with the shape of the area. Then you use the following formula to solve for area in square feet:

Formula for volume using Trapezoidal Rule

$Area\ (SF) = d \times [½ \times (y_0 + y_n) + y_1 + y_2 + y_3 + \ldots y_{n-1}]$

Where:

- d is the width of each piece
- y_0 is the length of the first line
- y_n is the length of the last line
- y_1 is the length of the second line
- y_2 is the length of the third line
- y_{n-1} is the length of the next-to-last line
- n is the number of pieces

Let's work through two examples to see how this works, beginning with the rectangle shown in Figure 7-12. First we'll divide the rectangle into four strips of equal width. Each strip is 15 feet wide. We won't measure the strips because we know they're all 10 feet long. So in Figure 7-12, n equals 4, d equals 15, and y equals 10:

$$\text{Area (SF)} = 15 \times [½ \times (10 + 10)] + 10 + 10 + 10]$$
$$= 15 \times [½ \times 20] + 10 + 10 + 10$$
$$= 15 \times [10 + 30]$$
$$= 15 \times 40$$
$$= 600 \text{ SF}$$

For the second example, we'll use a situation that's a bit more realistic. Take a look at Figure 7-13. This topographic map shows a small lake that's surrounded by a 4-foot-high berm. The lake has a uniform depth of 3 feet and the owner wants the lake filled in and leveled off. The owner wants the berm material used for the fill.

We'll have to do several calculations to find out if the berm contains enough material:

- the volume of the lake
- the volume of the earthen berm
- the difference between these two volumes

But before we start, there's a point I want to make about Figure 7-13. You'll notice that Figure 7-13 has two 100 foot contour lines and two 104 foot contour lines. It is very important for us to know which 100 foot or 104 foot contour line is which. In this example I've accomplished this by calling the

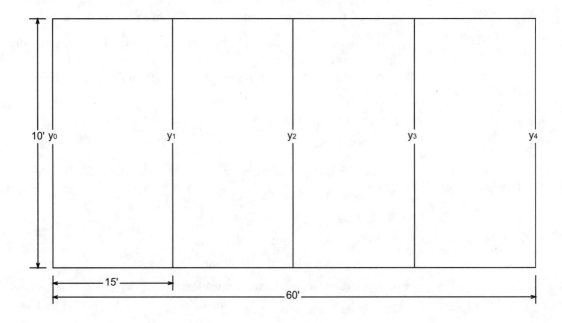

Figure 7-12
Using the Trapezoidal Rule

118 *Estimating Excavation*

inner 100 and 104 foot contour lines, the contour lines associated with the lake and located inside of the berm, inside contour lines. Meanwhile, the outer 100 and 104 foot contour lines, the contour lines associated with the berm and located outside of the lake, are outside contour lines. Clearly, there are many other ways to differentiate the two sets of contour lines from one another. Your main goal is knowing, at a glance and beyond any doubt, exactly which contour line you're working with in each worksheet or contour line tracing. Experiment and find a system that works for you and then stick with it.

Getting back to Figure 7-13, let's see how you use the Trapezoidal Rule to find the area each contour line encloses. You work with one contour line at a time, repeating these six steps for each contour line:

1) Trace the contour line, noting the scale used.
2) Divide the area into labeled strips of equal width.
3) Record the standard width you use for the strips.
4) Measure the length of each strip.
5) Record these lengths on the worksheet.
6) Use the Trapezoidal Rule to calculate the area.

Now let's try out the steps by finding the area of the inside 100 foot contour line. This contour line also represents the surface area of the lake. We'll use Figure 7-14, a traced copy of the inside 100 foot contour line, for steps 1 through 4. Then we'll use Figure 7-15, the inside 100 foot contour line worksheet, for steps 5 and 6.

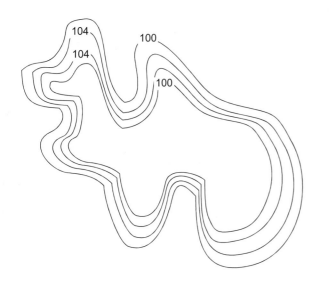

Figure 7-13
Topographic map of a small lake surrounded by a 4-foot-high berm

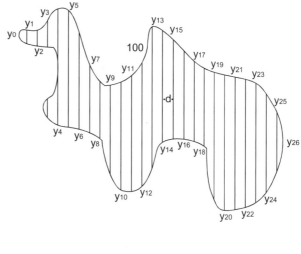

Figure 7-14
Inside 100 foot contour line with parallel lines for Trapezoidal Rule

Irregular Regions & Odd Areas **119**

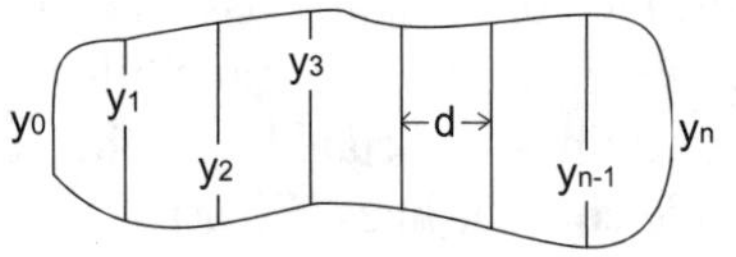

						yo = First line yn = Last line {y26} yn-1 = Next to last line {y25} d = Distance between lines {10'}		
Trapezoidal Rule								
Area = d x [½ x (y₀ + yₙ) + y₁ + y₂ + y₃ + ...yₙ₋₁]								
Line number		y_0	y_1	y_2	y_3		y_4	y_5
Scale distance		6	23	27	30 + 28 = 58		40 + 50 = 90	133
Line number		y_6	y_7	y_8	y_9		y_{10}	y_{11}
Scale distance		80	62	55	55		64	112
Line number		y_{12}	y_{13}	y_{14}	y_{15}		y_{16}	y_{17}
Scale distance		126	145	116	86		64	57
Line number		y_{18}	y_{19}	y_{20}	y_{21}		y_{22}	y_{23}
Scale distance		59	67	75	133		132	124
Line number		y_{24}	y_{25}	y_{26}				
Scale distance		110	83	22				
Calculations for: inside 100' contour line								

Area = 10 x [½ x (6 + 22) + 23 + 27 + 58 + 90 + 133 + 80 + 62 + 55 + 55 + 64 + 112 + 126 + 145 + 116 + 86 + 64 + 57 + 59 + 67 + 75 + 133 + 132 + 124 + 110 + 83]
= 10 x [½ x (6 + 22) + 2,113]
= 10 x [½ x 28 + 2,113]
= 10 x [14 + 2,113]
= 10 x 2,127
= 21,270 SF

Figure 7-15
Trapezoidal Rule worksheet and area calculations for the *inside* 100' contour line

Figure 7-14 is already divided into 27 labeled lines. At the lower left, you'll find the scale, the value of d and the name of the contour line. Now take a look at lines y₃ and y₄. Notice what makes these two lines different from the other lines? They both consist of two pieces. Whenever a line has multiple parts, you measure the length of each part. Record each part's length on your worksheet and find their sum. You'll use their total length in the equation and calculations.

How the Avoid the Trapezoidal Rule's Biggest Pitfall of All

The area calculations for the Trapezoidal Rule are all quite simple. But that doesn't mean it's foolproof. Be careful not to use the length of yₙ (the last line) twice in the area calculations. Back at the very beginning of the formula you averaged the lengths of the first and last lines. Don't forget that. It is surprisingly easy to forget and then here's what happens. You'll plug yₙ in, for the second time, at the very end of the equation. Fortunately, there's an easy way to check your work for this error and all it takes is one quick glance. Compare the last number in the addition string with the length you recorded for y_n. Are the two numbers the same? If y_n and y_{n-1} really are the same length, that's okay. Otherwise you just saved yourself from counting the same line twice.

Use Figures 7-16 and 7-17 to follow along with the area calculations for the inside 104 foot contour line. For the outside 104 foot contour use Figures 7-18 and 7-19. Then for the outside 100 foot contour use Figures 7-20 and 7-21.

Figure 7-22 shows all of the math used to find the lake and the berm volumes. In the last section of Figure 7-22, Fill volume excess (+)/shortfall (-), notice that there's a shortfall of 300 cubic yards. Obviously, you can't complete the job using only the material in the berm. Finding 300 cubic yards of compatible fill material and importing it takes time and costs you money. Be sure to consider and include costs like this shortfall in your estimates. That way you'll never end up holding the bag.

Coming up, in Chapter 8, how to use shrink/swell factors to make your earthwork estimates even more accurate.

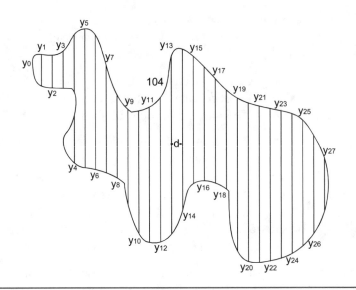

Figure 7-16
Inside 104 foot contour line with parallel lines for Trapezoidal Rule

y_0	= First line				
y_n	= Last line {y_{27}}				
y_{n-1}	= Next to last line {y_{26}}				
d	= Distance between lines {10'}				

Trapezoidal Rule

Area = d × [½ × (y_0 + y_n) + y_1 + y_2 + y_3 + ...y_{n-1}]

Line number	y_0	y_1	y_2	y_3	y_4	y_5
Scale distance	8	29	36	40 + 30 = 70	45 + 49 = 94	150
Line number	y_6	y_7	y_8	y_9	y_{10}	y_{11}
Scale distance	150	100	75	70	72	125
Line number	y_{12}	y_{13}	y_{14}	y_{15}	y_{16}	y_{17}
Scale distance	168	175	166	140	102	75
Line number	y_{18}	y_{19}	y_{20}	y_{21}	y_{22}	y_{23}
Scale distance	72	75	78	144	146	145
Line number	y_{24}	y_{25}	y_{26}	y_{27}		
Scale distance	137	124	101	39		

Calculations for: inside 104' contour line

Area = 10 × [½ × (8 + 39) + 29 + 36 + 70 + 94 + 150 + 150 + 100 + 75 + 70 + 72 + 125 + 168 + 175 + 166 + 140 + 102 + 75 + 72 + 75 + 78 + 144 + 146 + 145 + 137 + 124 + 101]
= 10 × [½ × (8 + 39) + 2,819]
= 10 × [½ × 47 + 2,819]
= 10 × [23.5 + 2,819]
= 10 × 2,842.5
= 28,425 SF

Figure 7-17
Trapezoidal Rule worksheet and area calculations for the *inside* 104' contour line

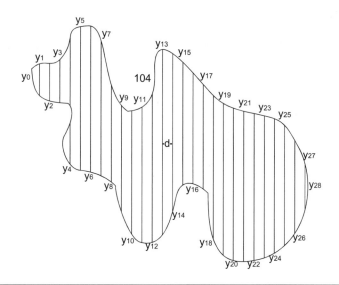

Figure 7-18
Outside 104' contour line with parallel lines for Trapezoidal Rule

		y_0 = First line				
		y_n = Last line {y_{28}}				
		y_{n-1} = Next to last line {y_{27}}				
		d = Distance between lines {10'}				
Trapezoidal Rule						
Area = d × [½ × (y_0 + y_n) + y_1 + y_2 + y_3 + ...y_{n-1}]						
Line number	y_0	y_1	y_2	y_3	y_4	y_5
Scale distance	8	34	42	50 + 35 = 85	123	159
Line number	y_6	y_7	y_8	y_9	y_{10}	y_{11}
Scale distance	164	158	109	82	86	140
Line number	y_{12}	y_{13}	y_{14}	y_{15}	y_{16}	y_{17}
Scale distance	179	188	186	172	144	108
Line number	y_{18}	y_{19}	y_{20}	y_{21}	y_{22}	y_{23}
Scale distance	89	88	90 + 34 = 124	155	158	157
Line number	y_{24}	y_{25}	y_{26}	y_{27}	y_{28}	
Scale distance	153	145	128	101	23	
Calculations for: outside 104' contour line						

Area = 10 × [½ × (8 + 23) + 34 + 42 + 42 + 85 + 123 + 159 + 164 + 158 + 109 + 82 + 86 + 140 + 179 + 188 + 186 + 172 + 144 + 108 + 89 + 88 + 124 + 155 + 158 + 157 + 153 + 145 + 128 + 101]
= 10 × [½ × (8 + 23) + 3,457]
= 10 × [½ × 31 + 3,457]
= 10 × [15.5 + 3,457]
= 10 × 3,472.5
= 34,725 SF

Figure 7-19
Trapezoidal Rule worksheet and area calculations for the *outside* 104' contour line

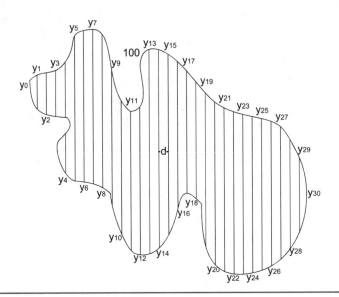

Figure 7-20
Outside 100' contour line with parallel lines for Trapezoidal Rule

y_0 = First line						
y_n = Last line {y_{30}}						
y_{n-1} = Next to last line {y_{29}}						
d = Distance between lines {10'}						

Trapezoidal Rule
Area = d x [½ x (y_0 + y_n) + y_1 + y_2 + y_3 + ...y_{n-1}]

Line number	y_0	y_1	y_2	y_3	y_4	y_5	y_6
Scale distance	11	40	50	58+32=90	129	165	174
Line number	y_7	y_8	y_9	y_{10}	y_{11}	y_{12}	y_{13}
Scale distance	175	168	142	109	151	171	177
Line number	y_{14}	y_{15}	y_{16}	y_{17}	y_{18}	y_{19}	y_{20}
Scale distance	200	197	185	168	145	117	106+30=136
Line number	y_{21}	y_{22}	y_{23}	y_{24}	y_{25}	y_{26}	y_{27}
Scale distance	164	168	172	171	168	160	145
Line number	y_{28}	y_{29}	y_{30}				
Scale distance	126	99	30				

Calculations for: outside 100' contour line

Area = 10 x [½ x (11 + 30) + 40 + 50 + 90 + 129 + 165 + 174 + 175 + 168 + 142 + 109 + 151 + 171 + 177 + 200 + 197 + 185 + 168 + 145 + 117 + 136 + 164 + 168 + 172 + 171 + 168 + 160 + 145 + 126 + 99]
= 10 x [½ x (11 + 30) + 4,262]
= 10 x [½ x 41 + 4,262]
= 10 x [20.5 + 4,262]
= 10 x 4,282.5
= 42,825 SF

Figure 7-21
Trapezoidal Rule worksheet and area calculations for the *outside* 100' contour line

Calculation Sheet

Project: Jones job   **Date:** 4-20

Areas:
Inside 100' contour (see Figure 7-15): 21,270 SF
Inside 104' contour (see Figure 7-17): 28,425 SF
Outside 100' contour (see Figure 7-19): 34,725 SF
Outside 104' contour (see Figure 7-21): 42,825 SF

Volume (CY)

Lake volume:
= (area inside 100' contour × depth) ÷ 27
= (21,270 × 3) ÷ 27
= 63,810 ÷ 27
= 2,363.3 CY

Volume (CY)

Berm volume:
= ({[(area 100_o + area 104_o) ÷ 2] × depth} ÷ 27) - ({[(area 100_i + area 104_i) ÷ 2] × depth} 27)

Where:
100_o = outside 100' contour
104_o = outside 104' contour
100_i = inside 100' contour
104_i = inside 104' contour

= ({[(42,825 + 34,725) ÷ 2] × 4} ÷ 27) - ({[(21,270 + 28,425) ÷ 2] × 4} ÷ 27)
= ({[77,550 ÷ 2] × 4} ÷ 27) - ({[49,695 ÷ 2] × 4} ÷ 27)
= ({38,775 × 4} ÷ 27) - ({24,847.5 × 4} ÷ 27)
= (155,100 ÷ 27) - (99,390 ÷ 27)
= 5,744.4 - 3,681.1
= 2,063.3 CY

Volume (CY)

Fill volume excess (+) / shortfall (-):
= berm volume - lake volume
= 2,063.3 - 2,363.3
= -300 CY

Conclusion:

Figure 7-22
Volume calculations for the small lake

8 Using Shrink and Swell Factors

At the end of Chapter 7, I mentioned that you need shrink and swell factors to make your estimates more accurate. That's because a given quantity of soil has no constant volume. Add moisture and the soil swells, expanding in volume. Soil volume also increases when it's loosened or disturbed by excavation. Conversely, the soil volume shrinks or contracts when you apply pressure to compact the fill. Actual shrink and swell factors consider the *combined* effect of:

- moisture content
- density (compact versus loose)
- soil type

We'll begin this chapter with a quick look at soil states and volume measurement. Then we'll move on to some practical examples of when and how to apply shrink/swell factors in your estimates. Next we'll look at how you link soil volume to equipment use and load factors. The chapter finishes up with a look at an alternate method of deriving shrink/swell factors from the soil's weight in different states.

Soil States and Their Units of Measure

It's understood within the construction field that soils have three distinct states. Let's take a look at the different soil states and how you measure them.

Bank material is undisturbed soil, that is, soil in its natural state and location. You measure the volume of bank material in units called *bank cubic yards*, or *BCY* for short.

Loose material is soil that's no longer in its natural state or location. This soil's loosened through digging, turning or some other excavation process. As a result, the soil volume expands. You measure the volume of loose material in *loose cubic yards*, or *LCY*.

Compacted material is soil that's removed from its natural state or location, placed elsewhere and then compacted. This process occurs in nature as well as a result of excavation. Compaction reduces the soil's volume relative to its loose state volume. You measure compacted state soil in units of *compacted cubic yards*, or *CCY*.

If you've ever dug a hole to plant a tree or shrub, you've already worked with soil in all three states. Before you began digging, you're standing on *bank state material*. First you dig the planting hole, piling the dirt off to one side as you dig. That mound of freshly turned earth is *loose state material*. Then you place the root ball in the hole and replace most of the soil you removed. Finally you finish the job by tamping down the earth or watering it in. That's compaction. All of the soil you put back into the hole is *compacted state material*.

There's one more point I want to make before ending the gardening lesson. Remember that pile of left-over dirt you still had after you filled the hole? You probably wrote it off as being replaced by the root ball, but that's not the whole story. What if you dug the hole but then you changed your mind. Instead of planting a tree, you filled the hole right back up again. You'd still have a pile of left-over dirt. Most soil won't compact back down to its bank state volume right away. It takes time and natural weathering to return soil to its original bank state.

Using Shrink/Swell Factors in Earthwork Estimates

Before you can figure out how much soil you'll need for a particular job, you need to know how much that soil will swell or shrink. The swell and shrink factors tell you this. For a specific soil, under specific circumstances, you multiply the volume of the soil by its shrink or swell factor to figure out how much you'll have. Here's an example. Say you have a trench to fill and a stockpile of damp earth. You want to know whether there's enough material in the stockpile to fill the trench. Start by calculating the volumes of both the stockpile and the trench. Then multiply the volume of the stockpile by the correct shrink factor. You need the shrink factor because you'll compact the fill and compaction reduces the volume of soil.

You'll find that every soil you work with is different. Sending a sample of the soil to a laboratory for testing is the only way to find exact shrink and swell factors. Such testing is standard on large projects and the results appear on the plans. On a smaller project, specific shrink/swell factors are rarely available. When that happens, use the approximate shrink/swell factors listed in Figure 8-1. Shrink/swell factors are based on ratios that compare soils' weights in each of the three states: bank, loose and compact. You'll notice that the soil types listed in Figure 8-1 are very general. That's why the factors are only approximate.

Soil type & moisture level	Swell factor	Shrink factor	Compaction requirements
Dry sand	1.13	1.00	BCY
Dry sand	1.32	0.83	95% S.P.
Dry sand	1.39	0.77	100% S.P.
Dry sand	1.38	0.78	95% M.P.
Dry sand	1.45	0.72	100% M.P.
Damp sand	1.13	1.00	BCY
Damp sand	1.16	0.98	95% S.P.
Damp sand	1.22	0.93	100% S.P.
Damp sand	1.21	0.94	95% M.P.
Damp sand	1.27	0.88	100% M.P.
Damp gravel	1.14	1.00	BCY
Damp gravel	1.23	0.93	95% S.P.
Damp gravel	1.29	0.87	100% S.P.
Damp gravel	1.32	0.84	95% M.P.
Damp gravel	1.39	0.78	100% M.P.
Dry clay	1.31	1.00	BCY
Dry clay	1.18	NA	85% S.P.
Dry clay	1.25	NA	90% S.P.
Dry clay	1.39	0.94	100% S.P.
Dry clay	1.39	0.94	90% M.P.
Dry clay	1.54	0.82	100% M.P.
Dry dirt	1.32	1.00	BCY
Dry dirt	1.31	1.00	85% S.P.
Dry dirt	1.39	0.95	90% S.P.
Dry dirt	1.54	0.83	100% S.P.
Dry dirt	1.45	0.90	90% M.P.
Dry dirt	1.61	0.78	100% M.P.
Damp dirt	1.28	1.00	BCY
Damp dirt	1.17	NA	85% S.P.
Damp dirt	1.23	NA	90% S.P.
Damp dirt	1.37	0.93	100% S.P.
Damp dirt	1.29	1.00	90% M.P.
Damp dirt	1.43	0.89	100% M.P.

BCY = bank cubic yards

S.P. = Standard Proctor

M.P. = Modified Proctor

NA = areas where the bank material has a greater density than required for the compacted material

Figure 8-1
Approximate conversion factors for soil swell and shrinkage

As we saw in Chapter 3, when we discussed the Proctor test, there are different degrees of compaction. Proctor requirements appear on either the plans or the job specifications. Figure 8-1 takes this into account and lists various levels of compaction in the far right column.

Let's try using Figure 8-1 to find out if there's enough material in that stockpile of damp earth to fill the trench. Here are the numbers:

- Stockpile volume = 200 CY
- Trench length = 900'
- Trench width = 3'
- Trench depth = 2'
- Compaction required = 100% Standard Proctor

We'll begin by calculating the trench volume in cubic yards. Here's the formula and all the math:

Formula for trench volume

Volume (CY)   = (length x width x depth) ÷ 27

Volume (CY)   = (900 x 3 x 2) ÷ 27

             = 5,400 ÷ 27

             = 200 CY

The stockpile and the trench have the very same volume — 200 cubic yards. So all's well, right? Not quite. You have 200 LCY of fill, but you need 200 CCY at 100 percent Standard Proctor. Now let's find out how much material you've really got in that stockpile.

Step 1

Find the shrink factor for damp dirt compacted to 100 percent Standard Proctor in Figure 8-1. The answer is 0.93.

Step 2

To find the post-compaction volume of the stockpile, multiply its loose volume (200 LCY) by the shrink factor (0.93):

200 x 0.93 = 186 CCY

There's not enough material in the stockpile to do the job. You're short by a total of 14 CCY. You'll need to bring that material in from elsewhere. Be sure that any bid you submit includes these costs.

Estimating the Number of Haul Trips

Shrink/swell factors have other useful applications. You use them, for example, to figure how many trips it takes to move a given amount of material from one location to another. Let's see how to do it. Here's what you know:

- Material type: sand
- Material condition: damp
- Total quantity: 1,000 CCY

- Required compaction: 95% Standard Proctor
- Per trip haulage capacity: 10 LCY

Note: Always measure hauling capacities in loose cubic yards. Also, this list omits two factors, time and resistance. We'll cover both factors later, in Chapter 13.

Step 1

Turn back to Figure 8-1 and find the correct swell factor for damp sand at 95 percent Standard Proctor. The answer is 1.16.

Step 2

To convert this volume into loose cubic yards, multiply the total quantity (1,000 CCY) by the swell factor (1.16):

1,000 CCY x 1.16 = 1,160 LCY

Step 3

To find the total number of trips, divide total volume in loose cubic yards by the vehicle's capacity:

1,160 LCY ÷ 10 LCY per trip = 116 trips

Using Material Weights to Customize Shrink/Swell Factors

As I mentioned earlier, shrink/swell factors are partly based on the weight of soils in different states expressed as a ratio. Figure 8-2 is a list of approximate weights for a variety of common materials with different moisture and compaction levels. Here are the formulas you use with the weights to find approximate shrink and swell factors.

Formulas for shrink and swell factors

Approximate shrink factor = loose state weight ÷ bank state weight

Approximate swell factor = bank state weight ÷ loose state weight

Remember, these are only approximate. If you need more precise data and it's not provided in the plans or specs, contact a soils engineer. If you choose to calculate your own shrink/swell factors, I strongly recommend obtaining accurate weights for local soils. You'll need to contact either the State Department of Transportation or a municipal, county or state planning or engineering agency.

Using Soil Weights to Calculate Equipment Load Factors

Let's look at the weight range you find in the far right column of Figure 8-2. At the high end is: "Sand, wet packed — 3,120 pounds/LCY." At the low end is: "Clay, dry — 1,940 pounds/LCY." A single loose cubic yard, depending on the material, differs in weight by as much as 1,180 pounds.

Equipment capacity charts simply state a volume in loose cubic yards. They don't take weight differences into account. That's your job. Fortunately, load factors make it easy. Suppose you need to move 43 LCY of natural clay and the stated capacity of the scraper you're using is 12 LCY. How many trips will it take? If you simply divide the total volume by the scraper's capacity, ignoring the load factor, you'll overload the scraper. First you need to find and apply the correct load factor, then calculate the number of trips you need to make.

Step 1

Find the load factor using this formula:

Formula for load factor

Load factor = pounds/LCY ÷ pounds/BCY

Step 2

Find these weights for natural clay in Figure 8-2, plug them into the formula and here's the result:

Load factor = 2,130 ÷ 2,960

= 0.72

Step 3

To apply the load factor, multiply the scraper's basic capacity (12 LCY) by the load factor (0.72). Here's the math:

12 LCY x 0.72

= 8.6 LCY

The result, 8.6 LCY, is the volume of natural clay that's equal in weight to the scraper's listed capacity of 12 LCY.

Step 4

Find the actual number of haul trips by dividing the total volume (43 LCY) by the scraper's modified capacity (8.6 LCY):

43 LCY ÷ 8.64 LCY

= 5

Material type & moisture level	Pounds per BCY	Pounds per LCY
Clay, natural	2,950	2,130
Clay, dry	2,290	1,940
Clay, wet	2,620	2,220
Common earth, dry	2,620	2,100
Common earth, wet	3,380	2,700
Limestone	4,400	2,620
Sand, dry and loose	2,690	2,400
Sand, wet and packed	3,490	3,120
Note: These weights assume that soils are monotypic, 100% one soil type. Accurate weights for actual soils (a mix of several different soil types) are determined through laboratory testing. The tests cover particle size and distribution, exact moisture content and level of compaction.		

Figure 8-2
Approximate weights of various materials in bank state and loose state

Pay Yards

It's important that you understand shrink/swell factors and the three soil states we've covered in this chapter. We'll refer to them throughout the rest of the book.

There's one more important use for this information and that's in calculating pay yards. That's the basis for your pay as an earthwork contractor. Pay yards are usually — but not always — measured in bank cubic yards. To be sure, check the plans or specifications. If the bid sheet doesn't show pay yards in BCY, it's up to you to make sure that your estimate does. A simple note can prevent a nasty surprise at the end of the job.

9

Topsoil, Slopes & Ditches

■ ■ ■ ■ ■ ■ ■ ■

In this chapter we'll focus on three important topics that every earthwork estimator must understand — the special requirements for topsoil, slopes, and ditches. Let's begin with topsoil. Topsoil is so important that you'll treat it differently from all the rest of the soil on a job site.

Dealing with Topsoil

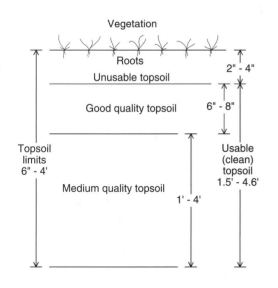

Figure 9-1
Topsoil layers

Topsoil is naturally rich in plant nutrients. New material (like leaves, grass and twigs) is always being added. Over time, this organic material decomposes. That adds nutrients to the topsoil, making it fertile. Topsoil is ideal for growing lawns and gardens.

Natural topsoil depths vary widely. In the U.S., for example, topsoil depths range from an inch or two in the arid Southwest to over 2 feet in nearby Midwestern states.

Figure 9-1 shows the layers that make up topsoil. At the very top is a layer with live vegetation. The next layer includes live root systems and the most recent organic material. This layer is between 4 and 6 inches deep and it's the layer that's richest in plant nutrients. Unfortunately, the roots, seeds, and debris that are also in this layer make it unsuitable for you to use either as topsoil or fill.

The next layer is the topsoil that's left after you strip off the root layer. This is clean topsoil — it doesn't contain any roots, seeds or debris. Clean topsoil is valuable. Job specs cover which parts of a site you strip and what you do with the stripped topsoil. Typically, clean topsoil is stockpiled on site and replaced after construction. Unusable topsoil is often disposed of off site. If the specs require hauling off site, find the nearest disposal location before working up an estimate. This may involve more mileage than you want to absorb as a cost.

An earthwork contractor's final task at most job sites is respreading topsoil. Make sure you have enough on hand when the time comes. Be aware that you may need to bring in additional topsoil from off site. Remember the top 2 to 4 inches of topsoil that you can't use? You may need to replace that now with clean topsoil. Don't get caught by surprise. Always read job specs carefully and completely, then plan ahead. Good topsoil is never cheap or easy to find.

From time to time, not often, you'll have more clean topsoil than you need. Offer to sell the excess to homeowners or small contractors. This can add up to a nice bonus on top of your normal profit on a job.

Take the time to choose your clean topsoil stockpile location carefully. Here are a few guidelines:

- Chose a site that's as far away from active work as possible. This helps keep the stockpile from getting scattered about and compacted by equipment.
- Avoid low-lying areas that puddle or may flood. You don't want it washed away.
- Control weeds by disking or with a short-term chemical herbicide.

Finding the Volume of Topsoil

Because you treat (and bid) topsoil separately, it's also smart to track topsoil quantities separately. Remember, most topsoil you remove gets replaced. You replace topsoil in its loose (not compacted) state. The only way to do that is by hand. That's why cubic yards of topsoil are often the most expensive cubic yards in an estimate.

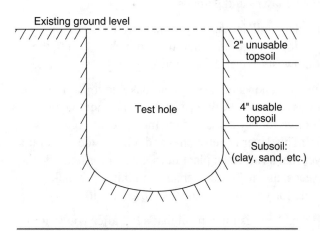

Figure 9-2
Typical topsoil sample test hole

The designer or engineer usually chooses the limits of topsoil work on a project. Standard practice is to strip topsoil to about 5 feet beyond the sides of a building and about 1 foot beyond the limits of roadways.

Look for a note on the plans that shows topsoil depth at the building site. If you don't find one, visit the site and make a personal inspection. Use a shovel to remove several plugs of soil, each from a different part of the job site. Figure 9-2 shows a profile view of a typical topsoil sample. The unusable topsoil zone extends from the surface down to the ends of the living root systems. Use the average length of all the roots, not the longest or shortest ones, to estimate how much soil you'll have to strip off. It's usually from 2 to 4 inches.

Here's a practical example of how you deal with topsoil, and how you calculate topsoil quantities. We'll use the site plan in Figure 9-3 and these two items from the job specs.

1) Earthwork contractor shall strip the top 6 inches of soil from the entire site. The earthwork contractor shall remove said topsoil in the two layers as described and disposition them as follows:

   Upper 2 inches of topsoil — remove from site for disposal

   Lower 4 inches of topsoil — stockpile on site

2) On completion of construction, the earthwork contractor shall respread topsoil to a depth of 6 inches over *all uncovered areas* of the site.

We need to know two volumes — stripped topsoil and replaced topsoil. Using these specs, the site plan (Figure 9-3) and basic math, here's how to find the answers.

Finding the Volume of the Stripped Topsoil

According to the specs, we strip topsoil from the entire site, so we'll begin by finding the area of the job site in square feet.

Formula for area of job site

$$Area\ (SF) = length \times width$$
$$= 120 \times 55$$
$$= 6,600\ SF$$

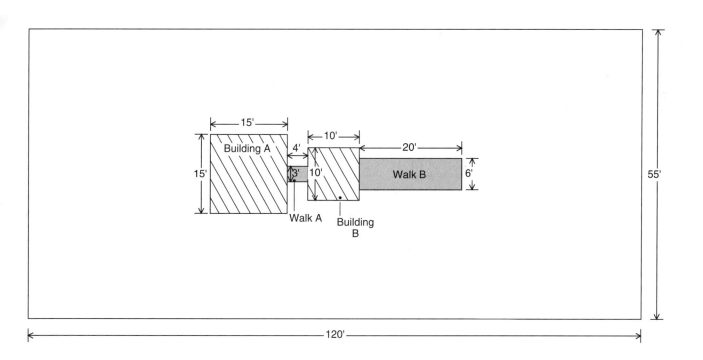

Figure 9-3
Site plan for sample topsoil calculations

Topsoil, Slopes & Ditches

You'll recall that the specs divided stripped topsoil into two categories. The upper 2 inches are discarded and the remaining 4 inches are stockpiled. We'll calculate their volumes separately in cubic yards using this formula:

Formula for volume in cubic yards

Volume (CY) = [area (SF) × depth (feet)] ÷ 27

Let's start with the discarded topsoil volume. Here's what we know:

- area = 6,600 SF
- depth = 2 inches, or 0.17 feet (2 ÷ 12 = 0.17)

And here's the rest of the math:

Volume (CY) = (6,600 × 0.17) ÷ 27
$$= 1,122 ÷ 27$$
$$= 41.55 \text{ CY}$$

We'll round that off to 42 CY.

You find stockpiled topsoil's volume exactly the same way. The area is the same as for the discarded topsoil but the depth is different. Also don't forget to convert the depth, 4 inches, into feet, 0.33 (4 ÷ 12 = 0.33). Here's the rest of this volume calculation:

Volume (CY) = (6,600 × 0.33) ÷ 27
$$= 2,178 ÷ 27$$
$$= 80.66 \text{ CY}$$

We'll call that 81 CY.

The total volume of stripped topsoil is the sum of the two volumes we just found.

Total volume of stripped topsoil (CY) = 42 + 81
$$= 123 \text{ CY}$$

Finding Replacement Topsoil Volume

First we need to know how many square feet of the site we'll be replacing topsoil on. That's equal to the difference between the site area and the sum of the area of the structures. We already know the site area is 6,600 SF. Now let's find the total area covered by structures, working structure by structure.

Building A

Area = 15 × 15
$$= 225 \text{ SF}$$

Building B

Area = 10 × 10
$$= 100 \text{ SF}$$

Walkway A

Area = 4 × 3

= 12 SF

Walkway B

Area = 20 × 6

= 120 SF

Total structure area (SF) = sum of all structure areas.

In our example that comes out like this:

Total structures area (SF) = 225 + 100 + 12 + 120

= 457 SF

The difference between the site area and the total structures' area is the total area to be respread with topsoil. Plug in the areas we found for Figure 9-3 and here's the math:

Area (SF) = 6,600 - 457

= 6,143 SF

Next we'll find the replaced topsoil's volume in cubic yards using this formula:

Volume (CY) = (area × depth) ÷ 27

From the job specs we know that the depth is 6 inches or 0.5 feet and here's the rest of the math:

Volume (CY) = (6,143 × 0.5) ÷ 27

= 3,071.5 ÷ 27

= 113.76 CY

After rounding off, to full cubic yards, it turns out that this job requires 114 cubic yards of replacement topsoil. Compare that volume with the volume in the stockpile and you'll see there's a shortfall of 33 CY (114 - 81 = 33).

Calculating Net Volumes for Earthwork

The existing ground level is your reference for measuring contour lines, plan lines, and typical profiles. That reference disappears along with the stripped topsoil and changes your working elevation. As a result, the original contour map no longer applies. The engineer often doesn't know the actual depth of the topsoil. You do, so the conversion's up to you. You can use the following formulas to find the actual depth and volume of earth moved.

Cut Areas

In areas involving cutting, you have to remove topsoil, cut the earth down to grade, and then replace topsoil to the specified depth. We'll use Figure 9-4 and this formula to find the total depth of cut:

Topsoil, Slopes & Ditches

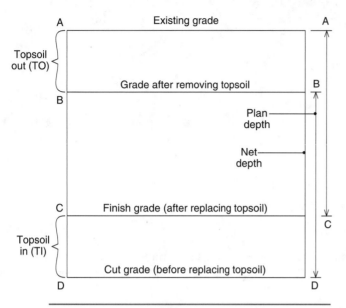

Figure 9-4
Calculating net cut depth

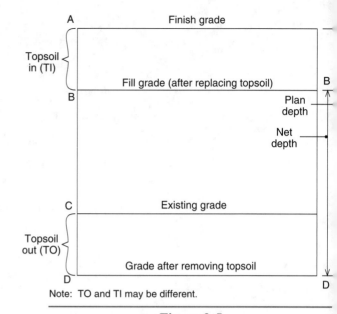

Figure 9-5
Calculating net fill depth

● ● ● ● ● ● ● ● ● ● ● ● ● ●
Formula for total cut depth

Net cut depth (feet) = plan depth - TO + TI

Where:

- line A-A = existing grade *before* removing topsoil
- line B-B = strip grade *after* removing topsoil
- line C-C = finished grade *after* replacing topsoil
- line D-D = cut grade *before* replacing topsoil
- plan depth = existing elevation (line A-A) - finished elevation (line C-C)
- TO = topsoil strip depth (topsoil out)
- TI = topsoil replace depth (topsoil in)

Most site plans you work with show only the existing grade (line A-A) and the finish grade (line C-C). That makes it easy to find plan depth. I strongly recommend that you always use net depth, instead of plan depth, in calculating earthwork volumes.

Fill Areas

You calculate total fill depths using almost the same steps as for total cut depth. We'll use the cross-section view of a fill area shown in Figure 9-5 as the sample job. And here's the formula we'll use to find total fill depth:

● ● ● ● ● ● ● ● ● ● ● ● ● ●
Formula for total fill depth

Net fill depth (feet) = plan depth + TO - TI

Where:

- line A-A = finished grade *after* replacing topsoil
- line B-B = fill grade *before* replacing topsoil

140 *Estimating Excavation*

- line C-C = existing grade *before* removing topsoil
- line D-D = cut grade *after* removing topsoil
- plan depth = finished elevation (line A-A) - existing elevation (line C-C)
- TO = topsoil strip depth (topsoil out)
- TI = topsoil replace depth (topsoil in)

Areas Combining Cut and Fill

On projects that combine cut and fill, I recommend making the calculations on two worksheets. Put all of your cut calculations on one worksheet and all of your fill calculations on a second worksheet. Separate worksheets are the best and easiest way to be sure you never mix up these two volumes.

Example 1

Let's try out these formulas using the cut area shown in Figure 9-6 as our example. We'll find the net cut depth using these specs for topsoil depths:

- At project's start — strip to a depth of 6 inches (0.5 feet).
- At project's completion — replace to a depth of 4 inches (0.33 feet).

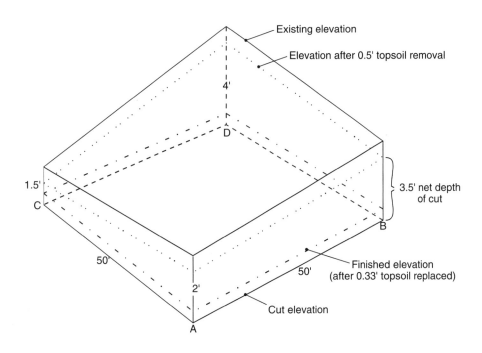

Figure 9-6
Calculating net cut area

Topsoil, Slopes & Ditches

Step 1 Find average cut

Average cut = (4 + 3.5 + 2 + 1.5) ÷ 4

= 11 ÷ 4

= 2.75 feet

Step 2 Calculate net cut depth

Net cut depth (feet) = 2.75 - 0.5 + 0.33

= 2.58 feet

Step 3 Calculate net cut volume

Net cut volume (CY) = (2.58 × 50 × 50) ÷ 27

= 6,450 ÷ 27

= 238.89 CY

Cut and Fill Areas Under Surface Structures

You use similar formulas to calculate the cut and fill under surface structures such as sidewalks, paved ditches, and roadways. The formulas you'll use look just the same. However, the meanings of *plan difference*, *TO*, and *TI*, are a little different here as you'll see. We'll start off with the formulas.

Formulas for total cut and fill under a structure

Formula for finding total cut under a surface structure

Total cut = Plan difference + TI - TO

Formula for finding total fill under a surface structure

Total fill = Plan difference + TO - TI

Where:

- plan difference = change in elevation between proposed and existing grades
- TO = topsoil strip depth (topsoil out)
- TI = construction material (concrete, asphalt, other) depth

Slopes and Slope Lines

A slope line is a straight line connecting two points. Engineers and designers use slope lines to indicate gradual slopes between finished elevations on site and existing elevations of undisturbed soil adjacent to the building site. The correct slope is very important to drainage.

Figure 9-7 shows a slope line connecting the top of a roadway to the bottom of a ditch. One name for the point where a slope and the roadway meet is *top of slope*. The slope meets the bottom of the ditch at the *toe of slope*. The slope here is 4:1. This is a simple ratio. It means that there's 1 foot of vertical climb included in 4 feet of horizontal distance. If the slope is 4:1 and the total vertical fall is 6 feet, then the toe of slope is 24 feet measured horizontally from the top of the slope.

You can also use the words run and rise to describe a slope. The *run* is the horizontal distance. The *rise* is the vertical distance. When the run is 4 feet and the rise is 1 foot, the slope is 4:1.

Here's a note of caution. When you're reading plans, you may see a slope labeled 1:4. Don't plow ahead assuming that the designer wants a slope with 4 feet of rise for every 1 foot of run. A slope of 1:4 is very steep — about the same as the roof on an A-frame house. Soil isn't stable at that angle. If you see a slope indicated as 1:4, it's a good bet the designer meant a 4:1 slope. It's very easy to reverse the numbers as you write them down. When you have any doubts, check with the designer.

Most slopes you work with range between 5:1 and 2:1. The steepest slope I've ever encountered was 1:1. That's 1 foot of run in 1 foot of rise, or equal to a 45-degree angle.

Designers have to consider several factors when planning slopes. First, the owners must be able to maintain the slope. Using a mowing machine isn't safe on any slope over 2:1. Second, drainage is easier to control with slopes in the 5:1 to 2:1 range. The water will run on these slopes uniformly. A good ground cover will protect 5:1 to 2:1 slopes. On steeper or flatter slopes, water will either run too fast and erode the slope, or won't drain fast enough. Water backs up and puddles on a slope that's less than 5:1.

You'll often work with drawings similar to the plan and profile sheet shown in Figure 9-8. The sheet includes a note from the project engineer that simply says the finish grade from the parking lot edge to the existing ground line is to be a 4:1 slope.

Let's move on now to your next task, calculating the topsoil volume you need to build that slope.

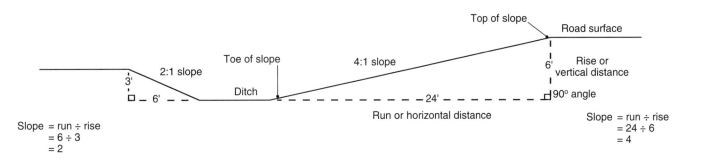

Figure 9-7
Calculating slope

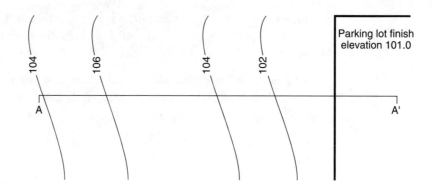

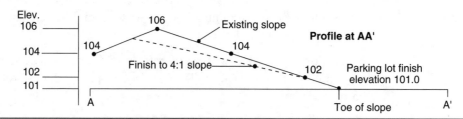

Figure 9-8
Calculating net cut depth

Calculating Topsoil Volume for a Sloped Area

A slope line like the one in Figure 9-9 connects two points with known elevations. Of course, the amount of slope affects how much topsoil you'll have to remove and replace. The slope line (field distance) is longer than the horizontal distance (plan distance) between the two points. You need to know the field distance before you can estimate the volume of topsoil.

You could calculate the actual length of the slope line, but there's a faster way. Professional estimators use charts like the one shown in Figure 9-10. To find the field distance for a known slope simply multiply the horizontal distance by the percentage you look up in Figure 9-10.

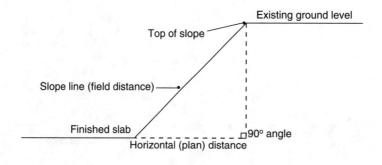

Figure 9-9
Side view of a sloped area

144   *Estimating Excavation*

Let's go through an example to show how this works. We'll use the retaining wall project shown in Figure 9-11. The specs call for finishing the slope shown in the section view with a 4-inch-thick layer of topsoil. We'll find the volume of topsoil needed in just four easy steps.

Step 1) Find the slope

Slope = run (horizontal distance) ÷ rise (vertical distance)

Slope = 20 ÷ 5

= 4

The slope in Figure 9-11 is 4:1.

Slope	Length of slope = horizontal distance x %
6:1	1%
5:1	2%
4:1	3%
3:1	5%
2:1	12%
1:1	41%

Figure 9-10
Table used to estimate the length of a slope line

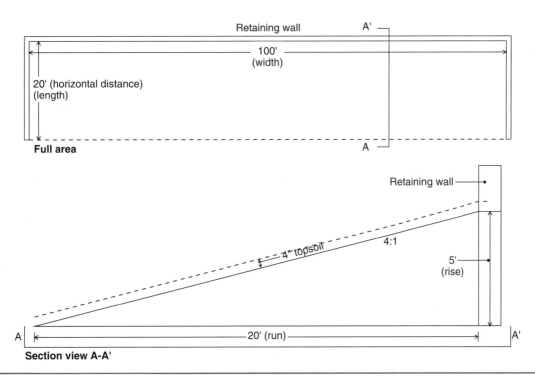

Figure 9-11
Calculating topsoil volume for a slope

Topsoil, Slopes & Ditches **145**

Step 2) Find the percentage increase for the slope

Using the table in Figure 9-10, find the percentage increase for this slope. The answer is 3 percent.

Step 3) Find the field distance or slope length

Field distance/slope length (feet) = horizontal/plan distance × increase factor

Field distance/slope length (feet) = 20 × 1.03

= 20.6 feet

Step 4) Calculate the topsoil volume in cubic yards

Volume (CY) = (slope length × slope width × topsoil depth) ÷ 27

Volume (CY) = (20.6 × 100 × 0.33) ÷ 27

= 679.8 ÷ 27

= 25.18 CY

Sloping Trenches for Safety

Years ago, contractors installing sewer, water, and other utility lines simply dug trenches the same width as their backhoe bucket, or only as wide as necessary. The trouble was the ditch walls had no support. The walls often caved in and workmen died. This led state and federal agencies to set safety standards. Any trench over 5 feet deep must have sloped sides or use trench boxes for support.

A trench box has two sides made of solid metal plate. You can rent or lease them in a variety of sizes. Select trench boxes with a width slightly less than the width of your trench and a bit longer than a standard section of the pipe you're installing.

Many utility line contractors prefer to slope the trench sides instead of using shoring boxes. The amount of slope needed for safety depends on the trench depth and soil conditions. Figure 9-12 lists soil types, with their average safe slopes.

Be very careful when estimating earthwork for trenches. Good judgment is essential if you're going to make any money on the job. Don't count on the plans or specs to mention sloped trenches. You probably won't find any guidelines on the degree of slope to use. However, your local OSHA (Occupational Safety and Health Administration) office should be able to supply the exact requirements.

Slope Lines for Drainage

The finished slope requirements for a roadway or a parking lot appear on the plans, but follow a different format. Instead of a ratio like 4:1, the slope is a fractional part of a foot per linear foot of surface. For example, most roads are higher (crowned) along their centerline. The usual slope is $^3/_8$ inch per foot.

Soil type	Slope
Sand	3:1
Loam	3:1
Sand / clay (mixture)	2:1
Clay	1:1
Warning: The slopes in this table are averages. Base all job site slope calculations on actual field conditions and the results of on-site soil testing.	

Figure 9-12
Soil types and their average maximum safe slope

Let's look at a standard two-lane road that's 12 feet wide. Along the centerline the elevation is higher than at the edges by 4½ inches. Other ways of measuring slopes like these include tenths of a foot, inches and fractions of inches. Other names for this kind of slope include: *drain slope*, *cross slope*, and *superelevation*.

Estimating Trenches

For estimating purposes, there are two types of trenches:

- Drainage channels that carry water away from buildings
- Utility line trenches for sewer, water, phone, and other utility lines

We'll look at both types of trench, starting with drainage channels.

Drainage Channels

The grade or incline of a drainage channel controls and conducts runoff. The channel shape and size depend on the expected runoff volume. Speed of runoff determines the type of lining material.

There are two common types of drainage channels, vee ditches and flat-bottom ditches. Figure 9-13 shows examples of both types. You use a vee ditch to supply drainage for relatively small volumes of water. If you expect a larger volume of water, you'd use a flat-bottom ditch. The estimated runoff volume determines the width of the ditch's flat bottom, typically 2 to 10 feet.

Fast runoff combined with heavy volume call for sealing or lining the trench with concrete so it's protected from erosion. Figure 9-14 shows a concrete-lined flat-bottom ditch. Notice the elevation note included in

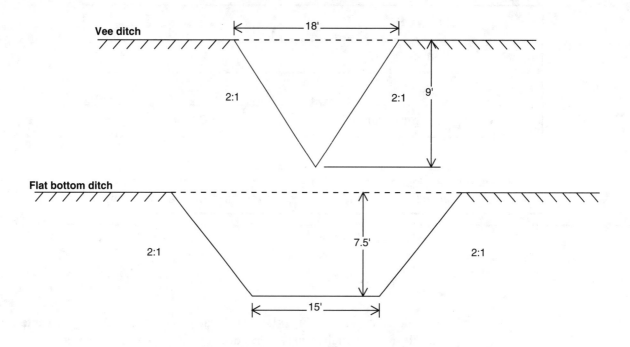

Figure 9-13
Section views of vee and flat-bottom ditches

Figure 9-14: *T.O.S. = 101.5*. The abbreviation *T.O.S.* in the figure stands for top of slab. This is often the only elevation that's supplied on the plans. There are two important points I want to make here. First, be careful not to confuse "top of slab" *T.O.S.* with "toe of slope" also *T.O.S.* Second, if the plans only give a *T.O.S.* elevation, it's up to you to calculate the excavation volumes from the top of slab down. For example, in Figure 9-14, total excavation depth includes, at the very least, the slab thickness, 6 inches, plus the depth of any bedding placed under the concrete.

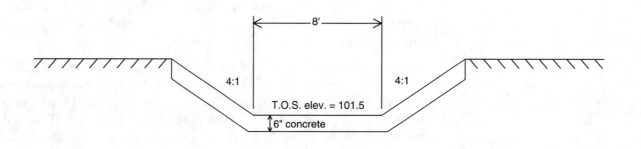

Figure 9-14
Concrete-lined flat-bottom ditch, section view

148 *Estimating Excavation*

Utility Trenches

Estimating excavation volumes for utility line trenches is a bit more complicated. We'll start by looking at the three factors that determine trench width:

- the material placed in the trench
- the excavation equipment
- the overcut required

Material Placed in the Trench

If you're trenching for 18-inch-diameter pipe, you don't need a 48-inch-wide trench. Here's the rule of thumb I use:

Trench width (feet) = pipe width + workspace to place bedding material and backfill

Most plans for utility trenches require bedding. Bedding is a layer of material that surrounds a pipe, cushioning it during laying and backfilling. Common bedding materials include sand, gravel, and concrete.

Now take a look at Figure 9-15. It shows an 18-inch-diameter water line laid in a trench. Notice that the figure also shows a 6-inch-thick bedding of sand surrounding the pipe on four sides. If you only consider the material placed in the trench, how wide is this trench? That's easy. Just add the pipe diameter to the depth of the bedding on both sides.

Trench width (inches) = 18 + 6 + 6

= 30 inches

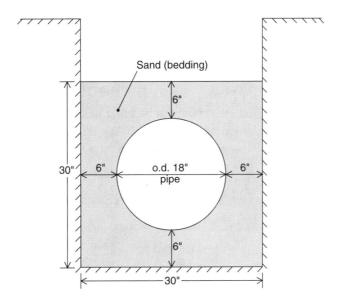

Figure 9-15
Excavation required for pipe and bedding

Remember that this width of 30 inches only considers what's placed in the trench, and that the depth of the material placed in the trench (depth of the bedding plus the pipe diameter) also equals 30 inches.

How much bedding material will you use per linear foot of pipe in Figure 9-15? Simply calculate the filled area (pipe plus bedding), then deduct the area of the pipe.

The pipe is a circle and here's the formula for the area of a circle:

Area (SF) = π × r²

equals *pi* (π) times the radius squared (r²).

Where:

- *pi* (π) = 3.1416
- radius (r) = 9 inches or 0.75 feet

Area (SF) = 3.1416 × 0.75²

= 3.1416 × 0.5625

= 1.77 SF

Find the area of the fill using this formula:

Area (SF) = width × depth

Plugging in the values from Figure 9-15 we find:

Area (SF) = 2.5 × 2.5

= 6.25 SF

To find the area of bedding material used per linear foot of pipe we'll use this formula:

Area (SF/foot of pipe) = filled area - pipe area

And here's the math for this example:

Area (SF/foot of pipe) = 6.25 - 1.77

= 4.48 SF/foot of pipe

To find the volume of bedding material you'll use for this job you need to know the length of the trench. We'll say it's 75 feet long. Now we'll calculate the bedding material volume in cubic yards. Here's the formula:

Volume (CY) = (area × length) ÷ 27

And here's the math for our example:

Volume (CY) = (4.48 × 75) ÷ 27

= 336 ÷ 27

= 12.44 CY

Here's a pitfall to watch out for in utility line trenching. Sometimes the plans only show the flow line elevation for pipe run. Any time you see only a flow line elevation on the plans, ask yourself this question: Does the elevation

• • • • • • • • • • • • • •
Formula for area of a circle

include the pipe wall thickness or not? Here's why that's important. Pipes have various wall thicknesses. Some have 2-inch-thick walls, some are as much as 6 inches thick. It makes a big difference in your estimate of total excavated volume. Here's a quick example that demonstrates my point.

Say the trench is 5 feet wide, 20,000 feet long and the pipe has walls 6-inches thick (0.5 foot). How big is the error (in cubic yards) when you don't consider pipe wall thickness?

$$\text{Volume (CY)} = (0.5 \times 5 \times 20,000) \div 27$$
$$= 50,000 \div 27$$
$$= 1,851.85 \text{ CY}$$

You wouldn't get paid for this work, since the pay items are figured on the flow line of the pipe.

Some utility lines are set in concrete. How do you find the volume of the concrete? You use the same formulas as we just used to find the volume for bedding material.

When utility lines cross heavily-traveled streets or highways, open trenches aren't allowed. Instead, you have to bore a passageway through the ground. Line the boring with encasement pipe and then run the utility line inside. Encasement pipes are typically much larger than what's run inside.

Utility lines inside an encasement pipe lack support. Filling the void between the two pipes with a dense material provides support. The most typical fillers are sand and concrete. How much sand or concrete does it take to do the job? Find the difference between the areas of the two pipes. Then multiply the result by the length of the encasement. Let's try an example, using Figure 9-16.

Figure 9-16 shows a 45-foot-long section of 18-inch-diameter waterline installed inside a 48-inch-diameter encasement pipe. Solve for the areas of both circles using the formula: Area = πr^2. The encasement pipe's area is 12.57 SF. The area of the waterline is 1.77 SF. Here's the rest of the math:

$$\text{Volume (CY)} = [(12.57 - 1.77) \times 45] \div 27$$
$$= [10.8 \times 45] \div 27$$
$$= 486 \div 27$$
$$= 18 \text{ CY}$$

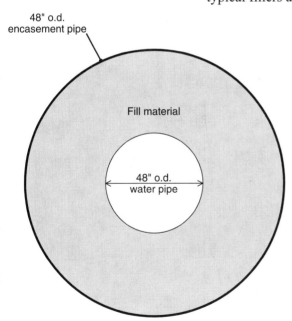

Figure 9-16
Calculating fill volume inside encasement pipe

Excavation Equipment

The excavation equipment you use is also a factor in determining trench width. Buckets on most excavating equipment have widths of 12, 18, 24, 30, 36, or 48 inches. For example, say the trench material width is 32 inches. Assume that the bucket's 36 inches wide.

Topsoil, Slopes & Ditches **151**

Overcut

The wider the trench is, the easier it is to work inside. Most earthwork contractors prefer to work with a wider trench, but extra excavation that's not in the specs or required for safety isn't always paid work. What does this have to do with your estimate? It means you have to track the overcut quantities separate from quantities that are pay items.

We'll use the manhole shown in Figure 9-17 as our example. According to the plans, this is a cast-in-place 48-inch-diameter 7-foot-deep manhole. The contract limits pay items to the same dimensions as the finished structure. That means you're paid to excavate a hole that's the same size as the finished manhole (48 inches across and 7 feet deep). Let's find the payable total excavated volume first.

Volume (CY) = (depth × area of one end of the manhole) ÷ 27

Where

- depth = 7 feet
- area one end of manhole = πr^2
- π = 3.1416
- r = 2 feet

Here's the math for the area calculation:

Area (SF) = 3.1416 × 22

= 12.57 SF

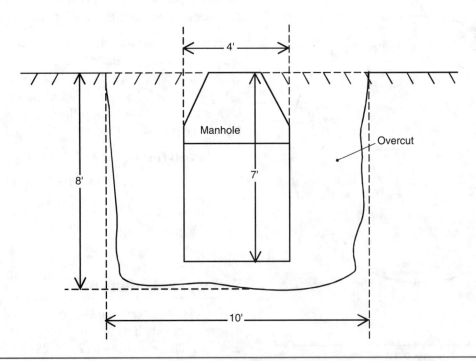

Figure 9-17
Estimating excavation and overcut volumes

Next, you calculate excavation volume in cubic yards:

Volume (CY) = (7 × 12.57) ÷ 27

= 87.99 ÷ 27

= 3.26 CY

That's the excavation volume you'll be paid for, but reality demands a lot more excavation. There's no room in this size excavation for the workers who form, pour and strip the forms from the manhole. In Figure 9-15, the dashed lines show the actual excavation outline and dimensions. You calculate this volume the same way as the paid volume.

Volume (CY) = (depth × area one end) ÷ 27

Where

- depth = 8 feet
- area one end = πr^2
- π = 3.1416
- r = 5 feet

And here's all the math

Area (SF) = 3.1416 × 5^2

= 78.54 SF

Next you calculate the excavation volume in cubic yards:

Volume (CY) = (8 × 78.54) ÷ 27

= 628.32 ÷ 27

= 23.27 CY

Now let's look at what this means in dollars and cents. The total excavated volume comes to about 23 CY. But in terms of pay items, the total is only 3 CY. The 20 CY difference is *overcut*.

Make it a point to include a clause in your standard contract allowing payment for overcut. Then be sure to calculate and include overcut in your bid. Sometimes you'll have to be flexible about this clause. You'll probably have to negotiate. Perhaps offer a lower rate for overcut if that's what it takes to get the job, but try not to work for free.

Get Organized, Stay Organized

Every estimate must be checked. You'll save time here by being organized. Clearly identify each calculation, keep separate steps separate, and always finish the task before moving ahead. Good organization and consistency are the key to accurate, professional estimating.

10 Basements, Footings, Grade Beams & Piers

■ ■ ■ ■ ■ ■ ■ ■ ■ ■

In this chapter I'll explain how to estimate excavation for basements, footings, and piers. We'll break the project down into logical steps to help eliminate (or at least reduce) the possibility of errors and omissions. The best estimators are both consistent and systematic. They follow the same procedures and use the same methods on every estimate. Develop good estimating habits and you'll produce more good estimates.

Estimating Basement Excavation Quantities

Although the examples and calculations in this chapter are for basement excavation, you use the same estimating procedure to estimate holding tanks, wells, lift stations, or any type of underground structure. Even though I refer to basements here, the concepts and processes also apply in many other types of work.

At first glance, it would seem that estimating basements should be a snap. Volume equals length times width times depth, right? Well, yes and no. That will get you the volume inside the basement walls. But there's more to excavation for basements than just the basement itself. How much extra working room do you need outside the basement walls? How much will you have to slope the side walls? How big will the equipment ramp have to be? Every basement excavation job will include complications like these.

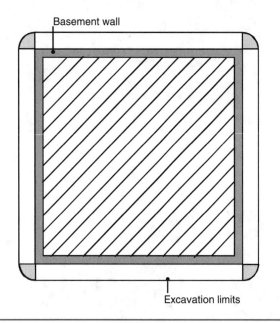

Figure 10-1
Calculating the volume of excavation

The Slope You Select

I've already explained that it's not safe to work in an excavation with vertical walls. Most types of soil are unstable at steep angles. Even firm soil can't be counted on to remain vertical in wet weather. So we'll nearly always have to slope the sides of basement excavations.

Figure 10-1 shows a square basement. The area beyond the basement walls that's not shaded has to be sloped back away from the pit during construction. When construction is finished and concrete in the basement walls has set, this sloped area will be backfilled and compacted.

Generally the excavation contractor determines the angle of slope at the basement perimeter. Steeper slopes require less excavation but may not be safe in some types of soil. More gradual slopes make it easier to get in and out of the pit, but require more excavation.

You'll recall from Chapter 9 that *run* is the horizontal dimension of a slope and *rise* the vertical dimension. The run and rise are usually expressed as a ratio. A ratio consists of two numbers separated by a colon. Slopes written as ratios list the run first and the rise second. The most common value for rise is 1. Typical values for run range from 1 to 4. Let's look at some examples. A 1:1 slope rises 1 foot in height for each foot of length. That works out to a 45-degree slope. A 4:1 slope is a very shallow slope. It rises only 1 foot in height over a 4-foot length. A 1:4 slope, meanwhile, is rather steep. It rises 4 feet in height for each foot of length. You'll rarely excavate a steeper slope than 1:4. A slope that's any steeper is unstable and unsafe.

The slope you select depends on the type of soil, safety considerations, the surrounding work space, and the construction methods being used. Figure 9-12 in Chapter 9 is a table listing average maximum safe slopes for different soil types. In most cases, a slope of 2:1 is safe.

If your job site includes more than one soil type, always use the slope, or *angle of repose,* that's recommended for the least-stable soil. For example, let's say the site has sand underneath a layer of clay. What slope do you use for the entire project? Sand is less stable than clay, and looking back to Figure 9-12, in Chapter 9, we find that a 3:1 slope is recommended for sand.

Here's something else to think about when you're choosing the slope for a basement excavation. What kind of concrete pour is planned for the footings, walls, and slab? The reach of the chute on a transit mix truck is between 15 and 18 feet. If the trucks can't get that close to the forms, they can't pour direct from the chute. The more gradual you make the slope, the further away the trucks have to park from the forms. For example, given a 2:1 slope and a depth of 10 feet, trucks won't be able to pour direct from the chute. The run of

156 *Estimating Excavation*

this slope comes to 20 feet (2 times 10) and that's too far to reach with most chutes. You'll have to build a ramp for the trucks or place the concrete by pump, or with a bucket. They're all expensive solutions.

Finding Volume — Outside Basement Walls

Once we know the run and rise of a slope, the total rise in feet, and the length of the slope, we can use triangles to calculate volumes. We'll start by finding the volume of the sloped area outside of the basement walls. In Figure 10-1 this area has no shading. Notice that the area with no shading means excluding all four of the corners. Later on in this chapter we'll cover how to find the volume of the corner areas.

In Figure 10-2 you see a cut-away partial view of a basement excavation. Notice that I said *partial*. Figure 10-2 shows only the excavation work done *outside of the basement walls*. The width of the footing (2 feet in Figure 10-2) is a dimension that appears on most plan sheets. However, the width of the work space (3 feet in Figure 10-2) isn't typically shown on plans. That's because the amount of work space is up to you. Workers from many different trades use this work space. First in line are the workers who build the forms for the basement walls. Later, workers installing DWV and HVAC and electrical lines also use this space. How much work space you allow depends on:

- the type of concrete forms
- the total excavation depth
- the number and type of utility lines
- the soil conditions

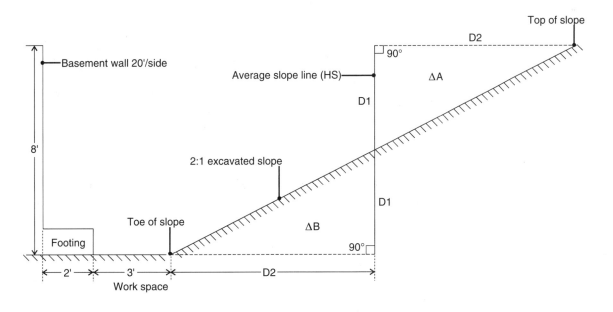

Figure 10-2
Using an average slope line

Basements, Footings, Grade Beams & Piers

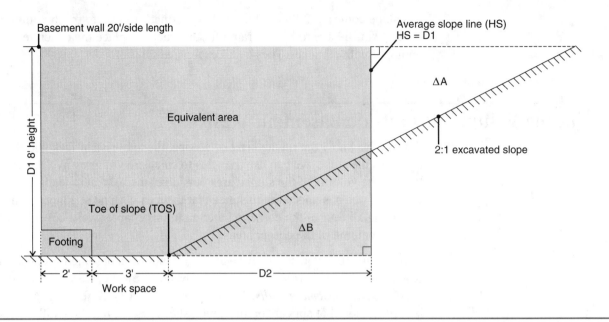

Figure 10-3
Using equivalent area

My Rule of Thumb is to allow no less than 3 feet and no more than 5 feet of work space. Of course, that's adjustable to fit the circumstances and the needs of your fellow subcontractors.

Before we start on the volume calculations I want to mention the basement wall footings. That excavation work isn't usually included as a part of basement excavation. I recommend that you follow the industry practice, and include footing excavation in the general excavation category. We'll assume that's been done here in our example too.

Using the Average or Half Slope Line and Equivalent Area to Calculate Basement Excavation Volume

To make slope calculations easier, estimators usually draw a vertical line down the midpoint of the slope. This is called the *average slope line* or *half slope line*, or *HS* for short. The average slope line in Figure 10-2 is the line labeled D1. Line D1 intersects the slope exactly at its horizontal midpoint.

The two dashed lines in Figure 10-2 marked D2 extend the existing and proposed ground levels to intersect with line D1. Notice that the lines D1, D2 and the excavated slope line form a pair of right triangles. We'll call them triangle A and triangle B.

The triangles also have the same length hypotenuse. That's the excavated slope line side of each triangle. Right triangles with the same hypotenuse are identical. That means that the area and volume of triangles A and B are also identical. This simple fact makes calculating the excavated area and volume much easier. I'll explain how it works using Figure 10-3. In Figure 10-3, triangles A and B are identical. Therefore, replacing triangle A with triangle B has no effect on the total volume. However, this substitution does change the shape of the region we're working with. Compare Figure 10-2 with Figure 10-3 and the effect is obvious. What was an irregular region in Figure 10-2 becomes, in Figure 10-3, a regular, shaded, rectangle labeled *equivalent area*.

You already know how to find the volume of a rectangle:

Volume = length × width × height

Here are the values for the variables in Figure 10-3:

- *Length* – one side of the basement = 20'
- *Height* – wall height, including footing = D1 or 8'
- *Width* – total horizontal distance from outside face of the basement wall to the average slope line, or the sum of the footing width (2 feet), the work space width, plus D2.

Obviously, we can't go any further without finding the length of line D2. D2 is the horizontal distance measured from the toe of the slope, or TOS, to the half slope line, or HS. Now let's see how you find its length.

Finding Toe of Slope to Half Slope — TOS to HS

This horizontal distance is equal to half the slope's total run, or:

Formula for TOS to HS

TOS to HS = total run ÷ 2

The total run of a slope equals:

Formula for total run

Total run = total rise × run ÷ rise

I believe in learning by doing, so let's try these formulas out with a couple of examples.

Example 1

The run to rise ratio is 1:1 and the total rise is 8 feet.

What is the TOS to HS distance?

Step 1 – Calculate total run

Total run = total rise × (run ÷ rise)

Total run = 8 × (1 ÷ 1)
$\quad\quad\quad$ = 8'

Step 2 – Calculate TOS to HS distance

TOS to HS = total run ÷ 2

TOS to HS = 8 ÷ 2
$\quad\quad\quad\quad$ = 4'

Example 2 (based on Figure 10-3)

The run to rise ratio is 2:1, the total rise is 8 feet, and *D2* = TOS to HS distance.

Basements, Footings, Grade Beams & Piers **159**

Repeat the steps used in Example 1 to find the length of D2.

Step 1 – Calculate total run

Total run = 8 × 2 ÷ 1
= 16'

Step 2 – Calculate length of D2

Remember, D2 = TOS to HS distance.

D2 = 16 ÷ 2
= 8'

Finding Width for the Equivalent Area

You recall that the width is the sum of: footing width, work space width and D2. Figure 10-3 supplies the first two values and we just found D2. Now find their sum.

Width = 2 + 3 + 8
= 13'

Calculating Volume by Equivalent Area

The dimensions of the *Equivalent area* in Figure 10-3 are:

- length = 20'
- width = 13'
- height = 8'

Using this formula, find its volume in cubic yards:

Volume (CY) = (length × width × height) ÷ 27

Volume (CY) = (20 × 13 × 8) ÷ 27
= 2,080 ÷ 27
= 77 CY

Basement Wall Dimensions

Before we go any further, let's stop for a bit to take a look at something that's tripped up many an earthwork estimator. Did you know that there are three ways to show basement wall dimensions? All three are legitimate systems. And all three are widely used in the construction business. The three systems are:

- inside wall line to inside wall line
- outside wall line to outside wall line
- center of wall line to center of wall line

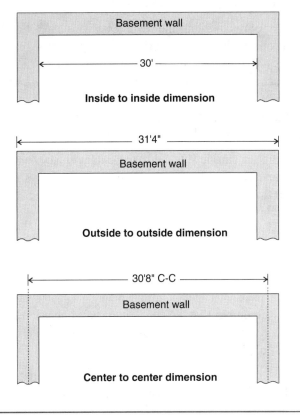

Figure 10-4
Dimensional systems

Figure 10-4 shows three plan views of the same part of a basement wall. In all three I've marked the same dimension, the length of the wall. But each time I get a different result. Why is this? Simple, I used a different dimensioning system each time.

It's up to you, as an earthwork estimator, to study the project plans and figure out the dimensioning system that's been used on the plans. So let's take a close look at each system, starting at the top of Figure 10-4.

Inside wall line to inside wall line

This system is typically used for *interior dimensions*. Measurements are from the inside face of one wall to the inside face of the opposite wall. In this dimensional system, wall thickness is not a factor that's considered at all.

Outside wall line to outside wall line

This system is typically used for *exterior dimensions*. Measurements are from the outside surface of one wall to the outside surface of the opposite wall. The resulting dimensions include the thickness of both walls.

Center of wall line to center of wall line

It's easy to tell when this system's used on a set of plans. You'll see *C-C* written alongside the measuring line. That stands for center to center. Measurements are from the center of one wall to the center of the opposite wall. The resulting dimension includes the width of one wall (half the width for each of two walls equals the width of one wall).

Unless basement plans use outside to outside dimensions, you'll need to adjust two dimensions, length and width, by the thickness of the wall or walls before you calculate the excavation volume. Think a few inches won't make much difference in an excavation that's 40 feet across? It adds up faster than you think. Assume you're estimating the excavation volume for a basement with these dimensions:

- area = 30' × 40'
- height = 8'
- wall thickness = 8"

These are inside to inside dimensions, but you assume they're outside to outside dimensions. Your estimate is short by almost 28 cubic yards. We'll assume a conservative cost of, say, $3 per yard. Multiply it out and you'll find that you've made an $84 mistake. So it really does pay to be sure that you're working with the right dimensions.

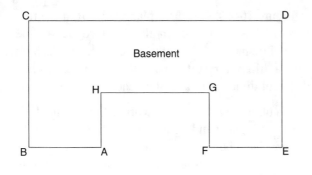

Figure 10-5
Plan view of a basement with two inside corners, G and H, and six outside corners, A, B, C, D, E and F

Inside vs. Outside Corners

Up to this point all the basements we've dealt with had four walls and four corners. As Figure 10-5 shows, basements sometimes have offsets and setbacks. Also note that Figure 10-5 has two types of corners — outside corners and inside corners. The outside corners are: A, B, C, D, E, and F. The inside corners are G and H.

Here's a simple rule for basements with square corners. The number of outside corners is always four more than the number of inside corners. Put that as a formula and it's:

Number outside corners – number inside corners = 4

In Figure 10-6 you'll find four basement plans with inside and outside corners that prove this formula.

Look back at Figure 10-1. Remember that we calculated the excavation in the sloped areas outside the walls, but ignored the volume in the shaded areas at the corners. Now let's pick up the volume at those corners.

Figure 10-7A is a plan view of a portion of a basement wall. Notice that one inside corner and one outside corner are shown. Arrows point down the slope toward the basement. Both the inside and outside corners are rounded and sloped toward the basement pit. It's easy to see that you'll do a lot more excavation for outside corners than you would for inside corners. How much more? Let's find out. You'll be glad to know that it won't take a separate calculation for each corner. There's an easier way.

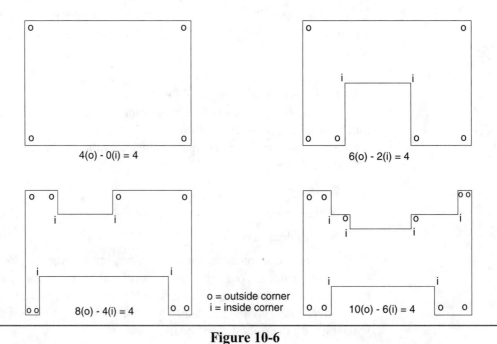

Figure 10-6
Basements with square corners always have four more outside corners than inside corners

162 *Estimating Excavation*

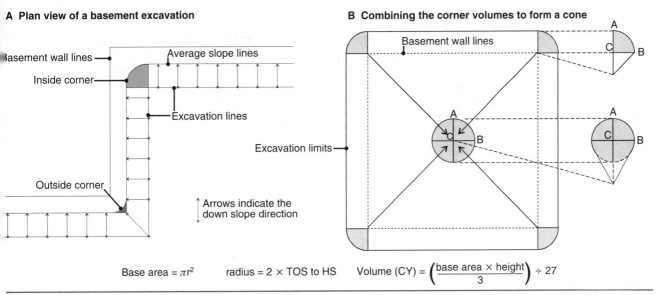

Figure 10-7
Calculating corner volume

Finding Corner Volumes — the Easy Way

At inside corners there's less material to remove. At outside corners, there's more material to remove. Luckily, these two amounts cancel out one for one. Remember the rule about corners? You always have four more outside corners than inside corners. You only need to calculate the volume of those four outside corners.

How do you go about finding the volume of four sloped outside corners? It's easier than you think. Suppose you joined all four corners together at the deepest point. What you'd have is an upside-down cone. The height of the cone is equal to the height of the basement wall. The formula for the volume of a cone is:

Formula for volume of a cone

Volume (CY) = [1/3(base area × height)] ÷ 27

The base of a cone is a circle, and the formula for the area of a circle is:

Base area = πr^2

But what is the radius of this circle? Take another look at Figure 10-7B. In the shaded corner areas a straight line drawn from point C out to the excavation limit line is a radius. It's equal to two times the TOS to HS distance.

Now let's find the volume of the cone shown in Figure 10-7B. Assuming that the walls are 8 feet high and the slope is 2:1, the radius represented by the line CB would be 16 feet. You can use the following formulas to find the radius, then solve for the volume of the cone:

TOS to HS = total run ÷ 2

Total run = total rise × run ÷ rise

Radius = TOS to HS × 2

Basements, Footings, Grade Beams & Piers

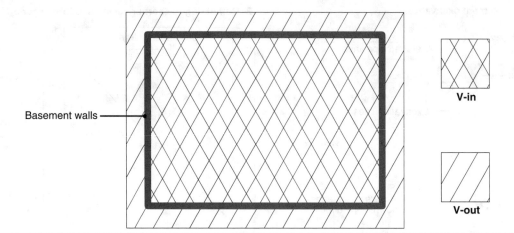

Figure 10-8
V-in and V-out portions of a basement excavation

Step 1 – Calculate total run

$$\text{Total run} = 8 \times 2 \div 1$$
$$= 16'$$

Step 2 – Calculate TOS to HS distance

$$\text{TOS to HS} = 16 \div 2$$
$$= 8'$$

Step 3 – Find the length of the radius CB

$$CB = 8 \times 2$$
$$= 16'$$

Step 4 – Find the volume of the cone in cubic yards

$$\text{Volume (CY)} = [1/3 \times 3.1416 \times 16^2 \times 8] \div 27$$
$$= (0.333 \times 3.1416 \times 256 \times 8) \div 27$$
$$= 2{,}142.5 \div 27$$
$$= 79 \text{ CY (rounded)}$$

Calculating the Total Volume for Basement Excavation

I recommend that you keep volume estimates inside the basement walls separate from volume estimates outside the basement walls. The soil you remove from inside the basement walls will have to be spread out over the site or hauled away. We'll call the material you remove from *inside the basement walls* the *V-in*. We'll call the material you remove from *outside the basement walls* the *V-out*. This material's stockpiled to be used as backfill when construction's complete. Figure 10-8 is a plan view of a basement excavation showing both the V-in and the V-out.

As we saw in Chapter 9, backfill work involves extra steps. Backfill requires not only compacting but also some hand work. If you don't want to do that extra work for free, you'll take my advice — calculate, estimate and bid V-out volumes separately.

Finding Volumes for Vertical Wall Basement Excavations

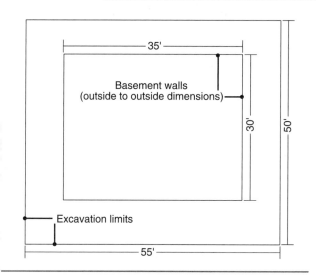

Figure 10-9
Site plan for a sample basement excavation

While most basement excavations have sloped sides, it's possible to find basement excavations with vertical walls. Some soils are capable of standing vertically, at least for a while. Suppose the job site's hemmed in with other buildings. The surrounding buildings all have basements that are just as deep as your basement. In a case like this you don't have any choice. You have to excavate vertical walls. You calculate total volume for vertical walls a bit differently.

Figuring Total volume, V-out and V-in

When the walls are vertical, you use this formula to find V-out in cubic yards:

V-out (CY) = Total volume (CY) – V-in (CY)

Here's the formula you use to find V-in volume in cubic yards — don't forget to use outside to outside dimensions:

V-in (CY) = ([basement] length × width × depth) ÷ 27

And the formula for total volume in cubic yards is:

Total volume (CY) = ([excavation] length × width × depth) ÷ 27

Let's do a simple example. Figure 10-9 is a site plan for a basement. The excavation depth is 8 feet and the specs list both the basement wall line and the excavation limit line as verticals. We'll use the formulas above to find three volumes for this excavation: V-in, total volume, and V-out. Figure 10-10 is a *Quantities Take-Off Sheet* for basement excavation volume. Notice that I round off both the total volume and the V-in volume before I subtract to find the V-out volume. This is a common practice in earthwork estimating. The small difference in volume isn't significant.

Finding Volumes for Sloping Wall Basement Excavations

Now we'll calculate the volumes for a basement excavation with sloping sides. Here's an important tip. Always find the V-out volume first, then figure the V-in volume, then add to find total volume. Make sure you follow this sequence for a basement including setbacks or offsets.

For this example we'll use Figure 10-11, the plan view, plus a detail view showing the equivalent area for a basement excavation.

Quantities Take-off Sheet

Project: Sample basement **Date:** _____

Quantities for: Excavation **Sheet** 1 **of** 1

By: DB   **Checked:** LL   **Misc:** _____

Volume type	Length (ft)	Width (ft)	Depth (ft)	Volume (CF)	Volume (CY)	Misc.
Total	55	50	8	22,000	814.81	
V-in	35	30	8	8,400	311.11	
V-out*	—	—	—	—	504.00	

Note:
*V-out volume = Total volume - V-in volume
　　　　　　= 815-311
　　　　　　= 504 CY

Figure 10-10
Worksheet for sample basement excavation volume calculations

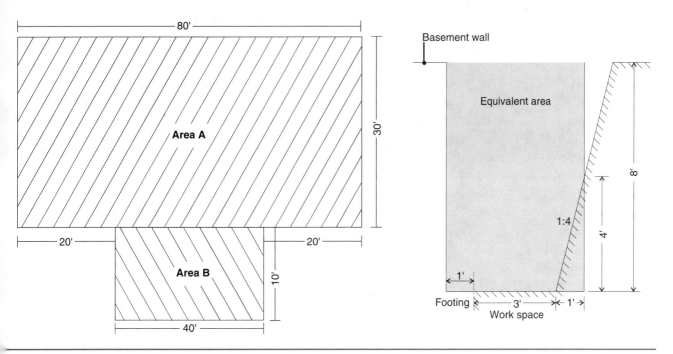

Figure 10-11
Calculating the volume of a sloping wall basement excavation

Let's start off with a few dimensions.

- Wall height = 8'
- Wall length = 240'
  Wall length is the sum of the lengths of all the basement walls.
  30 + 80 + 30 + 20 + 10 + 40 + 10 + 20 = 240
- Width of equivalent area = 5'

That's the sum of: outside footing width, work space and TOS to HS distance

1 + 3 + 1 = 5

Figuring total volume, V-out and V-in

Here are the formulas we'll use to find V-out volume in cubic yards:

V-out (CY) = (wall length × equivalent area) ÷ 27

Equivalent area (SF) = width × height

Equivalent area (SF) = 5 × 8
$\qquad$ = 40 SF

V-out (CY) = (240 × 40) ÷ 27
$\qquad$ = 9,600 ÷ 27
$\qquad$ = 355.56 CY

Basements, Footings, Grade Beams & Piers 167

Next we'll find the corner volume in cubic yards with this formula:

Corner volume (CY) = [(πr^2 × height) ÷ 3] ÷ 27

Where:

- r (the radius of the cone's base) = 1'
- π = 3.1416
- Height = 8'

Corner volume (CY) = [(3.1416 × 1^2 × 8) ÷ 3] ÷ 27

$\qquad\qquad\qquad$ = (25.13 ÷ 3) ÷ 27
$\qquad\qquad\qquad$ = 8.38 ÷ 27
$\qquad\qquad\qquad$ = 0.31 CY

You complete the V-out calculations using this formula:

V-out, total (CY) = V-out + corner volume

V-out, total (CY) = 355.56 + 0.31
$\qquad\qquad\qquad$ = 355.87 CY

To calculate the inside volume, the V-in, we first divide the basement space into two regular rectangles. In the plan view in Figure 10-11 they're labeled Area A and Area B. The dimensions and areas of these two rectangles are:

- Area A = 80 × 30 = 3,400 SF
- Area B = 10 × 40 = 400 SF

You use this formula to find V-in in cubic yards:

V-in (CY) = [(area A + area B) × depth] ÷ 27

V-in = [(3,400 + 400) × 8] ÷ 27
$\qquad$ = (3,800 × 8) ÷ 27
$\qquad$ = 30,400 ÷ 27
$\qquad$ = 1,125.93 CY

To find total excavation volume in cubic yards, round both the V-out and the V-in to full cubic yards and find their sum.

Total excavation volume (CY) = V-out (full CY) + V-in (full CY)

Total excavation volume (CY) = 356 + 1,126
$\qquad\qquad\qquad\qquad\qquad$ = 1,482 CY

Basement Excavation Depths

Be careful to determine the correct depth for basement excavations. Check the indicated depth of the basement on the plans, the existing ground lines, the topsoil depth (both in the surrounding area and under the basement slab), and the floor slab depth.

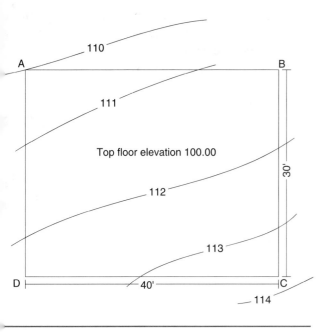

Figure 10-12
Finding the "real" excavation depth for a basement

Assume Figure 10-12 shows four basement corners, and the contours show existing elevations. Suppose the specs give a finished elevation for the basement slab's top of 100.00 feet. What's the excavation volume?

In Chapter 6 you learned to use a grid system in an area take-off from a topo map. Using that method, find the basement corner depths in Figure 10-12 and then subtract the top-of-slab elevation (100.00).

- Point A = Elevation 110.00 – 100.00 = 10.00 feet
- Point B = Elevation 111.35 – 100.00 = 11.35 feet
- Point C = Elevation 113.70 – 100.00 = 13.70 feet
- Point D = Elevation 112.20 – 100.00 = 12.20 feet

Find the average depth using the values you found above.

$$\text{Average depth} = (10.00 + 11.35 + 13.70 + 12.20) \div 4$$
$$= 11.81 \text{ feet}$$

We haven't made any allowance for topsoil that's to be stripped, or for topsoil that's replaced later, or for the thickness of the slab. Let's assign values to each of these factors and then see how to incorporate them into excavation depth.

Assume the following:

- Slab thickness = 6" (add)
- Stripped topsoil depth = 8" (subtract)
- Replaced topsoil depth = 4" (add)

We'll convert these values from inches to feet (divide by 12), work through the addition and subtraction, and apply the result to the average depth we already calculated. The result's the real depth of the excavation shown in Figure 10-12.

$$\text{Real depth} = 11.81 + [(6 + 4 - 8) \div 12]$$
$$= 11.81 + (2 \div 12)$$
$$= 11.81 + 0.17$$
$$= 11.98'$$

To find the excavation volume in cubic yards you multiply the real depth by the area of the basement and divide by 27. Here's how it works for this example:

$$\text{Area} = 30 \times 40$$
$$= 1{,}200 \text{ SF}$$

$$\text{Basement excavation volume} = (1{,}200 \times 11.98) \div 27$$
$$= 14{,}376 \div 27$$
$$= 532.44 \text{ CY}$$

Here's something else to remember as you work with basement excavation volumes. Many basements are built with their walls extending 1 to 2 feet above the finished grade. This reduces the excavation depth by the same amount. This information appears on the plans. Check your plans carefully.

Sample Basement Estimate

To test your understanding, figure the excavation volume for the basement shown in Figure 10-13. Figure 10-14 is a detail showing the V-out region with the average slope line. Base all your calculations on the data given in these two figures, and the job specifications that follow. I recommend that you work all the way through your estimate. Then check it against my work as shown in Figures 10-15A, 10-15B and 10-15C.

Here are the job specifications:

- Soil type is a sand/clay mix
- Strip topsoil depth = 10"
- Replace topsoil depth = 4"
- Wall height = 8'
- Basement slab thickness = 8"
- Elevation, at top of slab = 94.0'
- Footing width (outside the basement wall) = 1'
- Workspace = 4'

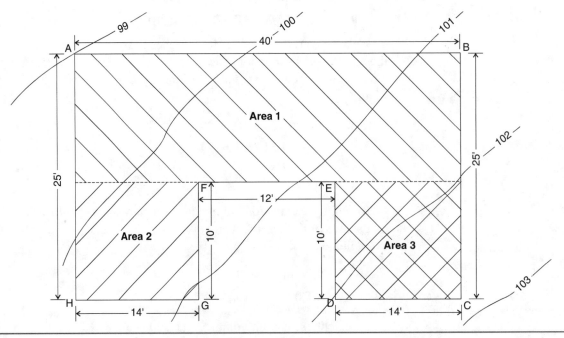

Figure 10-13
Plan view of the sample basement with existing contour lines

170   *Estimating Excavation*

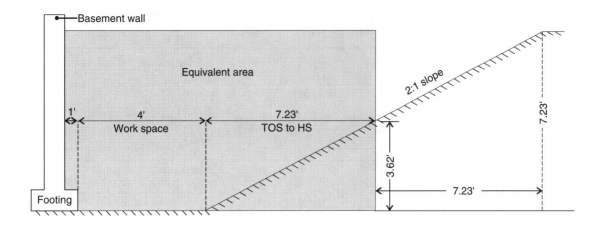

Figure 10-14
Average slope line detail for sample basement estimate

Step 1 – Determine the depth

Using the grid square take-off method, find the elevations for points A through H. Subtract the top of slab elevation (94.0 feet) from each to find the excavation depth. Then find the average depth for points A through H.

Calculate real depth from average depth by adding 12 inches (8 inches for the floor thickness and 4 inches for topsoil replaced) and subtracting 10 inches (for topsoil removal). The correct result for real depth is 7.23 feet. If you get a different number, check your figures against mine (Figure 10-15A).

Step 2 – Find the equivalent area

Let's find a safe slope for this job. Checking Figure 9-12 (see Chapter 9) we find that 2:1 is a safe slope for clay. Figure 10-14 shows slope profile at 2:1. We also know that the total rise is 7.23 feet. That means the total run equals 14.46 feet ([7.23 × 2] ÷ 1 = 14.46) and the TOS to HS distance is 7.23 feet (14.46 ÷ 2 = 7.23).

Before we can find the equivalent area, we need to know the total width. Figure 10-14 has all the data you need. The outside footing is 1 foot. The work space is 4 feet, and the TOS to HS is 7.23 feet. Add these together to arrive at a total width of 12.23 feet. Now let's calculate the equivalent area in square feet. Simply multiply the total width (12.23 feet) by the average depth (7.23 feet).

Equivalent area (SF) = 12.23 × 7.23
   = 88.4 SF

Step 3 – Calculate V-out

Figure 10-13 includes the lengths of all the basement walls. Their sum is the total wall length. You use total wall length and the equivalent area to find wall V-out in cubic yards. Next you find corner V-out in cubic yards. Finally, you find the sum of wall and corner V-outs, and that's your total V-out. The correct answer is 549.7 cubic yards. See Figure 10-15B to review the math.

Basements, Footings, Grade Beams & Piers

Quantities Take-off Sheet

Project: Sample estimate
Quantities for: Determine real depth
By: _____ **Checked:** _____

Date: _____
Sheet 1 **of** 3
Misc: _____

Point	1 Elev.	2 Elev.	C.I.	Measured distance	Point elevation	Depth (point elev. – top slab elev. 94.0')
A	99	100	1	0.0	99.0	5.0
B	101	102	1	0.2	101.2	7.2
C	102	103	1	0.85	102.85	8.85
D	102	103	1	0.0	102.0	8.0
E	101	102	1	0.3	101.3	7.3
F	100	101	1	0.5	100.5	6.5
G	101	102	1	0.15	101.15	7.15
H	100	101	1	0.45	100.45	6.45
					Total	56.45

Note:

$$\text{Average} = 56.45 \div 8 = 7.06$$

$$\text{Real depth} = 7.06 + \left(\frac{8 \text{ (slab)} + 4 \text{ (replace topsoil)} - 10 \text{ (strip topsoil)}}{12} \right)$$

$$= 7.06 + 0.17$$
$$= 7.23'$$

Figure 10-15A
Take-off sheet calculating average and real depth for sample project

Calculation Sheet

Project: Sample Estimate   **Date:** _____

V-out, walls (CY) = $\dfrac{\text{perimeter} \times \text{area}}{27}$

Perimeter = AB + BC + CD + DE + EF + FG
            + GH + HA (See Figure 10B)
         = 40 + 25 + 14 + 10 + 12 + 10 + 14 + 25
         = 150'

Area (SF) = width × depth

Depth = 7.23' (See Figure 10-15A)

Width = footing + work space + TOS to HS
        (see Figure 10-14)
      = 1 + 4 + 7.23
      = 12.23'

Area = 12.23 × 7.23
     = 88.4 SF

V-out, walls (CY) = $\dfrac{150 \times 88.4}{27}$

= $\dfrac{13{,}260}{}$
= 491.1 CY

V-out, corners (CY) = $\dfrac{1/3 \pi r^2 \times h}{27}$

π = 3.1416
r = 2 × TOS to HS (see Figure 10-14)
  = 2 × 7.23 = 14.46'

V-out, corners (CY) = $\dfrac{0.33 \times 3.1416 \times 14.462 \times 7.23}{27}$

= $\dfrac{0.333 \times 3.1416 \times 209.09 \times 7.23}{27}$

= $\dfrac{1{,}581.49}{27}$

= 58.6 CY

Total V-out (CY) = V-out, walls + V-out, corners
                 = 491.1 + 58.6
                 = 549.7 CY

Conclusion

Figure 10-15B
Calculating V-out for the sample project

Quantities Take-off Sheet

Project: Sample estimate **Date:** _____

Quantities for: V-in & Total **Sheet** 3 **of** 3

By: _____ **Checked:** _____ **Misc:** _____

Area	Length (ft)	Width (ft)	Depth (ft)	Volume CF	Volume CY	
1	40	15	7.23	4,338.0	160.67	
2	10	14	7.23	1,012.2	37.49	
3	10	14	7.23	1,012.2	37.49	
				Total V-in	235.65	

Note:
Total volume (CY) = Total V-in + Total V-out
= 235.65 + 549.7
= 785.35 CY

Figure 10-15C
Calculating V-in and total volume for the sample project

Step 4 – Determine V-in

First, you divide this irregular shape into three regular rectangles. They appear in Figure 10-13 labeled as Area 1, Area 2 and Area 3. Find the volume of each rectangle in cubic yards. The sum of these three volumes is the total V-in in cubic yards. The right answer is 235.65 cubic yards. Figure 10-15C shows my math.

Step 5 – Figure total volume

Find the total volume by adding V-out to V-in. Total excavation volume for this project is 785.35 cubic yards. My calculations appear in Figure 10-15C.

Estimating Ramps

On a basement job, you'll usually have to cut a ramp to move excavation equipment in and out of the pit. The location, size, and material of this ramp affect the excavation quantities. But most estimators don't actually estimate the volume of the ramp. The only purpose of a ramp is to provide temporary access to the pit. It increases efficiency and more than pays for itself in time saved.

However, you may have to estimate ramp excavation occasionally. So I'll explain the estimating procedure. The mathematics required to make a close technical estimate of an equipment ramp is beyond the scope of this book. And it's highly unlikely that you'll ever need to make exact calculations for a ramp anyway. Rather than provide details you'll never use, I'll explain an easy way to get results that are acceptable for most, if not all, purposes.

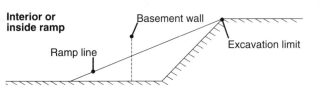
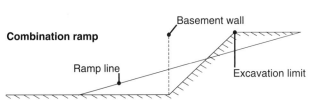
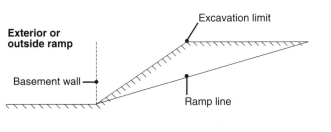

Figure 10-16
Three kinds of ramps

Ramps are classified by their location relative to the outside limits of the basement wall. They are either interior, exterior, or a combination of the two. Figure 10-16 shows each type of ramp. An interior ramp is totally within the limits of the excavation. An exterior ramp is located outside the basement wall and may be as much as 10 feet away from the wall. The combination ramp is both inside and outside the excavation area. The type of ramp determines the quantities and placement procedure — and, of course, the cost.

An interior ramp is the most expensive and the least desirable. The excavation equipment has to work around the ramp until all the wall work is done except the wall area that falls within the ramp. At that point, it takes hand work or a backhoe to remove the ramp. Don't use an interior ramp if space is available outside the basement.

An outside ramp is the least expensive because no part of it has to be removed before the walls are

Basements, Footings, Grade Beams & Piers **175**

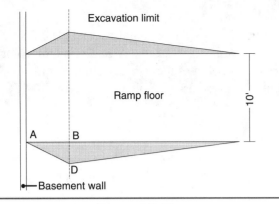

Figure 10-17
Plan view of outside ramp

completed. Of course, the ramp has to be backfilled and compacted when basement excavation is finished. Note the outside ramp shown in Figures 10-17 (plan view) and 10-18 (elevation view). We'll use triangle area formulas to calculate ramp volumes based on both right and oblique triangles.

The two shaded areas in Figure 10-17 are the sloping embankment along the sides of the ramp. I recommend that you ignore the soil volume moved in this area. The formulas for calculating this embankment are complex and the volumes are small. To compensate for ignoring the volume in the shaded areas, we'll be a little more generous in calculating volumes in other areas.

Look at Figure 10-18. The oblique triangle labeled V-ramp defines the volume of earth to be moved for this ramp. You use three formulas to find the area of an oblique triangle:

• • • • • • • • • • • • • •
Formula for volume of a ramp

$A = r \times s$

Where:

$r = \sqrt{[(s - a) \times (s - b) \times (s - c)] \div s}$

$s = \tfrac{1}{2} \times (a + b + c)$

To find the area of the V-ramp triangle, we need to know the lengths of the three sides: a, b, and c. Remember that the V-ramp triangle isn't a right triangle, so we can't use the Pythagorean theorem to find the lengths of the sides.

However, Figure 10-18 does include two right triangles, labeled RT1 and RT2. Notice that the hypotenuse of RT1 is the same as side c in the V-ramp triangle and that the hypotenuse of RT2 is the same as side b in the V-ramp triangle.

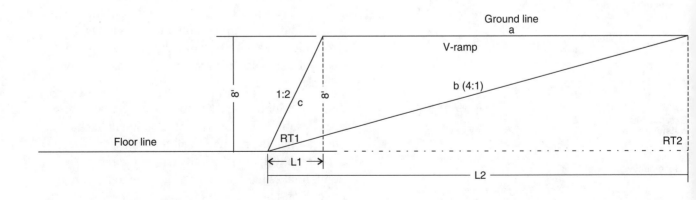

Figure 10-18
Elevation view of outside ramp

Now, what do we know about these triangles RT1 and RT2?

We know that the height of triangle RT1 is 8 feet. That's the same as the excavation depth. We also know that the slope of the hypotenuse of RT1 is 1:2. That means the rise is 2 feet for each 1 foot of run. So if the rise is 8 feet, the run must be 4 feet. The run is the same as length L1. So L1 is 4 feet. Write *4'* in beside L1 on Figure 10-18.

The Pythagorean theorem says:

Pythagorean theorem

The square of the hypotenuse of a right triangle equals the sum of the squares of the other two sides.

In this case, the square of side c equals the square of 8 feet plus the square of 4 feet. The actual length of side c is the square root of the result.

$$c^2 = 8^2 + 4^2$$
$$= 64 + 16$$
$$= 80$$

Find the square root of 80 to determine the length of side c. The result is 8.94', which we'll round to 9'. Pencil *9'* in by side c in Figure 10-18.

Now let's find the length of side b of the V-ramp triangle. This side is the same as the hypotenuse of triangle RT2. It's also the horizontal length of the ramp floor. You get to determine this length. It should be as short as possible to save on space and yardage, but long enough so the workers and their equipment can use it easily. Let's assume a slope of 4:1 for the ramp, 4 feet of run for each 1 foot of rise. So, for the rise of 8 feet, the run is 32 feet. Mark *32'* by L2 in Figure 10-18. Again, using the Pythagorean theorem:

$$b^2 = 32^2 + 8^2$$
$$b^2 = 1{,}024 + 64$$
$$b^2 = 1{,}088$$

Find the square root of 1,088 and that's the length of side b. The result is 32.99', rounded to 33'. Pencil *33'* in by side b in Figure 10-18.

The length of side a is L2 minus L1, or 29' (33 − 4 = 29).

Now you know the lengths of all three sides of the V-ramp oblique triangle:

- a = 29'
- b = 33'
- c = 9'

You figure its area using these formulas:

$$A = r \times s$$
$$r = \sqrt{[(s-a) \times (s-b) \times (s-c)] \div s}$$
$$s = 1/2 \times (a + b + c)$$
$$s = 1/2\,(29 + 33 + 9)$$
$$= 1/2 \times 71$$
$$= 35.5$$

$$r = \sqrt{[(35.5 - 29) \times (35.5 - 33) \times (35.5 - 9)] \div 35.5}$$
$$= \sqrt{[6.5 \times 2.5 \times 26.5] \div 35.5}$$
$$= \sqrt{430.6 \div 35.5}$$
$$= \sqrt{12.13}$$
$$= 3.5$$

Area (SF) = 3.5 × 35.5
= 124.3 SF

We'll round that off and call the area 124 square feet.

The area of the V-ramp oblique triangle in Figure 10-18 is 124 square feet. According to Figure 10-17, the ramp is 10 feet wide. Now find the ramp volume in cubic yards.

Volume (CY) = (124 × 10) ÷ 27
= 1,240 ÷ 27
= 46 CY (45.93 CY before rounding)

You don't have to estimate interior ramps because all the soil is within the excavation area. It's already calculated as part of the V-in and V-out. You can calculate combination ramps by constructing working triangles as we did in Figure 10-18. But only calculate the part of the ramp that's outside the excavation limit line.

Grade Beams and Piers

When difficult soils or load problems mean that normal footings can't support the foundation, grade beams (or grade beams and piers) provide the needed support. Piers are drilled into the ground below the footing and poured with concrete. The footing or grade beam floats on these piers, allowing the entire structure to move slightly without doing any structural damage. Sometimes a belled footing is needed at the bottom of the pier to distribute the weight over a broader area. Figure 10-19 shows a pier shaft and a bell.

Working with grade beams and piers is very specialized work. Most excavation contractors don't try to do it themselves. They hire subcontractors who have the special equipment required. But you may have to calculate the volume of the holes so the subcontractor bidding the job can figure the volume of soil he has to haul off the site. With this in mind, let's consider the following problem.

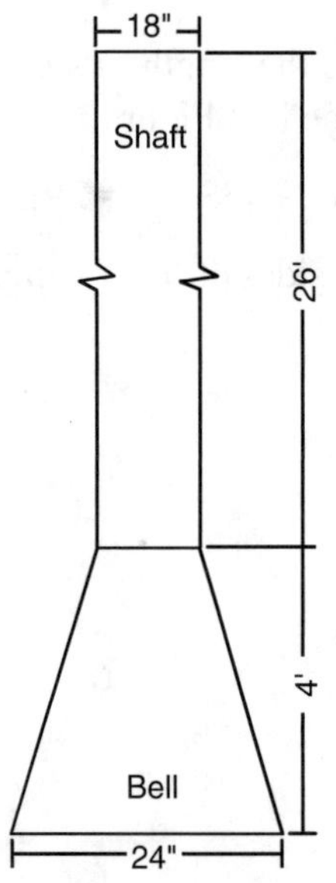

Suppose you need to figure out the volume of a pier shaft and bell, like the one in Figure 10-19. The shaft is 18 inches in diameter and 26 feet deep. The bell's base diameter is 24 inches and its depth is 4 feet.

Figure 10-19
Finding the volume of a pier shaft and bell

178 *Estimating Excavation*

Calculating the Shaft Volume

To calculate the volume of the shaft, multiply the area of an end of the shaft by its length. For a shaft diameter of 18 inches (or 1.5 feet), the radius is 0.75 feet.

Formula for volume of a shaft

$Area = \pi r^2$

$$\begin{aligned} Area\ (SF) &= 3.1416 \times 0.75^2 \\ &= 3.1416 \times 0.56 \\ &= 1.76\ SF \end{aligned}$$

Shaft volume (CY) = (area × depth) ÷ 27

$$\begin{aligned} Volume\ (CY) &= 1.76 \times 26 \div 27 \\ &= 45.76 \div 27 \\ &= 1.7\ CY \end{aligned}$$

Calculating the Bell Volume

To estimate the volume of the bell, calculate the area of the top and bottom circles of the bell. Add the two together and divide by 2 to find the average area. Then multiply by the depth of the bell. We already know the area of the top circle is 1.76 square feet. Let's figure the area of the bottom circle. The radius of the bottom circle is 1 foot (one-half of 24 inches).

$$\begin{aligned} Area\ of\ bottom\ circle &= 3.1416 \times 1^2 \\ &= 3.1416 \end{aligned}$$

$$\begin{aligned} Average\ area &= (3.1416 + 1.76) \div 2 \\ &= 4.9 \div 2 \\ &= 2.45\ SF \end{aligned}$$

$$\begin{aligned} Bell\ volume &= (2.45 \times 4) \div 27 \\ &= 9.8 \div 27 \\ &= 0.36\ CY \end{aligned}$$

Total volume of shaft and bell = 1.7 + 0.36 or 2.06 CY

Unless it's specified by the designer, the diameter and depth of the bell will usually be determined by the size of the contractor's drill rig.

Figuring the volume of piers, shafts, bells, grade beams, and footings is complex work. Make it easier and reduce the chance of errors by splitting the area into simple, regular parts. Then calculate each area and volume as a separate step. Finally, add the parts to find the whole. Work systematically. Be consistent. Be well organized. Keep your work neat and tidy so it's easy to check, both for you and for another estimator. That's the key to consistently accurate excavation estimates.

11
All About Spoil and Borrow

In this chapter we'll define spoil and borrow and learn how to calculate the volume of each. On many jobs you'll need an accurate estimate of how much soil has to be hauled in or hauled away and how much it'll cost. That makes this topic one that's very important to any excavation estimator.

Spoil is any excavated material that can't be used on the project. This is excavated material that you'll have to remove from the site. *Borrow* is material that you need to bring to the site in order to complete the job. Your source for borrow material is the *borrow pit*.

Obviously, you want to avoid spoil and borrow whenever possible. A balanced job has all the fill that's needed available on site. And when the job's complete, there's no spoil to haul away. The easiest, least expensive and most profitable excavation jobs involve neither borrow nor spoil.

Sometimes you can avoid borrow and spoil by temporarily stockpiling material on site or close by during construction. Use it later for backfill when construction is finished. Carefully calculate the amount of material that's stored and how much space you'll need for it. You'll find instructions for calculating stockpile area later in this chapter.

Underlying Costs of Spoil and Borrow

If you can't avoid importing soil, ask yourself the following three questions at the start of your estimate:

1) What borrow pit is closest to this job site?

2) Is the borrow pit material compatible with the on-site material?

3) What are the costs of moving the material?

Obviously, your costs depend on how you answer each of those questions. So we'll take the time for a closer look.

Locations – Borrow Pit vs. Job Site

The closer the borrow pit is to your job site, the better off you are. First, the closer your job site is to the borrow pit, the fewer miles of hauling you pay for. Second, the closer the job site and the borrow pit, the better your chances for a good material match.

Obviously, the fewer miles you haul borrow material, the lower your costs. The same is true when you're hauling spoil. Other cost factors include traffic loads, and street and bridge conditions.

Borrow material should always be as similar as possible to on-site material. On some projects a good match between the two soils is very important. You may need a test by a soils engineer to make sure that the borrow meets design standards. Check the plans and specifications to see who provides and pays for this testing. More often than not, tests like this are done at the expense of the excavation contractor.

Spoil Disposal

When you're dealing with spoil, there's an extra factor to consider — soil type. We'll look at the best possible case first. Suppose your spoil is rich, high-quality topsoil. Good topsoil's a valuable commodity, as we saw earlier in Chapter 9. Someone will want it. Not only that, they'll pay you for it and for your time too.

Unfortunately, most spoil isn't high-grade topsoil. Instead it's material such as rock, muck or clay, and miscellaneous debris. There just isn't a lot of demand for material like this. Sometimes it's a problem just finding a disposal site that's close enough to be practical. Here are a few tips to try if you run into trouble along these lines. Your local building department keeps public records listing all the excavation projects still in the approval stage. Check this list. Are any of the projects close to your job site? Does the paperwork show that they'll need fill? You're likely to have just what they need. Keep an eye out for private party ads looking for fill dirt or offering to accept fill. If none of these pan out, you'll have to use the nearest legal dump site that accepts spoil. Obviously, it's to your advantage to dispose of spoil as close to the job site as possible and reduce those hauling costs.

Interim Spoil

Not all spoil necessarily remains spoil. Material you remove from a site temporarily is called *interim spoil*. It's taken from a job site during construction and then brought back later to complete the project. You use interim spoil only when there's no other choice. For example, say the job site's very small, or has extreme topography such as steep slopes or deep ravines. In those cases interim spoil is the only answer. Interim spoil is unique, expensive, and something to avoid — because you handle it twice.

Spoil and Borrow Volume Calculations

Formula for total volume of spoil

You find the total volume of spoil on a project using this formula:

Total spoil vol. (CY) =
[total cut vol. – (total backfill vol. + total fill vol.)] × swell factor

Say that the spoil material is moist sand and you already know these volumes:

- Cut = 500 CY
- Fill = 200 CY
- Backfill = 150 CY

Here's the math for spoil volume in cubic yards:

Total spoil volume (CY) = 500 – (200 + 150)

= 500 – 350

= 150 CY

The swell factor for moist sand in Figure 8-1 is 1.13:

Total spoil volume (LCY) = 150 × 1.13

= 169.5 LCY

Before we move on, here are two excellent reasons for always using loose cubic yards for spoil volume.

1) It reminds you to use swell and shrink factors.
2) If you subcontract haulage, the bids are sure to be per LCY.

If you get a negative value for spoil, it means there's no spoil to haul off. In fact, you don't have enough material to do the backfill and fill work called for in the plans. That means it's time for you to locate a borrow pit. Project engineers and designers do their best to minimize borrow amounts. But despite all these efforts to avoid borrow, some jobs still require imported material. We'll talk more about balancing cut and fill volumes in the next chapter.

Turn back to Chapter 8 if you want to review the subject of shrink and swell factors. Two types of stockpiles require special consideration when it comes to applying shrink and swell factors: interim spoil stockpiles and topsoil stockpiles. We'll look at each of these special situations in depth.

Interim Spoil Stockpiles

Interim spoil, you recall, receives extra handling. First, you excavate it on-site. Second, you stockpile it off-site. Third, you bring it back on-site for use. At step one and step two you're working with loose state material so you apply the swell factor. Normally, the volume of an interim spoil stockpile doesn't change between the second and third steps. However, there are three exceptions. Here are the exceptions and how to deal with each of them:

1) Material stockpiled for more than three months. Settling changes the state of stockpiled material. You now have a stockpile of compact material. Recalculate the volume in compact cubic yards applying the correct *shrink* factor.

2) Material stockpiled unprotected through rainy season. You've lost an unknown quantity of material via water erosion and changed the moisture content of the material. Recalculate the volume and apply the right swell factor for the moisture level.

3) Material that's sold to another party. Don't cheat yourself; recalculate the stockpile's volume and apply the appropriate shrink or swell factor.

There are special rules for applying shrink and swell factors to stockpiles of topsoil.

- Apply only the swell factor to topsoil out volumes
- Apply only the shrink factor to topsoil in volumes

Calculating the Volume of a Stockpile

Figure 11-1 shows the same stockpile of loose soil in several different views. We'll use this stockpile to introduce the procedures you use to find stockpile volume. First, divide the stockpile into three sections, as shown in Figure 11-1A. The middle section is the prism shown in Figure 11-1C. The two end sections, meanwhile, combine to form the cone shown in Figure 11-1D.

Finding the Volume of the Middle Section

The middle section is the prism ABCDEF shown in Figure 11-1C. The volume of a prism equals its end area times its length. The ends of a prism are triangles, like triangle ACE in Figure 11-1C. Here's the formula you use to find its area:

Area = $\frac{1}{2}$ × (base × height)

Area triangle ACE = $\frac{1}{2}$ × (35 × 12)

$= \frac{1}{2}$ × 420

= 210 SF

Next we'll find the prism's volume in cubic feet. Multiply the end area (210 SF) by the prism's length (65 feet):

Prism volume (CF) = 210 × 65

= 13,650 CF

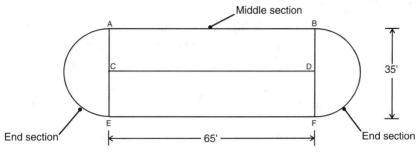

A. Top view

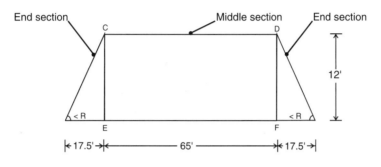

B. Front view

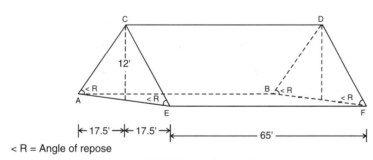

< R = Angle of repose

C. Middle section - prism

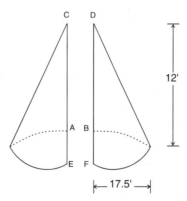

D. Combined end sections

Figure 11-1
Four views of a sample stockpile

All About Spoil and Borrow **185**

Finding the Volume of the Combined End Sections

Now let's find the volume in the two end sections. Remember, if we lump them together they form the cone in Figure 11-1D. We'll find the volume of the whole cone using a diameter of 35 feet and a height of 12 feet.

You recall that a circle's radius is half its diameter. But do you also recall the formulas to find the volume of an upside-down cone?

Volume (CF) = ⅓ × (base area × height)

Base area (SF) = πr^2

• • • • • • • • • • • • • •
Formulas for volume of an upside-down cone (end area)

Plug in the values for the cone shown in Figure 11-1D:

Base area = 3.1416 × 17.5²

= 3.1416 × 306.25

= 962.12 SF

Next, you find the volume in cubic feet:

Volume = ⅓ × (962.12 × 12)

= ⅓ × 11,545.44

= 3,848.48

Here's the formula you use to find the total volume of the stockpile in Figure 11-1A in cubic yards:

Volume (CY) = (prism volume + cone volume) ÷ 27

Plugging in the volumes for the prism and the cone you get:

Total volume (CY) = (13,650 + 3,848.48) ÷ 27

= 17,498.48 ÷ 27

= 648.09 CY

Finding the Volume of a Stockpile of Unknown Height

Suppose you have to estimate the volume of a stockpile and you don't know how high the stockpile is. The easy part of the job is to measure the width and length of the stockpile. The hard part is finding the height of the pile.

When you dump or pile loose soil, it forms a peak at the top and slopes outward on all sides. The angle between the side of a stockpile and the ground is called the *angle of repose*. It's usually between 20 degrees (for very loose material such as muck) and 40 degrees (for firm material such as dry loam). Add more material to a stockpile and the angle of repose remains the same. Most of the added material slides down the sides of the pile. The base grows broader, while the height increases only slightly. No matter how much material you add, the angle of repose stays the same. Because the angle of repose is constant, we'll use it to calculate the stockpile's height.

The Reverse Angle Method

We'll use the reverse angle method to find the stockpile height. To use this method you'll need a 100-foot tape, a standard carpenter's square and a plumb bob on a line. You use the 100-foot tape to measure the lengths of the stockpile's sides and ends. Then you use the carpenter's square and the plumb bob to determine angles.

Figure 11-2A is a top view of the sample stockpile. Figures 11-2B through 11-2D are detail views of the same stockpile. We'll go through this step by step.

Step 1: Setting up the carpenter's square

Stand the carpenter's square beside the stockpile, as shown in Figure 11-2, with the long leg horizontal and the short leg vertical. The bottom corner of the square's long leg should just touch the side of the stockpile, *point X* in Figure 11-2B. It's important for the long leg of the square to be parallel to the ground and as level as possible. You can check this by resting a bubble level along the top edge of the square's long leg. The square is level when you center the bubble.

Step 2: Setting up the plumb bob

Hold the plumb line in front of the long leg of the square. Suspend the plumb bob so that its tip just clears the ground. Position the tip right over the point where the stockpile meets the ground, *point V* in Figure 11-2B.

Step 3: Finding the tangent of the reverse angle

The plumb line crosses the long leg of the square at *point W* in Figure 11-2. Record the distance from the end of the leg to the plumb line. In Figure 11-2B this distance is *line XW* and it's 15 inches long.

The short leg on a standard carpenter's square is 12 inches long. Check your square just to be sure that it is a standard square. Remember to use the inside scale for this measurement, not the outside scale. In Figure 11-2B the short leg of the square and *line WV* both equal 12 inches.

You now know the lengths of two sides of triangle VWX in Figure 11-2B. The ratio of the lengths of these two sides, XW and WV, is a mathematical function. It's called the tangent of an angle. Here's the formula:

Formula for the tangent of an angle

Tangent of angle =
length of angle's opposite side ÷ length of angle's adjacent side

In Figure 11-2B, for Angle 1 the opposite side is line WV and the adjacent side is line XW.

Plug in the values we found for the two sides:

Tangent angle 1 = 12 ÷ 15

= 0.8

Now you can use that to find the angle of repose.

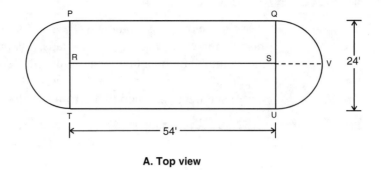

A. Top view

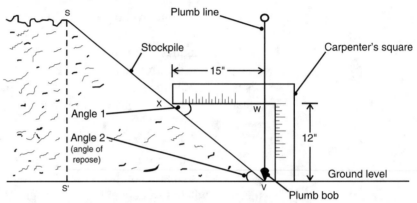

B. Setup for finding a reverse angle

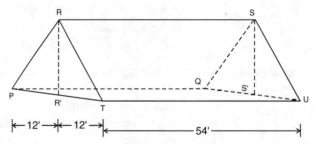

C. Prism - middle section

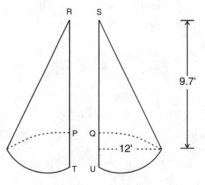

D. Combined end sections

Figure 11-2
Finding the volume of a stockpile of unknown height

188 *Estimating Excavation*

Step 4: Finding the angle of repose

Once you know the tangent of an angle, it's easy to find the angle that produces that tangent. Figure 11-3 is a list of tangents for angles ranging from 20 to 40 degrees. These are the angles you're most likely to need when calculating stockpile volumes. Here's how it works. You find the tangent listed in Figure 11-3 that's closest to the tangent value we just found for Angle 1 (0.8). The closest listed tangent in Figure 11-3 is 0.80978. To find the angle for this tangent, read across to the angle column. A tangent of 0.80978 is formed by a 39-degree angle. Therefore, Angle 1 measures about 39 degrees.

We know that the top of the square and the ground are parallel. So Angle 1 and Angle 2 are identical angles. We also know that Angle 2 is the stockpile's angle of repose. So the angle of repose is 39 degrees. But we're not finished yet. Remember we still need to find the stockpile height and volume.

Step 5: Calculating height from the angle of repose

Let's identify Angle 2's opposite and adjacent sides using Figure 11-2B. Angle 2's opposite side is the broken line *SS'*. Line SS' is also the stockpile height. Angle 2's adjacent side is *line S'V*. We know the length of S'V is half the width of the stockpile, or 12 feet (24 ÷ 2 = 12).

How do we find the length of SS'? We'll use the tangent function.

Tangent of angle =
length of angle's opposite side ÷ length of angle's adjacent side

Degrees	Tangent	Degrees	Tangent
20	.36397	31	.60086
21	.38386	32	.62487
22	.40403	33	.64941
23	.42447	34	.67451
24	.44523	35	.70021
25	.46631	36	.72654
26	.48773	37	.75355
27	.50953	38	.78129
28	.53171	39	.80978
29	.55431	40	.83910
30	.57735		

Figure 11-3
Tangents for angle of repose

All About Spoil and Borrow

We know the tangent for Angle 2 is 0.8.978. We also know that the length of Angle 2's adjacent side is 12 feet. What we want to find is the length of the opposite side so we'll rewrite the tangent function as follows:

*Length of angle's opposite side =
tangent of angle × length of angle's adjacent side*

Now you just plug in the values:

Length of SS' = 0.8.978 × 12

= 9.7'

Step 6: Calculating total volume

Here's your chance to try out the procedures and formulas introduced at the beginning of this chapter. Using the data from Figure 11-2 and the preceding five steps, find the volume of the stockpile in cubic yards. After you've finished, compare your result with that shown in Figure 11-4.

Calculating Volume for a Stockpile of Set Area

The space available for a stockpile location on most job sites is limited. That means you'll often want to know how much material you can expect to stockpile in that space. A stockpile that spills over into another contractor's workspace won't make you any friends on the job site. And don't forget to leave yourself the workspace you'll need. The amount of workspace you need depends on the type of equipment you use.

In Figure 11-5A you see a plan view for a stockpile. The dimensions, after allowing for workspace, are 30 feet wide by 70 feet long. Now, let's say that the angle of repose for the soil is 40 degrees. Find how much soil you can pile here using the tangent function and Figures 11-3, 11-5A and 11-5B.

Step 1: Calculating the height

Turn to Figure 11-3 and find the tangent for a 40-degree angle. The tangent is 0.83910. The length of the adjacent side is half the stockpile width (see Figure 11-5A). The adjacent side is 15 feet (30 ÷ 2 = 15). The side opposite this angle is also the height dimension for the stockpile. Rewrite the tangent function to solve for the length of the angle's opposite side:

*Length of angle's opposite side =
tangent of angle × length of angle's adjacent side*

Plug in the known values for the tangent of the angle and the length of the adjacent side. Here's the math:

Stockpile height (feet) = 0.83901 × 15

= 12.6 feet

\multicolumn{2}{c}{**Calculation Sheet**}
\multicolumn{2}{c}{Project:_____ Date:_____}

$$\text{Stockpile vol. (CY)} = (\text{prism vol.} + \text{cone vol.}) \div 27$$

Prism vol. (CF) = end area × length
End area (SF) = ½ (base × height)
Prism (see Figure 11-2C)
End area = area of triangle PRT (see Figure 11-2C)
Prism length = 54'
Base △PRT is side PT
PT = 24'
Height △PRT = 9.7'
End area = ½ (24 × 9.7)
         = ½ × 232.8
         = 116.4 SF
Prism vol. = 116.4 × 54
         = 6,285.6 CF
	Cone vol. (CF) = ⅓ (base area × height)
	Base area (SF) = πr^2
	Cone (see Figure 11-2D)
	Height = 9.7'
	π = 3.1416
	r = 12'
	Base area = 3.1416 × 12²
	         = 3.1416 × 144
	         = 452.4 SF
	Cone vol. = ⅓ (452.4 × 9.7)
	         = ⅓ × 4,388.28
	         = 1,462.76 CF

$$\text{Stockpile vol. (CY)} = \frac{6,285.6 + 1,462.76}{27}$$

$$= \frac{7,748.36}{27}$$

$$= 287 \text{ CY}$$

Conclusion

Figure 11-4
Calculations sheet for sample stockpile volume

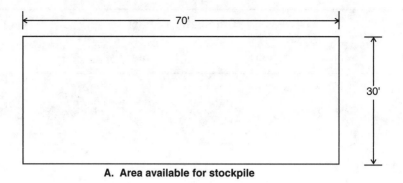

Figure 11-5
Calculating maximum stockpile volume in available area

Step 2: Calculating prism volume

You use the dimensions from Figure 11-5B, the height you found in Step 1, and these formulas:

Volume (CY) = (end area × length) ÷ 27

End area (SF) = ½ × (base × height)

The prism length is 40 feet. Find the area for one end of the prism. This is a triangle 12.6 feet high with a 30-foot base. So the end area is:

End area (SF) = ½ × (30 × 12.6)

$= ½ × 378$

$= 189 \text{ SF}$

Next you find the triangle's volume in cubic yards:

Volume (CY) = (189 × 40) ÷ 27

$= 7{,}560 ÷ 27$

$= 280 \text{ CY}$

Step 3: Calculating cone volume

Combine the two end sections of the stockpile and you have a cone. Use this formula to find the cone's volume in cubic yards:

Volume (CY) = [1/3 × (area of base × height)] ÷ 27

Height is 12.6 feet (see Step 1). The base of a cone is a circle and the area of a circle equals π (3.1416) times the radius squared (15^2 = 225). So the area of the base is 706.86 SF (3.1416 × 225 = 706.86). And the volume is:

Volume (CY) = [1/3 × (706.9 × 12.6)] ÷ 27

= [1/3 × 8,906.94] ÷ 27

= 2,968.98 ÷ 27

= 109.96 CY

Step 4: Calculating total volume

Just add the prism volume and the cone volume to find the total volume:

Volume (CY) = 280 + 109.96

= 389.96 CY

We'll round that off and call it 390 cubic yards. That's the maximum possible volume of this soil that you could stockpile in the area shown in Figure 11-5A. But there's one last calculation to make. You still need to convert the stockpile volume into loose cubic yards by applying the correct swell factor.

In the next chapter, I'll describe how engineers and estimators use balance points to "balance" the cut and fill. On an ideal job, you don't have either spoil or borrow — because cut and fill balance exactly. You won't see a job like that every day. But it's always the goal.

12 Balance Points, Centers of Mass & Haul Distances

■ ■ ■ ■ ■ ■ ■ ■ ■ ■

Both engineers and estimators use the words "balance point." But the words mean something entirely different to an engineer than they do to a dirt contractor. To an engineer, "balance point" is an imaginary line where the cut on one side of the line is equal to the fill on the other side of the line. Engineers try to plan earthwork so the volume of cut matches the fill volume. Of course, what's planned as a balanced job may not work out exactly that way. Cutting and filling soil isn't an exact science.

Balance Points to an Excavation Estimator

To the excavation estimator, the balance point is midway between the cut and the fill. A very simple example is shown along Profile 2 in Figure 12-1. The cut volume and the fill volume are identical. That's not going to happen in real life. Don't worry about that now. Later in this chapter we'll look at some more complex cut and fill jobs where the cut and fill volumes *aren't* equal. For now, I'll use Profile 2 in Figure 12-1 to define three important concepts for any earthwork estimator: center of mass, haul distance and balance point. We'll start with the center of mass.

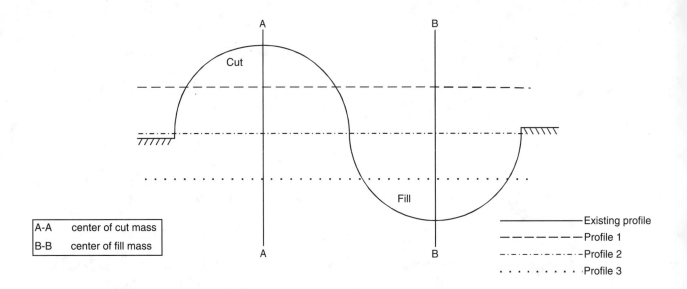

Figure 12-1
A simple example of balancing cut and fill

Figure 12-1 has *two centers of mass*. The center of mass for the cut area is line *A-A*. The center of mass for the fill area is line *B-B*. The distance separating these two centers of mass, measured along Profile 2, is the *haul distance*. The midpoint on Profile 2 between lines *A-A* and *B-B* is the *balance point*. Now let's look at why these are so important. You can't find your haul distance unless you know the locations of both centers of mass. The balance point tells you how far soil will be moved. And you use haul distance to determine your round trip, or cycle time, per load. The greater the haul distance, the greater the cost.

In this chapter, you'll learn how excavation estimators find balance points and practice the skill by calculating average haul distance for a sample project.

Balance Points to an Engineer

An engineer determines the finish grade, or grades, that appear on the project plans using arbitrary balance points. I'll show you how balance points work using a pair of examples. Let's start by taking another look at Figure 12-1.

This profile shows the existing profile plus three possible finish profiles for a project. Notice how much alike the cut and fill areas are in shape and size. Compare the three proposed profiles using their different proportions of cut and fill volume and here's what you find:

- *Profile 1:* small cut volume versus very large fill volume
- *Profile 2:* cut and fill volume are about equal
- *Profile 3:* very large cut volume versus small fill volume

Obviously the best finish grade for the project shown in Figure 12-1 is Profile 2. In this case that's a pretty simple choice. But in reality it's never that straightforward.

Here's a more realistic example. Take a look at Figure 12-2, a profile view of a one-mile section from a larger project. There are three profiles shown in Figure 12-2: the existing profile and two proposed finish profiles. The shapes of the cut and fill areas aren't symmetrical and the cut and fill volumes aren't equal for either finish profile.

Now let's compare the two proposed profiles. Profile 1 is a level surface with a single elevation. Profile 2 slopes in from both ends toward the center, providing drainage for the site. Profile 1 requires a much larger volume of fill material than you'll have on site from the cuts. Finishing the job would mean importing a large volume of fill from an off-site borrow pit. Profile 2 requires a smaller volume of imported fill, and with careful planning, comes close to balancing. Profile 2 has another advantage — a shorter haul distance. The lowest elevations for Profile 2 roughly match the lowest existing elevations.

Engineers understand how important carefully planned profiles are and do their best to balance the job and shorten haul distances. But no matter what the engineer has done or failed to do, make sure your plans use the minimum possible haul distance.

Reducing Haul Distances

There are two types of costs in every excavation job: the cost of loading soil and the cost of moving soil once it's loaded. The cost of loading soil will be about the same for every contractor using equipment appropriate for the job. But the cost of hauling soil will be lower for the contractor who reduces the

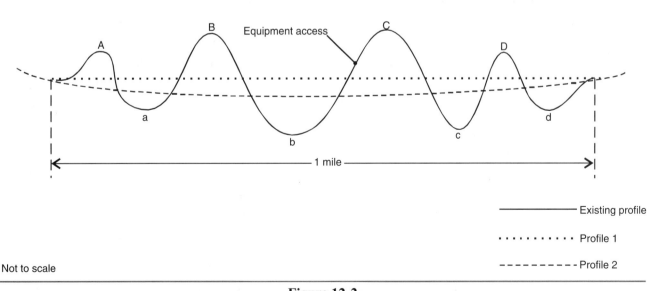

Figure 12-2
A more complex example of balancing cut and fill

average haul distance. Any money you save by reducing the average haul distance adds to your profit on the job. Good planning can maximize both equipment productivity and profit.

In Figure 12-2 I've used capital letters, *A, B, C,* and *D,* to label the cut areas and lower case letters, *a, b, c,* and *d,* for the fill areas. How would you plan this job? Well, you might start at cut *A* and move to fill *a.* Then you'd move to cut *B* and use part of this material to finish fill *a* and use the rest to start on fill *b.* Then you'd move on to cut area *C* and so on. I suppose that would work. And many contractors tackle the problem that way, working from left to right or north to south, finishing one area and then moving on to the next.

But suppose equipment access to the section shown in Figure 12-2 is at cut *C.* You have to begin working at the center and work toward the ends. Which way do you fill from cut *C*? Does the material go into fill *b* or fill *c*? And, if fill *c,* how much goes into *c,* before you start filling at *b*?

Here's what happens if you don't plan this or any job carefully. You end up hauling a lot of fill a mile or more from cut area *A* to fill *d.* That's the expensive way to do cut and fill work.

Let's work though two examples. The only difference between these two examples is the location of the borrow pit. Both assume the following:

- Finish profile is Profile 2 in Figure 12-2
- Volumes of cut and fill listed in Figure 12-3
- Swell/shrink factor is 1.14

Cut Areas	Volume (CY)
Area *A*	1,000
Area *B*	3,000
Area *C*	7,000
Area *D*	1,200
Total cut	**12,200**

Fill Areas	Volume (CY)
Area *a*	3,500
Area *b*	8,000
Area *c*	9,000
Area *d*	1,700
Total fill	**22,200**

Figure 12-3
Cut and fill volumes for the sample project shown in Figure 12-2

Example #1

Assume the borrow pit lies between cut areas *C* and *D* in Figure 12-2. Here's a five-step plan for this job.

Step 1: Fill area a

You need a total of 3,500 cubic yards of material here. Use 1,000 cubic yards from cut area *A* plus 2,500 cubic yards from cut area *B* for a total of 3,500 cubic yards. You have 500 cubic yards left from cut area *B*.

Step 2: Fill area b

Here you need a total of 8,000 cubic yards of material. Use the 500 cubic yards you have left from cut area *B*, plus all 7,000 cubic yards from cut area *C*, plus 500 cubic yards from cut area *D* for a total of 8,000 cubic yards. You have 700 cubic yards left from cut area *D*.

Step 3: Fill area d

This fill area requires a total of 1,700 cubic yards of material. You have 700 cubic yards of material from cut *D* left to use. Fill area *d* is short 1,000 cubic yards. This material comes from the borrow pit.

Step 4: Fill area c

The total volume of fill needed here is 9,000 cubic yards. All of this material comes from the borrow pit.

Step 5: Finishing fill areas c and d

You need another 10,000 cubic yards (1,000 + 9,000) of material. Apply the shrink factor, 1.14, to arrive at the total borrow volume of 11,400 compact cubic yards.

Under this plan, the fill areas where you need the borrow material are the closest to the borrow pit and the result is minimum haul distance.

Example #2

Assume the borrow pit lies to the left of cut area *A* in Figure 12-2. Here's our plan for this job.

Let's start by supposing that you ignore the change in the borrow pit's location. You decide you'll just use the same plan as for Example #1. Here's what happens. To finish the job in fill areas *d* and *c* you'll end up hauling borrow material from one end of the project to the other! Not a very efficient way to do the job, is it? There *is* a better way to plan this job. Let's see how using the volumes listed in Figure 12-3.

Step 1: Fill area a

We'll start by using all 1,200 cubic yards from cut area *D* plus 500 cubic yards from cut area *C* for fill area *d*.

Step 2: Fill area b

For fill area *c* you need 9,000 cubic yards. Combine the leftover 6,500 cubic yards from cut area *C* with 2,500 cubic yards from cut area *B*.

Step 3: Fill area d

That brings you to fill area *b* where you need 8,000 cubic yards. Use the 500 cubic yards you still have left from cut area *B* plus 1,000 cubic yards from cut area *A* to start on fill area *b*. The remaining 6,500 cubic yards you need to finish fill area *b* come from the borrow pit.

Step 4: Fill area c

And don't forget fill area *a*, where you need 3,500 cubic yards. This material will also come from the borrow pit.

Step 5: Finishing fill areas c and d

Total borrow, after applying the shrink factor of 1.14, is 11,400 compact cubic yards, just as in Example #1. But by carefully planning the cut and fill, you wind up with the two fill areas using borrowed material as close as possible to the borrow pit. That means the minimum haul distance — and minimum cost.

Always try to find a way to minimize haul distance. A bit of creative thinking sometimes helps. Here's an example of what I mean. Take another look at fill area *d* and cut area *D* in Example #1. Remember, the borrow pit is located between cut areas *C* and *D*. That also happens to be the location of fill area *c*. Here's what I'd try in a case like this. I'd go to the project engineer and ask for permission for a temporary overcut of 1,000 cubic yards in cut area *D*. With this additional material I'll complete not only fill area *b* but also all of fill area *d*. After that I'll bring cut area *D* back up to grade using material from the borrow pit. But never carry out a creative solution unless your plan's been approved by the project engineer or manager.

Calculating Haul Distances

To plan cut and fill work, you have to calculate not only the volume to be moved but also the center of the mass. Knowing the center of mass, or *CM*, you can set balance points. Your balance points don't need to be exact, but the more accurate they are, the more money you'll save on hauling.

To calculate the balance point, you have to know the haul distance. To know the haul distance, you have to know the center of mass. Let's begin by finding the center of mass.

Figure 12-4 shows a typical problem. The contractor will use material from the semicircular area on the north end of the job to build up the cul-de-sac circle on the south end. Your haul distance, measured from cul-de-sac edge to semicircle edge, is 1,300 feet. Obviously not all the soil is excavated at the edge of the semicircle and it's not all dumped at the edge of the cul-de-sac. Therefore, the actual haul distance on this job is more than 1,300 feet

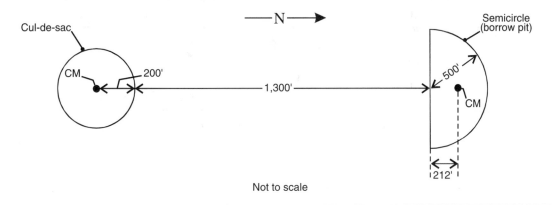

Figure 12-4
Haul distance

for nearly every load. We'll find the *average* haul distance for this project. That's a three step process. First, we'll find the distance between the south edge of the semicircle and the north edge of the cul-de-sac. Second, we'll find the distance from the edge to the center of mass, *CM* in Figure 12-4, for the semicircle and the circle. Third, the *average* haul distance equals the sum of three distances: fill area edge to center of mass + borrow pit edge to center of mass + edge to edge.

Finding the Distance from Edge to Center of Mass

Figure 12-5 lists formulas used to calculate the distance from the edge of a circle, semicircle, rectangle or a triangle to the center of mass. These formulas use the following abbreviations:

- *CM* = center of mass
- *x* = distance from center of mass to an edge
- *r* = radius
- *d* = diameter
- *h* = height
- *w* = width
- *b* = base
- *A-A* and *B-B* = axis drawn through the center of mass

Calculating the haul distance for a circle

Using the circle in Figure 12-4 and the first formula from Figure 12-5, let's find the average haul distance. Notice that the formula used for a circle is very simple.

$x_a = x_b = d \div 2 = r$

Balance Points, Centers of Mass & Haul Distances 201

Formula	Circle
$x_a = x_b = \dfrac{d}{2} = r$	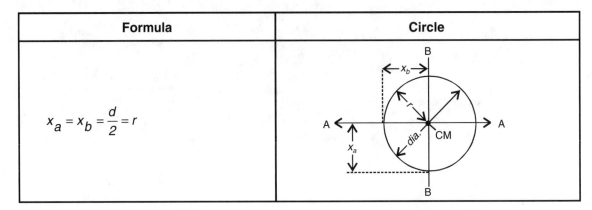

Formula	Semicircle
$x_a = \dfrac{d(3p - r)}{6p} = .288d = .576r$ $x_b = \dfrac{2d}{3p} = .212d = \dfrac{4r}{3p} = .424r$	

Formula	Rectangle
$x_a = \dfrac{h}{2}$ $x_b = \dfrac{b}{2}$	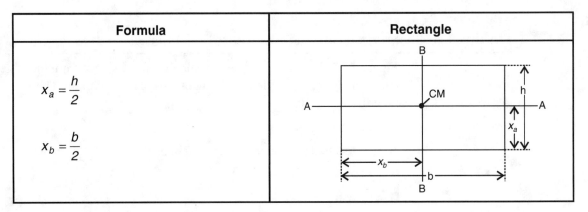

Formula	Triangle
$x_a = \dfrac{2}{3}h$ $x_b = \dfrac{1}{3}h$	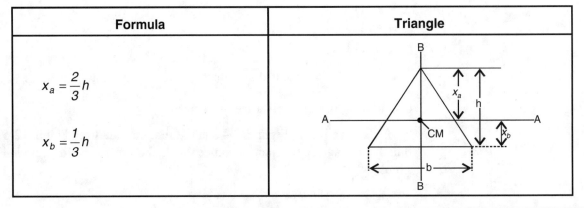

Figure 12-5
Formulas for calculating center of mass

x_a and x_b are exactly the same because they're radii for the same circle. From Figure 12-4 we know that the radius of the circle is 200 feet. So for the cul-de-sac end of the job we'll add 200 feet to the haul distance.

Calculating the haul distance for a semicircle

Now let's find the average haul distance for the semicircle in Figure 12-4. Here's the formula we'll use:

$x_b = 0.424 \times radius$

We know the radius for the semicircle from Figure 12-4 is 500 feet so:

$x_b = 0.424 \times 500$

$= 212'$

The semicircle adds 212 feet to the edge-to-edge haul distance.

The average haul distance is the sum of the two edge-to-center-of-mass distances plus the edge-to-edge haul distance given in Figure 12-4. Here's the math:

Average haul distance = 1,300 + 200 + 212

= 1,712'

Finding a *Vertical* Center of Mass

So far we've assumed that cut and fill depths are uniform throughout each cut and fill area. In reality that's seldom the case. Cuts and fills are deeper in some places and shallower in others. The result is a proportional shift in the location of the center of mass. But if the location of the center of mass changes, so does your average haul distance. This raises two big questions:

1) How do you find the center of mass for an area without a uniform depth?

2) How does it factor into your average haul distance calculations?

The best way to answer both questions is with an example. In this example we'll find two different average haul distances. The first average haul distance assumes an area with a uniform depth. The second average haul distance is for an area without a uniform depth. Otherwise it's identical to the first area.

Figure 12-6 is the plan and profile sheet for our sample project, a 680-foot-long section of a road project. Let's start by taking a close look at both parts of Figure 12-6. We'll start at the top of the sheet in Figure 12-6 with the plan view.

The plan view, you remember, is an overhead view of the job site. The main features of any plan view are the centerline and a series of measured distances, usually at 100-foot intervals, marked along the centerline's length. In Figure 12-6 the measured distances start on the left with Sta. 0+00 and end on the right with Sta. 6+80. The station-to-station interval is 100 feet, except for Stations 1+70, 6+30 and 6+80. Why are they different from the rest? Let's find out.

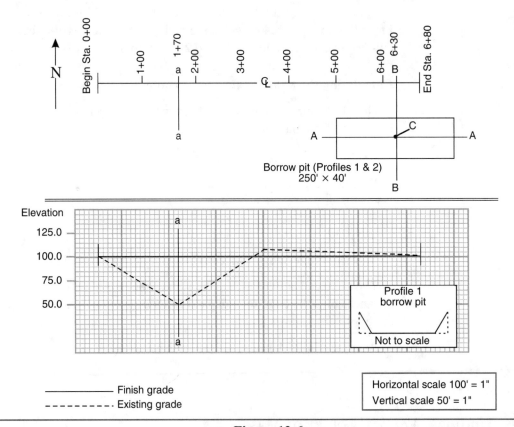

Figure 12-6
Plan and profile sheet for a section of a road project

The reason Station 6+80 doesn't follow the rule is easy to explain. Station 6+80 marks the east limits of this job site. This won't work as an explanation for either of the two remaining stations. Did you notice that Sta. 1+70 and Sta. 6+30, and none of the others, have alternative names? Station 1+70 is also line *a–a* and Sta. 6+30 is also line *B–B*. As you'll recall from Chapter 4, surveyors don't just mark off the standard 100-foot intervals on a centerline. They also stake important features and significant changes in elevation.

The reason for Sta. 1+70 is obvious as soon as you look at the profile view in Figure 12-6. Station 1+70, or line *a–a*, passes right through the lowest point found along the existing elevation, shown with a dashed line. But that's not all. Line *a–a* also divides this fill area, roughly an equilateral triangle, into two equal parts. So line *a–a* is the fill area's center of mass.

Sta. 6+30, or line *B–B*, also marks one axis for a center of mass. The mass, of course, is the rectangle just south of the centerline at Sta. 6+30 labeled *Borrow pit (Profiles 1 & 2)*. This 250- × 40-foot borrow pit is the source for all fill material we'll use on this project. Line *B–B* divides the borrow pit into two equal parts across its length and line *A–A* does the same along the borrow pit's length. The point where *B–B* and *A–A* cross is labeled *C*. Line *B–B* doesn't appear in the profile view in Figure 12-6 simply because the borrow pit's not located along the road centerline.

204 *Estimating Excavation*

Like most plan and profile sheets, Figure 12-6 uses one scale for horizontal dimensions and a different scale for vertical dimensions. Plan and profile sheets always list the scales used and so does Figure 12-6. We'll scale off the dimensions, using the borrow pit dimensions and these scales from Figure 12-6:

- horizontal scale: 100' = 1"
- vertical scale: 50' = 1"

Let's start off with the borrow pit width. From the plan view we know that the actual width is 40 feet. What's the equivalent measured plan dimension using the vertical scale? To find out you just divide 40 by 50; the answer is 0.8 inch.

Now let's do a reverse calculation. Take a measurement off of the plan sheet and change it back to an actual distance. Suppose you measured the borrow pit's length from the plan view in Figure 12-6 and it measured 2.5 inches long. Using the horizontal scale, what's the borrow pit's actual length in feet? To find out you just multiply 2.5 by 100. The result is 250 feet.

The distance from the road centerline to the north edge of the borrow pit is next. We measure 0.7 inch, then multiply by 50 to find an actual distance of 35 feet.

Now we'll find the average haul distance for the sample project shown in Figure 12-6 using two different borrow pit profiles: *Profile 1* and *Profile 2*.

Calculating the Average Haul Distance for Profile 1

I expect you already noticed the small profile view of the *Profile 1* borrow pit at the lower right of Figure 12-6. But be sure that you also notice the note that says: *Not to scale*. Why include this profile view? To show you that it's shape is symmetrical. And that's important because it makes finding its center of mass much easier. Here's how it works. The fact that the sides slope equally allows us to ignore the slope. Instead we'll proceed as if this borrow pit had a uniform depth and treat is like a simple rectangle. To find this rectangle's center of mass we'll use these formulas from Figure 12-5:

Formulas for center of mass

$x_a = width \div 2$

$x_b = length \div 2$

You pick up these dimensions from Figure 12-6:

- Length = 250'
- Width = 40'

And here's the math:

$x_a = 40 \div 2$

$= 20'$

$x_b = 250 \div 2$

$= 125'$

At the scale used in Figure 12-6, the lines would be 0.8 and 1.25 inch long. Line *A–A* divides the borrow pit into two equal parts from west to east. Line *B–B,* meanwhile, divides the borrow pit into two equal parts from north to south. The point labeled *C* marks the intersection of lines *A–A* and *B–B*.

The distance separating station 1+70 (fill area center) and station 6+30 (borrow pit center of *Profile 1*) is 460 feet (630 – 170 = 460). But this is only part of the average haul distance because it's measured along the centerline of the roadway. We need to add on two more distance measurements to find the average haul distance. First, add the distance from the roadway centerline to the edge of the borrow pit. That's 35 feet. Second, add the distance from the edge of the borrow pit to its center of mass. That distance is the same as the value we found earlier for x_a — 20 feet. Add all three distances and the result is the average haul distance for the *Profile 1* borrow pit.

Distance Sta. 1+70 to Sta. 6+30	= 460'
Distance roadway to edge of borrow pit =	35'
Distance edge to center of mass	= + 20'
Average haul distance	515'

That takes care of calculating average haul distances for all the symmetrical borrow pits out there. Unfortunately, there aren't many. Most borrow pits don't look like *Profile 1*. Instead, most borrow pits look a lot more like *Profile 2*. When you're working with a borrow pit shaped like *Profile 2*, most of the material comes from the far end of the pit. That means your average haul distance is longer. The big question is, how much longer?

Calculating the Average Haul Distance for Profile 2

Here's a rundown of what we know about this borrow pit (Figure 12-7):

- Length = 250'
- Width   = 40'
- Depth   = 10'

Profile 2 is an oblique triangle — it's not symmetrical. In the plan view *Profile 2* and *Profile 1* are identical. That means we can skip calculating the following distances for *Profile 2*:

- Distance roadway to pit edge   = 35'
- Distance pit edge to center of mass = 20'
- Distance Sta. 1+70 to Sta. 6+80   = 460'

The cross section's center point and the center of mass are not the same point. As you might expect, the center of mass for *Profile 2* is off-center and nearer to the pit's deepest point. But notice that I said the center of mass is *closer* to the deepest point.

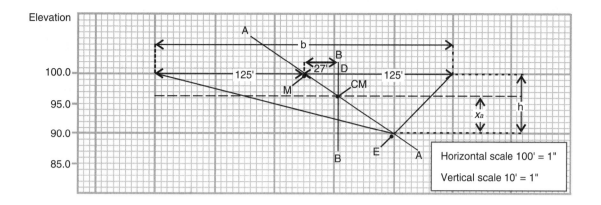

E	apex of triangle & point of lowest elevation = 90.0'	A-A	= axis passing through points E and M
b	triangle base & horizontal baseline, length = 250'	B-B	= axis perpendicular to b & passing through CM
		CM	center of mass
h	triangle height/depth = 10'	x_a	= $^2/_3 \times$ h = 0.667 × 10 = 6.67'
M	midpoint, horizontal baseline = b ÷ 2 = 250 ÷ 2 = 125'	x_b	= $^1/_3 \times$ h = 0.333 × 10 = 3.33'

Figure 12-7
Profile view of borrow pit *Profile 2*

No matter how irregular a shape is you can always find its center of mass by using sophisticated mathematics. But lucky for us there's an easier way. Your results won't be quite as accurate using this method, but they're more than adequate for an estimate.

In Figure 12-7, *Profile 2* is 250 feet long and 10 feet deep. Now let's find the center of mass for the *Profile 2* borrow pit.

Step 1: Find the horizontal midpoint

The base of the triangle that is the *Profile 2* borrow pit is also the *horizontal baseline.* You find the horizontal midpoint by dividing the length of the base, 250 feet, by 2. The result is 125 feet (250 ÷ 2 = 125). Using the horizontal scale from Figure 12-7, 50' = 1", so 125 feet scales off as 2.5 inches (125 ÷ 50 = 2.5). Measure 2.5 inches in from either end (west or east) along the *horizontal baseline* and mark this point. In Figure 12-7 I call this point *M*.

Step 2: Add line A–A

Line A–A is an axis that connects the *horizontal midpoint*, point *M*, and the deepest point in *Profile 2*, point *E*.

Step 3: Find the vertical midpoint and the vertical baseline

We'll use the center of mass formulas for right and oblique triangles from Figure 12-5 to locate the vertical midpoint for *Profile 2*. Here they are:

$x_a = {}^2/_3\, h$

$x_b = {}^1/_3\, h$

The depth of the borrow pit, *h*, is 10 feet.

$x_a = {}^2/_3 \times 10$

$ = 0.667 \times 10$

$ = 6.67'$

At the vertical scale of 10' = 1", that's 0.667 inches (6.67 ÷ 10 = 0.667). In Figure 12-7 the borrow pit's deepest point is point *E*. You'd measure 0.667 inches up from point *E* and make a light pencil mark. Then add a *vertical baseline* that passes through that point and is parallel to the *horizontal baseline*.

Step 4: Find the actual center of mass

Draw a straight line that connects points *E* and *M*. In Figure 12-7 this is line *A-A*. The point where line *A-A* intersects the *vertical baseline* is the actual center of mass for *Profile 2*. In Figure 12-7 this intersection is the point labeled *CM*.

Step 5: Find center-of-mass-to-edge distance

Draw a perpendicular line that passes through point *CM* and intersects the *horizontal baseline*. In Figure 12-7 this is line *B-B*. Point *D* marks the intersection of line *B-B* and the *horizontal baseline*. The distance between *CM* and point *D* is the distance from the edge to center of mass. In Figure 12-7 the distance between *CM* and point *D* measures 0.54 inches or 27 feet (0.54 × 50 = 27). So the center of mass of this borrow pit lies 27 feet east of its horizontal midpoint.

Step 6: Calculate the average haul distance

Simply add 27 feet to the 515 total haul distance, increasing it to 542 feet.

I recommend ignoring the vertical haul distance of 6.67 feet. Like most vertical haul distances this one's too small to be worth the bother.

In the next chapter we'll look at the costs of doing business as an excavation contractor. Those costs include machine production rates, operating expenses, and owning equipment.

13
Earthmoving Equipment: Productivity Rates and Owning & Operating Costs

Up to this point we've been concentrating on estimating volumes of earthwork. Volume is always important on an earthwork job and it's rarely easy to estimate. But there's more to estimating earthwork than calculating volumes. In this chapter we'll change our focus. We'll look at costs that good estimators never overlook in their estimates.

The costs we'll cover fall into three categories:

- Equipment — purchase or lease costs, maintenance and operation costs
- Labor — costs for wages, insurance, withholding and other taxes
- Overhead — costs for office space, equipment and supplies

Contracts for earthwork projects are awarded by the competitive bid process. General contractors invite bids from companies that specialize in excavation work. The excavation contractors submit bids based on the project plans and specifications provided by the general contractor. Each bid quotes a dollar cost per cubic yard of material moved. As an estimator it's your job to work up these bids. That means calculating *two* different cost totals in dollars per cubic yard for each job. One of these totals comes straight from the amount of material that's moved. The second total covers the costs of doing business as an excavation contractor.

As an excavation estimator you're a member of the contractor's planning team. You'll help decide what personnel and machines to use on each job. One construction company I worked for paired up a field superintendent and an estimator for each project. They worked as a team and developed a coordinated plan of attack for each project. This team not only organized the job but also scheduled the equipment. Projects ran smoothly and on schedule.

Good equipment cost estimates start with good equipment operating cost records. The more performance records you have, the better your chances of developing accurate cost figures to use in bids. If you don't have the records, then you'll have to estimate average annual operating costs. If you have equipment records from past jobs, I strongly recommend using them as the basis for your operating costs. Operating costs based on actual experience are much better and more realistic than any estimate.

Obviously, I don't have your actual costs for your equipment, so I've used my own data instead in this chapter. And I've included the formulas and factors you need in order to calculate *your* costs from your own data.

There are three major factors to consider when you develop machine ownership and operating costs: power, speed, and production. This chapter covers all three factors in depth. We'll begin with power.

Machine Power

Each earthmoving machine has only a certain amount of available power. An important part of your job as an earthwork estimator is matching your machine's usable power with your job's power requirements. Usable power is the available power limited by job conditions. Required power is the amount of power it takes to move not only the machine but also its load. The two most important factors that determine the amount of power you require are rolling resistance and grade resistance.

Rolling resistance is the force the ground exerts against the machine through its tires, measured in pounds of pull. The machine won't move without enough power to overcome this resistance.

Grade resistance is the force exerted by gravity on a machine as it moves uphill or downhill. This is also measured in pounds. Grade resistance is a dual factor — it acts positively under some conditions and negatively in others. The effect of grade resistance depends on the direction of travel relative to the grade. It's a negative when you travel uphill and a positive when you go downhill.

Rolling Resistance

Forces that affect rolling resistance include friction, maintenance, tire design and inflation. Here's a rule of thumb for roughly estimating rolling resistance:

Rolling resistance = 40 lb per ton of weight on wheels

This rule of thumb assumes:

- Normal maintenance
- Hard, level road surface
- Wheeled machines

Surface type & condition	Rolling resistance factors (lb/ton)*
Concrete or asphalt	40
Hard gravel surface	65
Packed snow	50
Loose snow	90
Packed dirt	100
Loose dirt	150
Loose sand or gravel	200
Soft, muddy dirt	320

*Rolling resistance factors are applicable only when calculating resistance for wheeled equipment.

Figure 13-1
Rolling resistance factors for wheeled machines

If either the road surface or equipment falls short of those assumptions, you'll need more power to move a ton.

I'm quite willing to assume that all of your equipment is well-maintained and in top-notch running condition. But the road conditions? That's another matter entirely! How many smooth, hard-surface, level roads have you seen lately on job sites, or anywhere else? There aren't many roads with ideal surfaces in the real world. And that's where rolling resistance factors come into play. You use RR factors to compensate for all the different variations on less-than-ideal road conditions. You'll find a table of RR factors, based on road surfaces, in Figure 13-1. Here's how you use RR factors in a formula for rolling resistance:

RR (lb) = weight on wheels (tons) × RR factor

Remember, you only use RR factors in resistance calculations for wheeled equipment. Why don't RR factors apply to track equipment? Track machines carry their own road surface with them and it's always the same. The machine's tracks *are* its road surface.

Now let's try out the formula using the RR factors from Figure 13-1. Suppose you have a 12-ton truck traveling on a level packed-dirt road. Find the rolling resistance in pounds. First, you need the RR factor for this road surface. Figure 13-1 shows that a surface of packed dirt has a RR factor of 100. So your values are 12 tons for the weight on wheels and 100 for the RR factor. Here's the calculation:

RR (lb) = 12 × 100

= 1,200 lb

Now let's change things around a bit. Say that you're running a pull-type scraper behind a wheeled tractor. What does this change mean to the way that you'll find the weight on wheels? It means weight on wheels *does not* include the tractor's weight. Weight on wheels here consists only of the scraper's weight. This is because the tractor furnishes all of the push or pull pounds needed here. If the tractor's attached to, or it's part of, the scraper, then weight on wheels is the sum of three weights: tractor, scraper and load.

How much resistance a machine works against makes a great deal of difference in how much power it takes to do a job. Here's a pair of examples that demonstrate my point.

Finding Rolling Resistance — Example 1

A wheel tractor attached to a fully-loaded scraper is moving on a level, packed-dirt road. The wheel tractor weighs 25,000 pounds. The scraper weighs 23,000 pounds and a full load of material weighs 21,000 pounds. Using the RR factors in Figure 13-1, find the rolling resistance.

Earthwork Equipment: Productivity Rates and Owning & Operating Costs **211**

Step 1: Find weight on wheels (tons)

You recall that the tractor's attached to the scraper, so weight on wheels equals the tractor weight plus the scraper weight plus the load weight. We know what all of these weights are in pounds, but we want the result in tons. Just find the sum of the weights and divide the result by 2,000:

$$\text{Weight on wheels (tons)} = (25{,}000 + 23{,}000 + 21{,}000) \div 2{,}000$$
$$= 69{,}000 \div 2{,}000$$
$$= 34.5 \text{ tons}$$

Step 2: Find rolling resistance (lb)

The RR factor for a packed-dirt road is 100, so here's the equation:

$$\text{Rolling resistance (lb)} = 34.5 \times 100$$
$$= 3{,}450 \text{ lb}$$

Finding Rolling Resistance — Example 2

We'll change just one variable; the type of road surface is sand. Everything else is the same as in Example 1. Check Figure 13-1 and you'll find the RR factor for level sand is 350:

$$\text{Rolling resistance (lb)} = 34.5 \times 350$$
$$= 12{,}075 \text{ lb}$$

That's a 250 percent increase in the rolling resistance! If I were you, I'd think seriously about using a larger tractor on this job.

Grade Resistance

Grade resistance is the force of gravity on any machine, wheel or track, that's moving on a grade. Let's take a look at an example to see what a 10 percent grade means. Suppose the grade is 10 percent and the horizontal distance you travel is 100 feet. By the time you travel 100 feet horizontally on a 10 percent grade, you'll also gain 10 feet in elevation.

A machine moving uphill must overcome not only grade but also rolling resistance. On level ground there's no grade resistance. When a machine moves downhill, the slope of the grade assists and partly cancels the effect of rolling resistance. You probably won't be surprised to learn that there are three formulas used to find total resistance. You use one formula to find RR traveling uphill. If you need to find RR traveling downhill, that's a different formula. Use the third RR formula for level travel. In these formulas *TR* is short for total resistance, *RR* is rolling resistance and *GR* is grade resistance. Here are the formulas:

Formulas for total resistance

$$TR \text{ (uphill travel)} = RR + GR$$

$$TR \text{ (level travel)} = RR$$

$$TR \text{ (downhill travel)} = RR - GR$$

Here's a rule of thumb for estimating grade resistance: For every 1 percent of grade, assume 20 pounds grade resistance per ton of vehicle weight. Turn that into a formula and here's what you get:

GR (lb) = weight on wheels (tons) × 20 (lb/ton) × % grade

Formula for grade resistance

For example, suppose a wheel scraper is traveling up a 6 percent grade on a hard gravel road. We'll find the rolling resistance, grade resistance and total resistance for this example assuming these weights:

- Wheel scraper weight is 60,000 pounds
- Load weight is 50,000 pounds

Step 1: Find weight on wheels (tons)

Weight on wheels (tons) = (60,000 + 50,000) ÷ 2,000
= 110,000 ÷ 2,000
= 55 tons

Step 2: Find grade resistance (lb)

Use the formula for grade resistance, keeping in mind that this is a 6 percent grade.

Grade resistance (lb) = 55 × 20 × 6
= 6,600 lb

Step 3: Find rolling resistance (lb)

Use the RR factors from Figure 13-1 and this formula:

Rolling resistance (lb) = weight on wheels (tons) × RR factor

Rolling resistance (lb) = 55 × 65
= 3,575 lb

Step 4: Find total resistance (lb)

Using the formula for uphill travel:

TR (lb) = RR + GR

TR (lb) = 3,575 + 6,600
= 10,175 lb

What's the total resistance if we change only the direction of the loaded scraper's travel? Same scraper, same load, same road and same 6 percent grade, but we'll use the formula for downhill travel:

TR (lb) = RR − GR

TR (lb) = 3,575 − 6,600
= −3,025 lb

A negative result means that this is a grade assistance, or pushing force, equal to 3,025 pounds acting on the scraper. To operate this scraper safely you need braking force at least equal to the 3,025 pounds grade assistance.

		Rimpull (lbs)	
Gear	Speed (mph)	Rated	Maximum
1	2.0	35,000	45,000
2	5.0	18,000	23,000
3	7.0	11,000	18,000
4	8.5	9,000	13,000
5	10.0	7,500	11,000
6	11.0	5,500	9,000
7	11.9	3,500	7,000
8	12.5	1,500	5,000

Figure 13-2
Rimpull chart for a wheeled tractor

Available Power

The power available from a machine depends on two factors: horsepower and operating gear speeds. Once you define the conditions it's easy to find the average operating speed for a machine. Start by calculating resistance using the formulas for total resistance. Then turn to the specifications sheet or operating manual for the machine. There you'll find tables and charts listing pulling power for selected gear ranges, ground speed and breaking forces.

Wheel and track machines are both rated in pounds of pull, but there are two different kinds of pull. Track machine pull ratings are in units of *drawbar* pounds of pull. This means that under certain specific conditions (operating gear, rpm and speed) the machine is able to pull the specified number of pounds on a drawbar. Wheel machine ratings are in *rim* pounds of pull. This is the number of pounds of pull that the wheel rims are designed to withstand before they break traction and slip while propelling the machine forward. Figure 13-2 shows an example of a rimpull chart. Drawbar pull charts, for track equipment, look much the same as Figure 13-2 and you use them the same way.

Machine Speed

This is the second of the three main factors you use in determining operating costs. Machine speed is simply how fast the machine can pull a load of a specified size under certain job site conditions. The faster a machine moves, the more material it can move per day. Machine speed depends on two factors, the gear ratio and the number of pounds of pull provided by each gear. To find machine speed you need to know the machine's weight and the total resistance. You use the weight and resistance data with the machine's specification chart to determine machine speed.

Here's how it works. We'll say that the machine is a wheel-type scraper and the total resistance is 8,500 pounds. Using Figure 13-2 as your machine specifications' chart, find the machine speed. It says fourth gear provides 9,000 pounds of pull at 8.5 mph. Fifth gear provides 7,500 pounds of pull at 10 mph. Clearly, fifth gear doesn't supply enough pull, so your best choice is to use fourth gear and run the loaded scraper at 8.5 mph. We'll see how you find empty machine speeds and total travel times a little later in this chapter. But first let's look at *usable power*. Not only is usable power related to available power, it also affects machine speed.

Usable Power

Usable power is simply available power less the power you lose either because of problems with traction or altitude. We'll take a close look at both factors, starting with traction.

Traction

Earthwork estimators define traction as a machine's ability to continue moving forward without the wheels or tracks slipping. When either tracks or tires slip, you lose speed. That's why traction is always a factor whenever you're figuring a machine's speed or efficiency.

You measure traction in either pounds of pull or pounds of push. There's a direct link between the weight on a machine's drive wheels and the amount of traction. It's physically impossible for a machine to exert a force greater than the weight on its drive wheels. Clearly, it's important to know not only the weight of a machine but also which are the drive wheels. For example, the drive wheels for a wheel tractor pulling a wheel scraper are on the tractor, not on the scraper.

To find out how many pounds of pull really are available (from a specific machine, operating on a specific type of surface) earthwork estimators use coefficients of traction. Figure 13-3 compares the coefficients of traction of tires and tracks on various types of surfaces. In Figure 13-3 you'll see that on a concrete or asphalt surface tires are the best choice. They operate at about 90 percent efficiency. Tracks slip easily on surfaces like concrete and asphalt. According to Figure 13-3, efficiency drops by more than half and track machines operate at only about 45 percent of total traction.

Now compare tracks to tires when the surface is dry dirt. This time the track machine has the advantage, operating at about 90 percent efficiency. Tires slip easily in dry dirt and their operating efficiency drops to about 55 percent.

Surface type and condition	Traction factors	
	Tires	Tracks
Concrete/asphalt	0.90	0.45
Normal dirt, dry	0.55	0.90
Normal dirt, wet	0.45	0.70
Sand, dry	0.20	0.30
Sand, wet	0.40	0.50
Gravel road	0.36	0.50
Snow, packed	0.20	0.27
Ice	0.12	0.12

Figure 13-3
Coefficients of traction

The percentage of gross vehicle weight (GVW) on the drive wheels appears on each machine's spec sheet. Use the data from the manufacturer if it's available. Otherwise, use one of the following formulas to calculate weight on the drive wheels.

Formula for weight on the drive wheels

Track machine pulling wheel scraper: weight on drive wheels = 100% GVW

Four-wheel tractor with attached scraper: weight on drive wheels = 40% GVW

Two-wheel tractor with attached scraper: weight on drive wheels = 60% GVW

Let's try pulling all of these factors and formulas together by finding the rimpull available in pounds. Here's what we know:

- Machine = a two-wheel tractor with an attached scraper
- Gross vehicle weight = 125,000 pounds
- Travel surface = hard gravel road

Step 1: Calculate the weight on the drive wheels

For a two-wheel tractor with attached scraper, the weight on the drive wheels equals 60 percent of the gross vehicle weight. Here's the math:

Weight on drive wheels (tons) = (125,000 × 0.6) ÷ 2,000

= 75,000 ÷ 2,000

= 37.5 tons

Step 2: Find the coefficient of traction (see Figure 13-3)

This is a wheeled machine and the travel surface is a hard gravel road. According to Figure 13-3 the coefficient of traction for this combination is 0.36.

Step 3: Calculate rimpull in pounds

Here's the formula and then the math.

Formula for rimpull

Rimpull (lb) = weight on drive wheels × coefficient of traction

Rimpull (lb) = 75,000 × 0.36

= 27,000 lb

But traction isn't the only factor that determines machine speed. The other factor is altitude.

Altitude

Altitude is a measurement of height above sea level. As altitude increases, atmospheric pressure decreases. And the lower the atmospheric pressure, the less horsepower a machine has. A naturally-aspirated engine (any engine not equipped with a turbocharger) loses about 3 percent horsepower for every 1,000 foot gain in elevation above 3,000 feet. This is a general value and may not be the value for your machines. Refer to your machine's owner manual or your equipment dealer for the individual machine values.

Here's how it works. Say a machine has a drawbar pull of 8,500 pounds at or below 3,000 feet and the job site elevation is 7,000 feet. What's is the actual drawbar pull for this machine at this elevation?

Step 1: Find the percentage of lost horsepower

As you now know, you lose 3 percent horsepower per 1,000 feet above 3,000 feet. Here's the math for our example:

$$\text{Lost horsepower (\%)} = (7,000 - 3,000) \times 3\%$$
$$= 4,000 \times 3\%$$
$$= 12\%$$

Step 2: Calculate actual drawbar pull in pounds

Actual drawbar pull in pounds equals the rated drawbar pull less 12 percent. For our example the math works out like this:

$$\text{Actual drawbar pull (lb)} = 8,500 - (8,500 \times 12\%)$$
$$= 8,500 - 1,020$$
$$= 7,480 \text{ lb}$$

Cycle Time

Cycle time is the measurement of how long it takes for a machine to pick up a load, travel to the dump site, dump the load, and make the return trip to the excavation site. One of the primary goals for excavation estimators is finding the shortest possible cycle times for equipment. Here's why. The shorter the cycle time, the more trips made per hour, the more material moved, the more money you make.

Cycle times are the product of two kinds of time: fixed time and variable time. Let's start by defining these two types of time.

Fixed Time

Fixed time refers to a group of operations including loading, dumping and maneuvering that, assuming similar conditions, take the same amount of time to accomplish from one job to the next. Manufacturers of earthmoving equipment often include estimates for fixed time in their equipment manuals. But, as you might expect, their estimates tend to be optimistic. Perhaps on a perfect job site they are accurate. I wouldn't know — I've never been on a perfect job site. So instead of using this somewhat unrealistic data, keep your own records. The data you compile by tracking real cycle times are far more meaningful. After all, it applies to your machines, your type of work and your conditions. Cycle times based on your customized fixed times are sure to be more accurate.

Variable Time

Variable time is the amount of time that a machine spends in transit between loading site and dumping site. Obviously this changes from job to job. The main factors in determining variable time are:

- Distance, by haul road, separating the loading site from the dumping site
- Percentage of grade
- Condition of the haul road

To find variable time for a job, clock several runs with a stopwatch and then find their average.

Cycle times vary from one type of equipment to another and even from machine to machine and operator to operator. Your selection of machines for a project depends on the job site conditions as well as the distance to travel between load and unload. As you saw earlier, sometimes wheel loaders are the best choice. Under different conditions the best machine for the job may be a track loader.

The sum of the fixed and the variable times is your estimated cycle time for a project. After work starts on a project, make several comparisons by clocking real cycle times and thinking of your estimate as a goal. Say that you notice that the cycle times on a project keep rising. That's a tip-off that there's a problem. Cycle times don't rise except when equipment is used inefficiently. Double-check the haul road's layout and condition. It's a good bet they need maintenance. Many excavation contractors find that it pays, in shorter cycle times, to keep and run a motor grader on site. The main job of the motor grader? Maintaining the condition of the haul road surfaces. Here are three goals to keep in mind from start to finish on every job:

- All machines working at full capacity and top efficiency
- Best possible haul road surface
- Use grades as productively as possible

Now that you know how to find cycle times, it's time we moved on to take a look at machine production and how to calculate productivity.

Machine Production

This is the third and final factor you use to determine your owning and operating costs. Here's a simple definition of machine production: *The quantity of material transferred between two locations within a specified period of time.* Three major factors determine machine production:

- Material
- Time
- Efficiency

Let me explain how this works with an example.

Finding a Production Rate in Cubic Yards per Hour

Suppose that you've signed a contract that requires moving 25,000 cubic yards of material in two weeks. To finish on schedule, how many cubic yards of material must you move per hour?

Before we start running any numbers, assume you have just one machine available for this job, and you'll run it eight hours a day, five days a week. Now it's time for some math.

Step 1: Finding the total hours

Eight hours per day, five days a week, for two weeks comes to a total of how many hours?

Total hours = 8 × 5 × 2

= 80 hours

Step 2: Finding the production rate in cubic yards per hour

You want to move 25,000 cubic yards of material in a total of 80 hours. So how many cubic yards must you move per hour?

Production rate (CY/hr) = 25,000 ÷ 80

= 312.5 CY/hr

Where do you go from here? The next step is to figure out how many machines and men it's going to take to achieve that production rate. But isn't there something wrong here? Ask yourself this question: Would you sign a contract without first doing the math so you knew what you were promising? I sure wouldn't, and neither should you! I buffaloed you into starting from the wrong end of that job, but in the process you learned something about production rates.

What Production Rates Tell You

Let's take a quick look at what you've learned so far. You know how to calculate material quantities. You also know how to find how long it takes to move a given quantity of material with a specific machine assuming ideal job site conditions.

In the real world perfect conditions are something you'll never find. Don't forget that you need to allow for that fact in your estimates. On any excavation job, no matter what, you always lose some time and capacity. A good estimate includes an allowance for this fact of life. How? My preferred solution is to always make a slight reduction to my productivity estimate.

Good production data is vital to estimators. It tells you how many machines you need to move the most material, in the least amount of time, for the least cost and therefore, the greatest profit.

	Actual productivity	
	Minutes per hour	Factor
Normal day operations		
Track equipment	50	0.83
Wheel equipment	45	0.75
Normal night operations		
Track equipment	45	0.75
Wheel equipment	40	0.67

Figure 13-4
Efficiency factors

Productivity is easy to calculate, but beware of this trap: "If two machines are good, then four machines are twice as good." It's just not true. More machines don't always equal more productivity. The opposite result is really more likely. Too many machines in too little space can reduce productivity, simply because they're in each other's way.

To find a machine's production rate you need to know its cycle time. Then you calculate production in trips per hour using the following formula:

• • • • • • • • • • • • • •
Formula for machine production

Machine production (trips/hour) = 60 minutes ÷ cycle time in minutes

Suppose your cycle time is 6.5 minutes. What's your machine production in trips per hour?

Machine production (trips/hour) = 60 ÷ 6.5
= 9.23 trips/hour

This formula assumes ideal conditions, but we know that's not realistic. Fortunately, the excavation industry recognizes that neither people nor their machines are 100 percent efficient. They analyzed data from thousands of jobs and developed factors that everyone uses.

Job Efficiency Factors and How to Use Them

An operator stops for a break or a drink of water. A machine breaks down or stalls without warning. Minor delays add up and they happen for all sorts of reasons. That's why people and their machines are never 100 percent efficient. The *job efficiency factors* in Figure 13-4 take this fact into account. Equipment manufacturers, engineers, designers and construction organizations have developed this data using information gathered over the years. You may wish to develop you own factors.

Efficiency factors make it easy for you to bring your productivity estimates into line with what's realistic and achievable. Let's try out the efficiency factors from Figure 13-4. We'll use the machine production rate of 9.23 trips per hour that we found earlier and assume all work is done during the daytime. If the machine is track-driven, our efficiency factor adjusted production rate is 7.66 trips per hour (9.23 × 0.83 = 7.66). We'll round that off and call it 8 trips/hour. If the machine has rubber tires, then the adjusted production rate is 6.92 trips per hour (9.23 × 0.75 = 6.92) or, rounded, 7 trips per hour.

Now let's see how these figures relate to material volumes. Suppose the capacity of your track machine works out to 100 cubic yards per hour. During the daytime this machine actually moves a total of 83 cubic yards per hour and at night 75 cubic yards per hour.

The job efficiency factors in Figure 13-4 are averages. They're fairly accurate for most types of equipment. But if the machine's spec sheet includes job efficiency data, use it rather than Figure 13-4.

Productivity Calculations for a Simple Dirt Job

Start by taking a look at Figure 13-5. This is the site plan for our project. This is a simple dirt job where we'll move material from a borrow pit to a fill area using the two haul roads *Road A* and *Road B*. The length of each road and the percent grade also appear in Figure 13-5. This time we'll skip calculating the centers of mass for both the fill area and the borrow pit. You'll find them marked on the site plan with the abbreviation *CM*. Instead we'll begin by taking a close look at the haul roads.

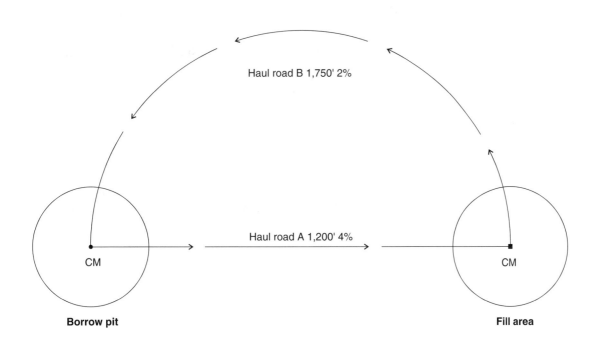

Figure 13-5
Layout of a simple dirt job

The Haul Roads

Figure 13-5 tells us that haul road A is 1,200 feet long and has a 4 percent grade. Haul road B is 1,750 feet long and has a 2 percent grade. We'll assume that both haul roads are only wide enough for one-way traffic. Now, which haul road should the loaded machines use? Don't make the mistake of thinking it won't make much difference. It makes a lot of difference and that difference shows in the profit/loss column.

Your best choice for the project in Figure 13-5 is to run the loaded machines on road A and the empty machines on road B. Why? For a start, loaded machines are heavier and travel more slowly than empty machines. Road A is 550 feet shorter than road B. That's one reason to run the loaded machines on road A. The second reason is even better. Running the loaded machines on road A is making the most productive use of that 4 percent grade by using it to your advantage. Here's how this works. When you run the loaded machines on road A, the entire trip is downhill. That gives you a positive grade resistance. Now let's get back to setting up that sample project.

Sample Project Machine & Haul Road Specifications

Here's all the data we'll need for the sample project calculations.

- Machine type = two-wheel-drive scraper
- GVW (gross vehicular weight) or empty weight = 45,000 lb (22.5 tons)
- Rated payload weight = 65,000 lb (32.5 tons)
- Loaded weight (GVW + payload weight) = 110,000 lb (55 tons)
- Rimpull chart = use Figure 13-2
- Length haul road A (measured from the borrow pit center of mass to the fill area center of mass) = 1,200 feet
- Grade, haul road A = 4%
- Grade, haul road B = 2%
- Haul road surface type and condition = hard-packed dirt, recently graded
- Rolling resistance factor = 100 (from Figure 13-1)

Let's start with the series of calculations that add up to total cycle time. We'll do each of these calculations twice; first for the loaded machines, traveling on road A, and then for the empty machines using road B.

Resistance Calculations

We'll make two separate sets of resistance calculations for this sample project, one for loaded machines and one for the empty machines. Remember these formulas?

GR (lb) = weight on wheels (tons) × 20 (lb/ton) × % grade

TR (lb) for uphill travel = RR + GR

TR (lb) for level travel = RR

TR (lb) for downhill travel = RR − GR

Where:

TR = total resistance

RR = rolling resistance

GR = grade resistance

Rolling Resistance

We'll start with rolling resistance, using the same formula for both calculations:

RR = weight on wheels × RR factor

There is a difference, of course, in how you define "weight on wheels." For the loaded machines "weight on wheels" is the gross vehicular weight plus the payload weight. But for the empty machines "weight on wheels" is just GVW.

Loaded:

RR = (22.5 + 32.5) × 100

    = 55 × 100

    = 5,500 lb

Empty:

RR = 22.5 × 100

    = 2,250 lb

Grade Resistance

Here's the formula for grade resistance:

GR (lb) = weight (tons) × 20 (lb/ton) × % grade

The loaded machines on road A are assisted by the 4 percent grade. Here's the math:

GR (lb) = 55 × 20 × 4

        = 4,400 lb

Empty machines on road B travel uphill on a 2 percent grade. The grade resistance equals:

GR (lb) = 22.5 × 20 × 2

        = 900 lb

Total Resistance

For the loaded machines in this example you find total resistance using this formula:

TR (lb) = RR − GR

$$TR\ (lb) = 5{,}500 - 4{,}400$$
$$= 1{,}100\ lb$$

To find total resistance for the empty machines you use this formula:

TR (lb) = RR + GR

$$TR\ (lb) = 2{,}250 + 900$$
$$= 3{,}150\ lb$$

Total resistance and rimpull are the same for a wheeled machine like the scraper we're using for this sample job.

Finding Operating Speed and Gear Using a Rimpull Chart

Let's review our rimpull needs for the sample job:

- For the loaded machines we need 1,100 pounds of rimpull
- For the empty machines we need 3,150 pounds of rimpull

We'll use the rimpull chart shown in Figure 13-2 to find the operating speed and gear for the loaded and the empty machines.

Now let's see how this works. In the *Rated* column, find the number closest to, but not less than, the 1,100 pounds of rimpull we know we need for the loaded machines. That's 1,500 at the bottom of the *Rated* column. Stay in the bottom row and move across to the *Gear* column to find the operating gear we'll use. The answer is eighth gear. Now move to the *Speed* (mph) column to find the machine's speed in miles per hour. The answer is 12.5 miles per hour.

Repeat these steps for the empty machines and check your results against the following list.

- Required rimpull = 3,150 pounds
- Gear = seventh
- Speed = 11.9 mph

Notice anything odd here? Most haul roads, like the roads in Figure 13-5, are less than a mile long. But most specification sheets list speeds in miles per hour, just like Figure 13-2. Let's make these speeds easier to work with by converting them into feet per minute. All you do is multiply the miles per hour by a constant 88. Here's how it looks as a formula:

Feet per minute = miles per hour × 88

Try it out converting the speeds we just found:

$$\text{Loaded speed (feet per minute)} = 12.5 \times 88$$
$$= 1{,}100\ \text{feet per minute}$$

$$\text{Empty speed (feet per minute)} = 11.9 \times 88$$
$$= 1{,}047\ \text{feet per minute}$$

Calculating Travel Times

We know that haul road A is 1,200 feet long. We also know the loaded machine's speed is 1,100 feet per minute. To find the travel time, you just divide distance by the rate of speed.

Loaded travel time (minutes) = 1,200 ÷ 1,100

= 1.10 minutes

Haul road B is 1,750 feet long and the empty machine's speed is 1,047 feet per minute. Empty travel time equals 1.67 minutes (1,750 ÷ 1,047 = 1.67)

The manufacturer's handbooks that come with most loaders include graphs of estimated loaded and unloaded travel times under a variety of conditions. There's no point in duplicating the work, so if the travel time data's included, use it.

Calculating Cycle Time

Cycle time, remember, is variable time plus fixed time. Variable time for the sample project is the sum of loaded travel time plus empty travel time. That comes to 2.77 minutes (1.10 + 1.67 = 2.77). Fixed time for the two-wheel-drive scraper used in the example equals load time (we'll say that's 0.60 minutes) plus dump and maneuver time (0.50 minutes). So fixed time equals 0.60 + 0.50 = 1.10 minutes, and cycle time equals 0.60 + 0.50 + 1.10 + 1.67 = 3.87 minutes.

Calculating Production Rates

Formulas for production rates

The first production rate we'll find for the sample project measures the number of trips per hour. Here's the formula and the math:

Production (trips/hr) = 60 ÷ minutes per trip

Production (trips/hr) = 60 ÷ 3.87

= 15.50 trips per hour

To find a more realistic production rate we'll apply an efficiency factor as the next step.

We'll use the efficiency factors from Figure 13-4. The machine's wheeled and we'll assume this is a normal day operation. That combination makes the efficiency factor 0.75, or 45 minutes of actual production per hour. The actual production rate is 11.62 trips per hour (15.50 × 0.75 = 11.62). We'll round that off and call it 11.5 trips per hour.

You already know the payload size for the machine is 32.5 tons. Now you also know a machine makes 11.5 trips per hour. So how many tons of material will a machine that's filled to capacity move in an hour? To find out, use this formula:

Production rate (tons/hr) = payload size × trips per hour

Here's the math for the sample project:

Production rate (tons/hr) = 11.5 × 32.5

= 373.75 tons per hour

If this were an actual job, you would know a lot more than this about the material you're moving. Make it your business to know at least this much about any material:

- Type
- Moisture content
- Swell factor

You need that information to calculate production rates, either in cubic yards per hour (CY/hr) or cubic yards per day (CY/day). Then you use the production rates to figure out how many machines you'll need on the job to move the material efficiently and on schedule. The other factors you'll want to consider include:

- Availability of machines
- Size of the operation
- Number of working days

On most earthmoving projects the focal point is some type of scraper. But scrapers don't work alone. Other machines either support your scraper units or do other types of work. So now we'll see how to calculate production rates for other common earthmoving machines.

Pusher Units

Wheel units that aren't self-loaders need to work with a pusher unit. But how many haul units can one pusher unit efficiently handle? You don't want either the haul units or the pusher unit to sit idle. In this business, wait time is wasted time. Strike a balance between your pusher unit cycle time and the number of haul units per pusher unit. The result is minimum wait time and maximum productivity.

First, you find the cycle time for your pusher unit using this basic formula:

Formulas for pusher units

Pusher unit cycle time (minutes) =
boost time + transfer time + return time + load time

Use the following industry standardized times for pusher units:

- Boost time + transfer time = 0.25 minutes
- Return time = 40% of load time

Now, simplify the formula as follows:

Pusher unit cycle time (minutes) = 0.25 + (1.4 × load time)

Next, you find the haul unit's cycle time using the same steps as earlier in this chapter. Once you know both the pusher and the haul unit cycle times, you use this formula to find the number of haul units you'll need per pusher unit:

Haul units per pusher unit = hauling unit cycle time ÷ pusher unit cycle time

Let's do an example. Suppose the cycle time for the hauling units is 4 minutes and load time for the pusher unit is 0.5 minutes. Here's the math:

Haul units per pusher unit = 4 ÷ [0.25 + (1.4 × 0.5)]
= 4 ÷ [0.25 + 0.7]
= 4 ÷ 0.95
= 4.21 haul units per pusher unit

Haul units are pieces of equipment, so a decimal number doesn't make much sense. We'll round this answer off to the nearest whole number, or four haul units per pusher unit.

Bulldozer Production Rates

Production rates for bulldozers are complex and involve many variables. Most excavation contractors don't calculate production rates for their bulldozers. Standard practice among excavation contractors is to simply consider the costs of keeping a bulldozer on site as part of job overhead.

Compactors with front blades now keep the unloading areas smooth on most job sites. But there's still plenty of dozer work on any job site, especially in finish work — working slopes down and dressing them up.

Compactor Production Rates

There are two types of compactors: sheepsfoot rollers and pneumatic rollers. Compactors of either type both deposit and compact fill material using a standard 6-inch lift. There are many variables to consider in finding compactor production rates. They include:

- Compactor type
- Material type and moisture content
- Compaction requirements
- Area involved
- Machine speed
- Required number of passes

The type of compactor, for instance, determines rolling resistance. Let's compare rolling resistance for the two types of compactor, assuming normal material and 6-inch lifts. For a sheepsfoot roller, rolling resistance is about 500 pounds per ton. For a pneumatic roller, add 10 pounds per ton per inch of compacted material.

Fortunately, there's a shortcut. Just use the following simplified formula:

Compactor production rate (CY/hr) = (w × s × l × 16.3) ÷ p

Where:

- w = width, compacted area (feet)
- s = speed (mph)
- l = lift thickness (compact inches)

Compactor type	Average speed (mph)
Self-propelled sheepsfoot roller	5
Self-propelled tamper unit	6
Self-propelled pneumatic roller	7
Sheepsfoot roller pulled by wheeled tractor	7
Sheepsfoot roller pulled by track tractor	4
Sheepsfoot roller pulled by motor grader	12

Figure 13-6
Compactor's average operating speeds

- 16.3 = mathematical constant
- p = passes, machine (number required)

Obviously *speed (mph)* is the shortcut here, but that leaves you with a big question. What is the source for this data? The best source is the specifications sheet for your specific machine. If you don't have that, use Figure 13-6.

Sample Calculation of Compactor Productivity

Now let's try out that formula on a sample problem. Here's all the data we need:

- Compactor type: 15-ton, self-propelled, 12-foot-wide sheepsfoot roller
- Lift thickness: 6-inch lifts
- Number of passes required: 5
- Machine speed: see Figure 13-6

Find the production rate for this compactor using this formula:

Compactor production rate (CY/hr) = (w × s × l × 16.3) ÷ p

We'll start by matching the formula variables to their values in this example. Here's the result:

- w = 12 feet
- p = 5
- s = 5 mph
- l = 6 inches

Next, simply plug these values into the formula and do the math:

Compactor production rate (CY/hr) = (12 × 5 × 6 × 16.3) ÷ 5
= 5,868 ÷ 5
= 1,173.6 CY/hour

Motor Grader Production Rates

Keeping a motor grader on-site is standard practice in the excavation business for all but the smallest projects. On any project, motor graders have two main jobs.

1) Keeping haul roads smooth

2) Leveling all finish grades

Both jobs are critical. But you don't measure or estimate these jobs as volumes. Production rates for motor grader work use the amount of time it takes to complete a task, rather than the volume of material moved.

Typically a grader makes the same number of passes regardless of the task being done. Here's the standard sequence. The first pass is a cutting pass with the blade set to cut to the depth of the bottom of the deepest ruts or hollows. It uses the lowest gear and speed of all the passes. The second pass, in a higher gear and speed, smoothes out small irregularities and any blade spill left behind by the first pass. The third pass is the finish pass. The gear and speed are either the same or slightly higher than in the second pass.

Calculating Motor Grader Task Times

Let's try a sample problem. Assume the following:

- Total number of passes: 3
- First pass gear: second
- Speed in second gear: 3.1 mph
- Second and third passes gear: third
- Speed in third gear: 4.2 mph
- Length of haul road: 4.1 miles
- Productive minutes per hour: 50
- Efficiency factor: 83 percent

What's the total time for this job? It's the sum of the times for each pass. The time for each pass equals the road length divided by grader speed. Here's the math for the first pass:

Pass 1 (hours) = 4.1 ÷ 3.1

= 1.32 hours

The second and third passes were identical, so we'll combine them here:

Pass 2 and 3 (hours) = (4.1 ÷ 4.2) × 2

= 0.976 × 2

= 1.95 hours

Next add these times together to find total time:

Total time (hours) = 1.32 + 1.95

= 3.27 hours

Actual total time equals total time divided by the efficiency factor. Here's the math:

Actual total time (hours) = 3.27 ÷ 0.83
= 3.94 hours

We'll round that off and call it 4 hours.

That's the end of our discussion of production rates for earthmoving equipment. I've limited this discussion to the standard machines that you'll find on almost every earthmoving job. There are other machines, of course, and we'll look at several and see how to calculate their production rates in the sample bid in Chapter 15.

Owning and Operating Costs

When you've figured the material quantities and machine time required, it's time to consider the hourly cost of owning and operating the equipment. And there's more to these costs than just the purchase price of the machine. Hourly cost has to include the ownership cost, operating cost, and the operator's wages. The wages are easy to calculate — just multiply the hourly wages times the number of operators. Ownership and operating costs are a little more complicated.

Ownership Costs

Ownership costs are a fixed cost because they continue whether or not a machine's working. These are the four components to ownership costs: depreciation, interest, insurance and taxes. We'll take a closer look at them, starting at the top of the list.

Depreciation

Let me make one thing clear at the start. I'm no tax expert. What I call depreciation is *not* the same as depreciation in the tax code. That type of depreciation is very complex and beyond my experience. Depreciation, as I use the word in this book, refers *only* to equipment ownership costs.

You'll agree, I'm sure, that the older a machine is, the less it's worth. Depreciation, to simplify greatly, is a way for you to spread out the purchase price of a machine over its useful lifetime. The useful life span of a machine depends on the working conditions, machine type and the skill of the operator.

The most important is operating conditions. For most earthmoving equipment, the standard useful life span is:

- Excellent conditions = 12,000 hours
- Average conditions = 10,000 hours
- Severe conditions = 8,000 hours

To establish a realistic value for a machine at the end of its useful life, I recommend contacting local equipment dealerships. They can quote Blue Book values. Some contractors I know assume the value of a machine at the end of its useful life is zero. This practice artificially inflates hourly operating costs. I recommend using only actual values to calculate hourly operating costs for equipment. Then calculate hourly depreciation using this formula:

Formula for depreciation

Depreciation per hour =
purchase price – tire value ÷ estimated useful life span

Most of those variables are familiar. Purchase price is clear enough and we just covered depreciation and estimated useful life span. But what about *tire value*? Tires on earthmoving equipment wear out rapidly. Over the useful lifetime of a machine you'll replace the tires many times. Because tire costs are an operating cost, not an ownership cost, deduct their value from the purchase price. When you buy a new machine, obtain the value of all tires from the equipment company and deduct it from the purchase price. For a used machine, either the equipment company or local tire supplier can provide you with the estimated value of the tires.

Interest, Insurance and Taxes

Each machine accounts for a specific percentage of your total costs in each of these three categories. We'll find the sum of the three percentages and use that to calculate ownership costs. In the rest of this chapter I refer to this group as *IIT* (my shorthand for interest, insurance and taxes). To calculate these hourly costs, find the sum of the percentages of your total annual costs and divide the result by the estimated hours of use per year. Remember that these are yearly costs. That means you should recalculate them annually, using the actual costs and the actual hours of usage. Use the actual costs for the past year to help you estimate the costs for next year.

You'll find hourly cost factors helpful in estimating your hourly operating costs. Figure 13-7 is a typical chart. Your equipment company should have a comparable chart for your equipment. To use Figure 13-7 you need to know the following:

- Machine's average hours of usage per year
- Interest rate percentage on machine's purchase loan
- Percentage of annual total insurance costs assigned to the machine
- Your per machine percentage of total tax paid

Find the sum of the three percentages. Find your result in the vertical scale on the left side of Figure 13-7. Read across and to the right until this horizontal line meets the diagonal line matching your average hours per year usage. Go straight down from that point and read the factor off of the horizontal scale that runs across the bottom of Figure 13-7.

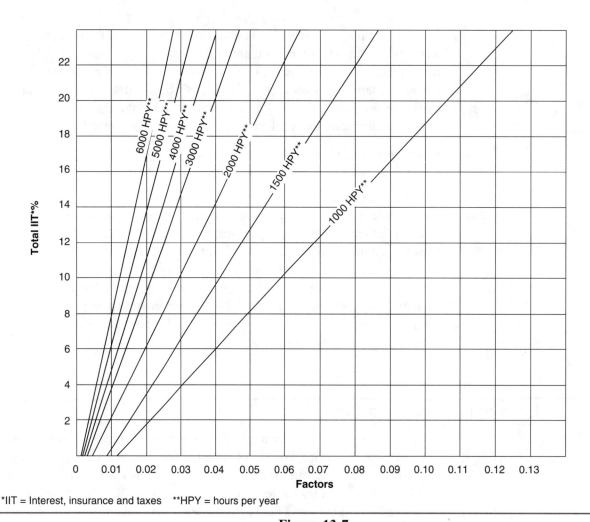

Figure 13-7
Graph of hourly cost factors for interest, insurance and taxes

Confused? This example should help. Assume the delivered purchase price for your machine was $75,000 and you use it an average of 3,000 hours per year (HPY for short). These are your rates:

- Interest = 7%
- Insurance = 5%
- Taxes = 4%

Step 1) Find the sum of the yearly percentages:
7 + 5 + 4 = 16

Step 2) Find 16% on the left vertical scale in Figure 13-7.

Step 3) From 16%, read across to the right, until this line intersects the 3,000 HPY diagonal.

232 *Estimating Excavation*

Step 4) From that intersection, read straight down to the factors scale across the bottom of Figure 13-7.

Step 5) From that scale, read your IIT factor.

You need the IIT factor to estimate your hourly cost using this formula:

Approximate hourly cost = IIT factor × delivered price ÷ 1000

Approximate hourly cost = 0.032 × 75,000 ÷ 1000
= $2.40 per hour

Operating Costs

Operating costs include fuel and lubricants, tires, and repairs. These all vary depending on how much use the machine gets.

Fuel and Lubricants

I recommend that you keep up-to-date and accurate records of your own costs for fuel and lubricants for each machine. If you don't have this information, use the manufacturer's estimates from your spec sheet or owner's manual. But start keeping records of your own right away. Typically, manufacturer data is overly optimistic. It's better than no data at all, but use it only if you have to.

Tires

Calculating the hourly cost of tires depends on several variables, including type of tire, site conditions and upkeep. Many tire manufacturers can provide estimated service life for their products. Again, you'll have more accurate estimates if you keep good records of tire costs for each machine. Either way, when you've estimated service life hours, here's how to figure estimated hourly cost.

Hourly tire cost = replacement cost of tires ÷ estimated service life (hours)

Remember that replacement cost of tires includes all of the costs for tire replacement. Besides the price of the tires themselves, you also pay for mounting, tubes, and taxes. Another cost you need to consider is the relatively new tire disposal cost. Many states now charge the tire company to dispose of old tires, and they pass this cost on to the consumer. Make sure you know what the fee is and include it in your calculations.

Repairs

Over the life of a machine, the repair cost will probably be higher than the fuel or tire cost. Earthmoving machines work hard in dirty and difficult conditions. Hard usage takes a toll on every machine. Keeping accurate

	Operating conditions		
Machine type	Excellent Repair factor	Average Repair factor	Severe Repair factor
Wheeled tractor scrapers	0.07	0.09	0.13
Off-highway trucks	0.06	0.08	0.11
Track-drive tractors	0.07	0.09	0.13
Wheeled tractors	0.04	0.06	0.09
Track loaders	0.07	0.09	0.13
Wheeled loaders	0.04	0.06	0.09
Motor graders	0.03	0.05	0.07

Figure 13-8
Repair factors for earthmoving equipment

records on individual machines will reveal the machine cost, and provide a guide for deciding when to trade it in for a new model or a bigger unit. If you don't have detailed records, calculate repair costs using this formula:

Repair cost per hour = Repair factor × (delivered price − tire cost) ÷ 1000

- The repair factor depends on the operating conditions. See Figure 13-8. Figure 13-9 shows a completed ownership and operating cost form for a typical motor grader.

Calculating the Overhead

Up to now, we've looked at costs directly attributable to a particular machine. But your bid must also cover fixed costs, usually called company overhead. In a larger company, the overhead may be considerable.

Overhead costs include:

- Building and grounds for your shop, storage yard, and office, including all taxes, insurance, and upkeep
- Utilities such as phone, heat, and electricity
- Legal and accounting fees, advertising, office supplies
- Management and support payroll, including withholding taxes, insurance and fringe benefits, for office and shop
- Superintendents and the pickups they use
- Lowboys and tractors for moving equipment

The machines that do the actual dirt moving must make enough money per hour to cover company overhead as well as the direct costs of operating the machines.

Machine Ownership and Operating Cost Summary

Machine type_____ Company number_____
Purchase date_____ Delivered price_____

Depreciation
Tire replacement cost

Location	Size	Quanitity	Cost
Front	_____	_____	_____
Rear	_____	_____	_____
Drive	_____	_____	_____

Delivered price - tire cost _____
Minus resale or trade-in value _____
Net depreciation value _____

Hourly ownership cost

$$\frac{\text{Net depreciated value}}{\text{Depreciation hours}} = \underline{\hspace{2cm}} = $$

Interest (_____%), Insurance (_____%), Taxes (_____%)

Estimated yearly use in hours_____

$$\frac{\text{Factor x price}}{1000} = \frac{\text{x}}{1000} =$$

Total ownership cost _____

Hourly operating costs

Fuel price (_____) x use/hour (_____) = _____

Lubricants and filters

Item	Unit price	Used/hour	Cost
Engine	_____	_____	_____
Transmission	_____	_____	_____
Finals	_____	_____	_____
Hyd	_____	_____	_____
Grease	_____	_____	_____
Filters	_____	_____	_____
Other	_____	_____	_____

Total fuel, lubricants and filters _____

Tires

$$\frac{\text{Replacement cost}}{\text{Estimated hours}} = \underline{\hspace{2cm}} =$$

Repairs

$$\frac{\text{Factor x (cost - tires)}}{1,000} = \frac{\text{x}}{1000} =$$

Other

Total operating cost _____
Operator wages _____
Total owning and operating cost _____

Figure 13-9
Sample machine ownership and operating cost summary

Machine number	O&O cost per hour	Average hours per year	O&O cost per year	Percentage overhead	Term overhead	Overhead cost per hour	Total hourly cost
1	29.82	1,500	44,730.00	30%	33,000.00	22.00	51.82
2	35.01	1,300	45,513.00	30%	33,000.00	25.38	60.39
3	36.22	500	18,110.00	12%	13,200.00	26.40	62.62
4	21.15	2,000	42,300.00	28%	30,800.00	15.40	36.55
		Totals	150,653.00	100%	110,000.00		

Figure 13-10
Total ownership and operation machine costs

Begin your estimate of overhead for the coming year by totaling actual overhead for the prior year. Be sure to include all company expenses that weren't the result of taking some particular job. That's your company overhead.

Next, make an estimate of whether overhead will increase or decrease during the coming year. Adjust actual overhead for last year by the estimated percentage of increase or decrease in overhead during the coming year.

Then convert estimated total overhead for the year into a cost per hour for each machine you'll be using during the coming year. There are two accepted methods of doing this.

First, you can total the average annual hours for all production machines. Then divide the total overhead cost by the total number of hours. The answer is your average hourly operating cost. This is the easiest method, but not the best. You probably won't use every machine on every project. That's why I recommend the second method — calculating an average hourly operating cost for each individual machine.

Figure 13-10 is the chart I use to compile the cost for each machine. Here's what the column headings mean:

Machine number – is a number you assign to a specific machine. Any numbering system will work.

O&O cost per hour – is owning and operating cost per hour.

Average hours per year – is either the actual total number of hours that you ran the machine in the previous year, or your estimate of use for the coming year.

O&O cost per year – is the owning and operating cost per hour times the estimated number of hours of use.

Percentage overhead – is the percentage of total machine hours assigned to this specific machine.

Term overhead – is the percent of overhead times the total yearly overhead cost.

Overhead cost per hour – is annual overhead divided by the average hours per year.

Total hourly cost – is the sum of owning and operating cost per hour and overhead cost per hour.

The numbers in Figure 13-10 are based on my estimate of total overhead of $110,000 per year.

Adding the Profit

The last dollar amount on your estimate — and probably the most important — is profit. There are almost as many ways to figure profit as there are earthwork contractors. There are two general rules, however.

- First, add the profit to the hourly cost for all "cost plus" bids. That's a project where the owner pays directly for all materials, and pays for the machine on an hourly basis.

- Second, the profit should always be the last item you add to the estimate. Keep that figure separate until the last minute.

Suppliers or subs may change their bid at the last minute, requiring an immediate change in your bid. Also, some area of your bid may be overstated while others may be below actual cost. (This is called loading the bid.)

Bid Price per Cubic Yard

At last we've almost arrived at the figure we've done all these calculations to find — the bid price. This is really a simple formula:

Bid price ($/CY) = total hourly cost of all machines ÷ cubic yards in project

In this chapter I've tried to tie together the methods for finding quantities, choosing the type and number of machines, how long the project will last, what it will cost, and how to add overhead and profit. We'll carry this information even farther in the next two chapters, with a practice estimate and a sample bid.

14 A Sample Take-off

If you've followed the instructions provided in the first thirteen chapters of this book, you should be able to estimate most common excavation work. Now, just to test your understanding, I'm going to give you a final exam for this course. But don't worry. This is an open book exam. In fact, to make your task easier, I'm going to offer hints on where to look if you need a little more information on how to do some calculations.

This final exam is broken into two parts. Part one is in this chapter. We'll take off quantities for a complete project — from start to finish. Figure 14-1 is the site plan. To make it easier for you, I've put all the figures together at the end of this chapter. In the next chapter, Chapter 15, we'll use the quantities calculated in this chapter to price the work.

This sample estimate is intended to be as realistic as possible. It's also intended to test your understanding of everything covered in this manual. Figure the quantities. Then compare your answers with my answers. If the answers agree, you're ready to begin estimating your own jobs.

The plan sheets in this chapter are much smaller than the plans you'll have for most jobs. That's because the plans in this chapter have to fit on a book page. That's a disadvantage, of course. And there's another disadvantage: The plans you see here may have been reduced or enlarged slightly when the book was printed. That's inherent in the book printing process and something we have to accept. So the dimensions you scale off the plans may not agree exactly with the dimensions in my examples. My advice is to ignore these differences. They'll be small, in most cases. The important thing is that you understand the procedure, not that we arrive at exactly the same figures.

For most estimates you'll list each excavation category separately. For example, your estimate will show the number of units and the unit costs for dirt excavation, for rock excavation, for utility line excavation, for backfill, and so on. Since your estimates have to show estimate details, my sample estimate will show all the details.

We'll use several different types of take-off forms to do the calculations. I introduced specialized take-off forms in many of the earlier chapters. We'll use some of these forms in this chapter. But some of the work is best done on plain ruled paper. The paper or form you use is a matter of personal preference, in my opinion. Most professional excavation estimators use printed forms for at least some of their work. But you can do an accurate take-off on any type of paper.

No matter what type of form you use, always show your calculations. That makes checking easier. Every professional estimator I know shows the calculation details on estimates. I recommend that you do the same. If you need to know where some figure came from, all the information should be there on the sheet, readily available for inspection. If you have to make revisions in the estimate, changes are easy if all the calculations are shown. If the calculations are too long to include on the take-off sheet, attach a separate piece of paper.

Nearly every set of plans and specs will include a section titled "General Specifications." That's where you should start reading. So that's where we'll begin this sample estimate.

General Specifications

General specifications will be very similar on most of the jobs you bid. But don't fall asleep while reading the general specs. Any unusual work conditions on a job are described in the "General Specifications." For example, if you have to maintain traffic movement during excavation or observe special precautions for environmental reasons, you'll discover that in the general specs section of the bid documents. A few words in the general specs can double your cost of doing the work. So stay alert and read every word.

Here are the general specifications for our sample project:

Item 1: Clearing and Grubbing

Due to the small amount of trees and undergrowth, clearing and grubbing will not be a pay item per unit. They will be on a lump sum basis.

Item 2: Soil Testing

The results of the borings and other miscellaneous soil information will be found in the plans. The contractor shall be responsible for any deviation from the results shown.

Topsoil tests show a 2-inch-thick layer of deleterious material unsuitable for use. This material must be removed completely from the job site. The usable portion of material averages 4 inches. Payment will be based on this figure. Any excess topsoil material will become the property of the contractor, and will be removed from the site.

Item 3: Service Road

The road on the back of the property shall be considered a service road only. Due to the lay of the land and drainage needs, no ditches will be required. The road shall be constructed of 6 inches of crushed stone of Type AB-3.

Item 4: Entrance Drive

The entrance drive shall be constructed of asphaltic concrete over a crushed stone base. Grade and line shall be according to the standard drawings for this area.

Item 5: Parking Lot

The parking lot shall be built to the size and specifications as shown on the general plan and as denoted here. The surface shall be 4 inches of asphaltic concrete over 6 inches of crushed stone base of Type AB-3. The crown of the lot shall be built into the lot using the surface asphalt. The grading shall be within plus or minus 0.10 foot. Contractor shall make the surface drain regardless of finished grade or plan notes.

Item 6: Buildings

Both the office building and the shop building require excavation work. Details for the shop building footing and the office building basement are covered in standard drawings, and in the plans.

Item 7: Rock Excavation

The contractor may encounter rock during excavation work at certain areas designated on the plans. In such case, the rock shall be excavated to a depth of at least 1 foot below grade line in areas of cut. Suitable material shall then be placed and compacted to bring area back up to grade. The contractor shall bid this item as anticipated only, at a unit price. Quantities shall be determined as work progresses.

Item 8: Utility Work

Both sanitary and storm sewer lines shall terminate in manholes at the southern property line. All sanitary sewer lines shall be 8-inch cast iron pipe. All storm sewer lines shall be 12-inch corrugated metal pipe. Trench shoring shall be used in excavation of all trenches with a depth of 3 feet or more. Bedding for all lines shall be as per standard drawings.

Doing the Take-off

We'll break the take-off into logical steps, figuring work in the same order it will be done in the field. That makes topsoil excavation our first topic.

Figuring the Topsoil

First, we'll find the total amount of topsoil to strip from the project to get down to workable material. Second, we'll calculate the amount of topsoil we can store and replace on the site. Third, we'll identify the areas where topsoil will be replaced. Finally, we'll find a place to store topsoil while doing other site work. If you have trouble with any of these calculations, review Chapter 9.

According to the general specifications, the existing site has 6 inches of topsoil. But the top 2 inches includes debris and vegetation that make it unsuitable for use as replacement soil. Figure 14-2 shows the parking lot, two roadways and the location for two buildings. Everything else on site will be covered by topsoil.

Check my calculations in Figure 14-3 to see if you agree. The site measures 340 feet by 240 feet. The 2 inches of unsuitable material total 503.70 CY. This volume has to be hauled away. The next 4 inches of reusable material total 1,007.41 CY. That's the amount of material we'll store, either on or off the site. I know from making the site visit and talking to the owner that this spoil can be stored in the northeast corner of the property during construction. So that's my plan for this job.

How much topsoil will it take to place a 6-inch layer over the area not covered by structures? See my calculations in Figure 14-3. First, I found the site's area, 81,600 SF. Then I found the total area covered by buildings, roads and parking lots, 22,050 SF. Next I'll subtract the covered area from the total area, and multiply the result by 0.5. That gives me the answer in cubic feet, so I'll divide it by 27 to get cubic yards. I get a result of 1,102.77 CY. We have 1,007.41 CY in the stockpile, so we'll need to truck in an additional 95.36 CY of topsoil. Let's round that off to a total of 96 CY. Figure 14-4 is a summary of my topsoil figures.

Rock Service Road

The specifications call for 6 inches of rock on the service road. Take another look at Figure 14-2. The service road's broken down into three parts, labeled Service road A, Service road B, Service road C. We already found the area for each of these parts in Figure 14-3. The service road won't need any extra compaction because it's all cut. That means you won't need to calculate any extra yardage for undercutting. Now let's find out how many cubic yards of rock we'll need. First, add the three areas together (see Figure 14-3). Second, multiply the sum of the areas by 0.5. Third, divide that result by 27. I get 120.37 CY of rock for an answer, and Figure 14-5 shows my calculations.

Bidding Rock: Tons or Cubic Yards?

Some estimators bid rock by the ton. Others prefer to estimate rock by the cubic yard. If you want to bid rock by the ton, you'll need to know how much a cubic yard of the rock weighs. The weight varies depending on the kind of rock. For this estimate, we'll use loose limestone rock. One cubic yard of loose limestone weighs 2,800 pounds, or 1.4 tons. So 120.37 CY of rock at 1.4 tons per cubic yard works out to about 168.5 tons of rock.

Grid Square Take-off

Figure 14-6 is a contour map of the building site. Dashed lines show existing elevations and solid lines show proposed elevations. Figure 14-7 shows a grid system laid over the contour map. Grid squares are numbered from A to I_1. Each 1¼-inch grid square represents an area 50 feet by 50 feet on the ground. That makes the map scale 1" to 40'. Check the scale by measuring for yourself. Each grid (except partial squares along the right and lower edge) should measure 1¼ inches by 1¼ inches.

This estimate is fairly simple because nearly all the grids require cut. We'll start by finding the existing elevation for each corner of each grid square. That's a three-step process. First, you find and record the elevation of the contour lines on either side of the corner. Second, measure the distance between the two contour lines and from the higher contour line to the corner. Third, using the measurements as a ratio, solve to find the corner elevation. If you need to review all or part of this procedure in greater detail, turn back to Chapter 6.

Scaling Existing Elevations

Let's start by finding the existing elevation at point A-1. That's corner 1 (the northeast corner) of grid square A in Figure 14-7. Notice that corner A-1 is between existing contour lines 100 and 102. So we know that the existing elevation at A-1 is more than 100 and less than 102. How much is corner A-1 above elevation 100? To answer that question we'll measure the distance between contour lines 100 and 102 along a line that passes through corner A-1.

Here's the estimating procedure:

1) Lay your measuring scale on Figure 14-7 so the point of origin (zero) is on contour line 100. The scale should run perpendicular to contour line 100.

2) Slide the end of the scale along contour line 100 until the scale edge runs over the top of corner A-1.

3) Note the distance between contour line 100 and contour line 102.

4) On my scale, the distance between 100 and 102 is 40 units. If you're measuring with an ordinary ruler, you'll get almost exactly 1 inch. (The unit of measure isn't important. Use any measuring system you like. The estimated elevation at A-1 will be nearly the same for any unit of measure.)

5) Then, without moving the scale, note the distance from contour line 100 to corner A-1. I get 7 units. With an ordinary ruler, it's about 3/16 inch.

Recording Existing Elevations

Now take a look at Figure 14-8. That's my worksheet for existing corner elevations from Figure 14-7. Let's use corner A-1 as an example, and see what sort of data's here and where it comes from:

Location – is the name of a specific grid square corner: A-1.

Low elevation – for A-1 is 100.

High elevation – for A-1 is 102.

Scale distance – for A-1 is 7/40, or the measured distance *Low elevation* contour to A-1 (7 units) over *Low elevation* contour to *High elevation* contour (40 units).

Contour interval – is 2 units.

Add elevation – is *Scale distance*, converted to a decimal, times *Contour interval*. For A-1 I get:

$7 \div 40 \times 2 = 0.35$

If you measure in fractions of an inch, the calculation looks like this:

$3/16" \div 1" \times 2.0 = 0.375"$

The two answers (0.35 and 0.375) aren't a perfect match, but they're close enough for this type of work.

Point elevation – is *Low elevation* (100.0) plus *Add elevation* (0.35), or for A-1:

$100 + 0.35 = 100.35$

I've used the same procedure to find the existing elevations for the four corners of all 35 grid squares, and recorded the data in Figure 14-8. Work out the rest of the corner elevations on your own. When you compare your results with mine they should be about the same.

Scaling and Recording Proposed Elevations

Repeat the same process using the proposed elevations, this time comparing your work with my data shown in Figure 14-9.

Calculating Excavation Volumes from the Grid Square Take-off

Once you know existing and proposed elevations at each corner, it's easy to find excavation quantities. Turn to Figure 14-10. This worksheet records the data used to calculate an *Average depth* for each grid square. Here's how it works using grid square A and A-1 for our example:

Proposed – elevations for each *Element*, or corner, come from the *Point elevation* column in Figure 14-9. For A-1 it's 100.

Existing – elevations for each *Element* come from the *Point elevation* column in Figure 14-8. For A-1 it's 100.35.

Depth – is *Existing* minus *Proposed*, for A-1:

$100.35 - 100 = 0.35$

Average depth – appears right below the grid name. This is a simple average, the sum of the four corner depths divided by the number of corners. In the case of our example, Grid A, I get:

$(0.35 + 2.14 + 0.91 + 0) \div 4 = 0.85$

Before we move on to the next step, take a look at the data and calculations for Grid I_1, and especially note the data in the *Depth* row. Three of these values are negative numbers. The proposed elevation is greater than the existing elevation, so the depths are negative. Furthermore, the *Average depth* is also negative. That tells us that Grid I_1 is fill, and the other 34 grid squares are all cuts.

Calculating Average Depth

Turn to Figure 14-11 to see how I use the average depth figures to find excavation volumes for either cut or fill.

Location – is the grid square name.

Length and Width – dimensions for each grid square aren't identical. For example, in Figure 14-7, eleven of the grid squares do not measure 50 by 50. That's a difference that counts. Always record data carefully.

Ave. depth – comes from Figure 14-10.

Vol. cut (CY) and Vol. fill (CY) – are *Length* times *Width* times *Ave. depth*, divided by 27. You enter the result in the *Vol. cut* column if it's positive, and in the *Vol. fill* column if it's negative. Using grid square A, here's how it works:

$50 \times 50 \times 0.85 \div 27 = 78.70$ CY

Notice that there are three totals listed at the end of Figure 14-11. *Total cut* = 13,062.06 CY, *Total fill* = 53.93 CY and, in the *Note* box, *Total spoil* = 17,040.65 LCY.

What About Topsoil?

You may have noticed that so far we've ignored topsoil depth in our calculations. Topsoil is removed before any other excavation begins, and it's not replaced until the construction is complete. Most professional excavation estimators work from existing and proposed grade elevations and ignore depth of the topsoil. Doing that on this job will result in building elevations that are 6 inches lower than you might expect. But it won't change the grid square excavation volumes. Here's why:

Look at Figure 14-12. This is a cut-away view of a grid square. The specs on this job require stripping off 6 inches of topsoil. Assume the finish elevation is 100.0. Now let's calculate the depth of cut two ways. The first time (or Case 1) we'll ignore the topsoil. The second time (Case 2) we'll include the topsoil in our calculations by using the elevations identified as *Alternate*.

Case 1, Ignoring the Topsoil – Depth of the cut at corner 4 equals existing elevation minus proposed elevation, or 102.22 – 100 = 2.22 feet.

Case 2, Considering the Topsoil – Existing depth is 102.22. Removing 6 inches leaves the depth at 101.72. Remember that number. In order to replace 6 inches of topsoil and have the finished grade be 100 we must start by excavating to a depth that's 6 inches below the finished grade. That's 100 minus 0.5, or 99.5. Therefore, cut at corner 4 equals existing elevation: 101.72, minus proposed elevation: 99.5, and the result is 2.22 feet.

Case 2 takes more time and math to arrive at the same exact answer as we found in Case 1. Don't you agree now that it's easier to ignore the topsoil?

On most jobs where you aren't concerned with final building elevations, my advice is to ignore topsoil removed and replaced. That's the procedure we'll follow in this chapter. Make an exception for large areas where topsoil will be removed but not replaced, such as a parking lot.

Entrance Road

A 28-foot-wide driveway runs 125 feet north from the south property line. Figure 14-13 shows this entrance road and proposed contour lines at the site. Figure 14-13a shows the entrance road in plan and profile views. Figure 14-14 shows a typical section of this entrance road.

Figure 14-15 gives the excavation limits for the entrance road. Using these excavation limits, I created the excavation template shown in Figure 14-16.

Calculating Average End Areas

Take another look at Figures 14-13 and 14-13a, and compare them. You'll notice that finished roadway elevations follow the profile of the finished contour lines. On most roadway jobs you'll have to average the end areas at each section to figure the excavation volume. On this job, we're in luck because the end areas are nearly identical. As a result we'll only need to calculate the end area once. For the other stations we'll use the same end area figure because it is the average. That saves a whole lot of figuring time.

If necessary, review Chapter 4 to refresh your memory on how to do average end area calculations. No matter which method you use (measuring strip, planimeter or arc section), your answer should match mine. As Figure 14-15 shows, average end area equals 34.1 SF.

To find the excavation volume:

1) Calculate the end area in two places.
2) Add the two end areas together.
3) Divide the sum by 2 to find the *average* end area.
4) Multiply the result by the distance between the two end areas and you'll have the excavation volume in cubic feet.
5) Divide the result by 27 to change cubic feet to cubic yards.

Let's see how this works with Sta. 0+00 and Sta. 0+18 in Figure 14-15. At Sta. 0+00 we only need to meet the existing grade. There's no excavation required, so the end area equals 0. At Sta. 0+18, however, using the full template, the end area equals 34.1. That takes care of step 1. Now for the rest of the math:

$0 + 34.1 = 34.1$

$34.1 \div 2 = 17.05$

$17.05 \times 18 = 306.9 \text{ CF}$

$306.9 \div 27 = 11.37 \text{ CY}$

Figure 14-17 summarizes my excavation volume calculations for the five stations (Sta. 0+00, 0+18, 0+50, 1+00 and 1+25) along the entrance road. My rounded off total for excavation volume, including a 31 percent swell factor, is 192 CY.

Remember, you can use any reasonable distance between these end sections. Once you're out in the field you'll find 100-foot intervals used in very flat areas. However, areas with larger or more frequent changes in elevation use 50- or 25-foot intervals.

Parking Lot

The general specifications require a parking lot made of 4 inches of asphaltic concrete laid over a 6-inch crushed stone base. Add these together to find the total excavation depth, 10 inches.

You'll remember that this site had 6 inches of topsoil. The first 2 inches have been stripped and hauled away. The remaining 4 inches were also stripped, and stockpiled. In the grid square take-off and entrance road calculations, we ignored the topsoil. Anytime you remove and replace topsoil it has no effect on excavation volumes. But we won't be replacing any topsoil in the parking lot. To include the topsoil in these excavation calculations I subtract topsoil depth from excavation depth to find the additional excavation depth. My answer comes out to 4 inches.

Take a close look at the parking lot area in Figure 14-2. You'll see that I've broken it down into two rectangular parts labeled *Parking lot A* and *Parking lot B*. We already know their areas from the calculations shown in Figure 14-3.

A Sample Take-off **247**

A's area is 1,750 SF and *B's* area is 4,350 SF. Add them together to find the total area, 6,100 SF. Next we'll find total excavation volume for a 4-inch-deep cut over an area measuring 6,100 SF.

(4" ÷ 12) × 6,100 SF = 2,033.33 CF
2,033.33 CF ÷ 27 = 75.31 CY
75.31 CY × 1.31 = 98.66 LCY

What About Drainage?

Figure 14-18 shows the parking lot with a break point represented by line AB. This line is the highest point on the lot. Water will drain from this line toward the service road and down to the catch basins. It's hard to excavate such a small slope. Building this sort of slope into the asphaltic surface, however, is easy. So we'll use 4 inches as the excavation depth throughout the parking lot. Later on we'll build the right drainage slope into the asphalt surface.

Sanitary Sewer Lines

Figure 14-19 shows the sanitary sewer system. From the general specifications and Figure 14-20 you can see that the sanitary sewer lines are 8-inch cast iron pipe. According to the general specifications, we'll have to use a trench box during excavation because the trench is 3 feet deep. So we'll calculate trench volumes using straight sides instead of sloping sides.

Our first job here is to find the volume of material we'll remove from the trenches and manhole areas. Then we'll calculate the volume of rock and concrete needed in the trench and around manholes and the backfill needed to finish the job.

Let's begin by examining the plan and profile sheets for this system, Figures 14-21 to 14-24. Each plan and profile sheet has two parts. The top part is a plan view showing general layout and alignment of the line or lines. The bottom part of each sheet contains a profile view of the line. This is a cutaway view of the piping and the manholes.

You'll also find the scale listed near the top of each sheet. If you're new to the business, the scale may look confusing at first glance. This is what's called a dual scale, and it's easy to use once you know how it works. In Figures 14-21 through 14-24 I've used the following scale: *Horizontal: 5 squares = 20'*, and *Vertical: 6 squares = 5'*. Why the difference between horizontal and vertical? It allows me to enlarge the sewer lines and related structures, such as manholes, without changing other dimensions. Your designer needs the extra space here to record inlet and outlet elevations and other data. One last note about dual scales. The plan view scale and the horizontal scale in the profile view should be (but aren't always) the same. Double check, just to be sure.

Sanitary Sewer Line Excavation

The method for finding cubic yards in utility line excavation is quite simple. At each structure, find the difference between the finished contour line at the top and the flowline of the pipe. Add a few inches for undercut of the pipe. Multiply the average depth by the length between the two structures. Multiply by the trench width. Finally, convert the answer to cubic yards.

We'll use a special calculation for the undercut. Most designers illustrate sewer lines as shown in Figures 14-21 through 14-24. They show all lines between manholes with an inlet elevation where the lines enter a manhole, and an outlet elevation where they exit. This elevation is the actual flowline inside the pipe. It doesn't allow for the thickness of the pipe or for 6 inches of bedding material under the pipe. In the example, I've added 6 inches to all depths to allow for bedding material. I've elected to ignore the pipe thickness, since an 8-inch-diameter pipe only adds about ½ inch. Where pipe diameters are larger, you'll probably want to consider the thickness of the pipe walls.

Figures 14-25 through 14-28 show how I calculated average depths for the lines between the manholes. Note the additional 6 inches (+0.5 foot) for the undercut.

Figure 14-29 summarizes the excavation work required for the four lines. The total is 369.86 LCY.

Sanitary Sewer Line Backfill

Figure 14-30 is a section view showing rock backfill to be placed around the pipe. Notice that 6 inches of rock are required over and under the pipe. Add the 8-inch diameter of the pipe and you get a total depth of 20 inches for rock backfill. That's the same as 1.67 feet, a figure we'll use to calculate dirt backfill quantities.

Dirt Backfill

Figure 14-31 shows my calculations for dirt backfill. Numbers in the length, width and average depth columns are the same as for trench excavation and come from Figure 14-29. Rock backfill and the pipe together reduce trench depth a total of 1.67 feet. Here are my calculations for Line 1:

Average depth of 4.70 feet is reduced by 1.67 feet, leaving 3.03 feet. Find the volume by multiplying length times width times depth.

$145' \times 2' \times 3.03' = 878.7$ CF

To find volume in compact cubic yards, you multiply volume by 1.25 (25 percent shrinkage) and divide by 27.

878.7 CF $\times 1.25 \div 27 = 40.68$ CCY

Rock Backfill

Figure 14-32 shows how I calculated rock backfill quantities. Numbers in the length and width columns are the same as for trench excavation and come from Figure 14-29. Depth is 20 inches or 1.67 feet. Of course, pipe fills a portion of this volume. How much of the 24-inch by 20-inch cross section area does the pipe actually occupy? The pipe's diameter is 8 inches, so the radius is 4 inches. The area of a circle equals pi (3.14) times the radius (4") squared. The result here is square inches. Square feet are easier to work with, so divide by 144 to convert to square feet. Plug in the numbers:

4" × 4" = 16 square inches (SI)
3.14 × 16 SI = 50.24 SI
50.24 ÷ 144 = 0.35 SF

Look at Figure 14-32 and you'll see that same number in the *Area of pipe* column.

Here's the rock volume calculation for Line 1:

Multiply *Width* (2 feet) times *Depth* (1.67 feet), and I get 3.34 SF. Deduct the *Area of pipe* (0.35 SF) and the result is 2.99 SF. Multiply 2.99 SF by *Length*, (145 feet), divide by 27, and I get 16.06 CY, the same number you'll see for Line 1 under *Rock fill vol.* in Figure 14-31. Calculate rock fill volumes for lines 2, 3, and 4 the same way. Finally, total the *Rock fill vol.* column and round the result to full cubic yards. My total rock fill volume for four sanitary sewer lines comes to 39 CY.

Developing Manhole Excavation Constants

Next we'll calculate excavation volumes for the three manholes in the sanitary sewer system. Let's start by developing two "SF per foot of depth" area constants. These constants are:

1) The excavation area constant
2) The manhole area constant

Figure 14-33 supplies all the data we need — two diameters and two formulas:

- Pipe diameter is 4 feet
- Excavation diameter is 8 feet
- Formula to find the radius of a circle is: r = diameter ÷ 2
- Formula to find the area of a circle is: area = πr^2
- The value of pi (π) is: 3.14

You can't find the area without knowing the radius, so we'll start there and then move on to calculate the areas. For the manhole I get:

4' ÷ 2 = 2' radius
3.14 × 2^2 = 12.6 SF area

For the excavation I get:

8' ÷ 2 = 4' radius

3.14 × 4² = 50.24 SF area

Calculating Manhole Excavation Volumes

Now let's see how you use these constants to calculate the manholes' excavation and backfill volumes. You find each manhole's excavation volume by repeating the same six steps. Using manhole No. 1, let's walk through the steps. Figure 14-34 is the worksheet with my results. We'll begin with finding the manhole depth, and by finding the correct plan and profile sheet:

1) Record the outlet, and the finish elevations for the manhole.

2) Subtract the outlet elevation from the finished elevation.
   Elevation difference   for No. 1 is 100 – 89.7 = 10.3

3) *Elevation difference* plus 1 foot is the manhole depth:
   Depth   for No. 1 is 11.3

4) Multiply *Depth* by the excavation area constant:
   Excavation area (SF)   or the constant, for No. 1 is 50.24

5) Divide the result by 27:
   Exc. vol. (CY)   for No. 1 is 21.03

6) Multiply **Exc. vol. (CY)** by 1.31:
   Shrink (–) or swell (+) factor   for No. 1 is + 1.31

The result for No. 1 is 27.55, the same value you see in the *Actual vol. (CY)* column in Figure 14-34.

Repeat the same steps for manholes 2 and 3. Total the *Actual vol.* column and round the result to full cubic yards as shown in Figure 14-34. I get a total excavation volume for the manholes of 108 LCY.

Calculating Manhole Backfill Volumes

After installing the concrete manholes, the next task is backfill around the structures, and that means another round of volume calculations. My backfill volume calculations are shown in Figure 14-35. Here's an example, using manhole No. 1 and following the column headings shown in Figure 14-35.

Depth   There's no change, so I reuse the data from Figure 14-34. For No. 1 this is 11.3 feet.

Excavation area   is the constant from Figure 14-33, 50.24 SF.

Manhole area   is the constant from Figure 14-33, 12.6 SF.

Exc. area – MH area   Subtract the manhole area from the excavation area to find the backfill area per foot of depth. For No. 1:

50.24 SF – 12.6 SF = 37.64 SF

A Sample Take-off

Backfill vol. (CF) Multiply backfill area by *Depth* and the result is backfill volume in cubic feet. For No. 1:

11.3' × 37.64 SF = 425.33 CF

Shrink (−) or swell (+) factor Backfill takes the shrink factor − 1.25. So for No. 1 I get:

425.33 CF × 1.25 = 531.66 CCF

Actual vol. Divide cubic feet of backfill by 27 to convert it to cubic yards. For No. 1:

531.66 CCF ÷ 27 = 19.69 CCY

Repeat the same calculations for the remaining two manholes. Total the *Actual vol.* column, and round the result to full cubic yards. I found that the total volume of backfill around the manholes came to 77 CCY. Before we move on, take a look at the *Note* section at the bottom of Figure 14-35. It says:

108 exc − 77 bkfill = 31 CY spoil

That's my shorthand for 108 CY excavated, less 77 CY replaced as backfill, leaves me a total of 31 CY of spoil.

Storm Sewer Lines

Enclosed storm sewer systems are an efficient way to provide a concealed drainage system for runoff water in urban areas. In rural areas, open ditch systems serve the same need. Figure 14-36 is a plan view of the enclosed storm sewer system for our sample project. We'll calculate excavation volumes for the storm sewer system by following the same steps we used earlier for the sanitary sewer system.

Storm sewer systems include structures called catch basins. These are precast concrete boxes that have an opening at the top, on at least one side. These openings allow surface runoff water to enter the storm sewer system. One of these openings, called a drop, or curb inlet, appears in Figure 14-37. There are two more standard drawings for the storm sewer system. Figure 14-38 shows the excavation limits for the catch basins. Figure 14-39 shows the backfill requirements. Further information about the storm sewer system comes from the General Specifications, including:

1) All lines are 12-inch corrugated metal pipe (CMP)
2) City line tie-in via manhole at south project limit line
3) Trench box required, so trench walls are vertical

Catch Basin Excavation

The plan and profile sheets for the three catch basins, the manhole, and their connecting lines appear in Figures 14-40 through 14-42. Notice that the profile views give inlet and outlet elevations for each catch basin. There's one dimension missing: the catch basin depth. Assume that the bottom of each catch basin is 1 foot lower than its outlet.

Figure 14-43 shows and summarizes excavation calculations for the three catch basins and manhole No. 1.

Elev. diff. is finished elevation minus outlet elevation. Both elevations appear in Figure 14-40. For catch basin 1 that's:

$100 - 98 = 2$

Depth is the Elev. diff. plus 1 foot. For catch basin 1:

$2 + 1 = 3$

Struct. exc. area is the constant we found in Figure 14-38, 36 SF.

Volume is *Depth* times *Struct. exc. area*, the result divided by 27 to convert it to cubic yards. Plug in the numbers for catch basin 1, and I get:

$3' \times 36 \text{ SF} = 108 \text{ CF}$
$108 \text{ CF} \div 27 = 4 \text{ CY}$

Actual vol. is *Volume* times the *Swell (+) factor*, 1.31. For catch basin 1:

$4 \text{ CY} \times 1.31 = 5.24 \text{ CY}$

Catch Basin Backfill

Figure 14-44, meanwhile, shows my backfill calculations for the three catch basins and manhole No. 1.

Depth repeats the data from Figure 14-43.

Struct. area is the area of the catch basin (see Figure 14-38) or manhole (see Figure 14-33).

Exc. area repeats the data from Figure 14-43.

Backfill area is *Exc. area* minus *Struct. area*. By plugging in the numbers for catch basin 1 I get:

$36 \text{ SF} - 4 \text{ SF} = 32 \text{ SF}$

Volume (CY) is *Backfill area* times *Depth*, result divided by 27. For catch basin 1 that's:

$32 \text{ SF} \times 3' \div 27 = 3.56 \text{ CY}$

Actual vol. is *Volume (CY)* times the *Shrink (–) factor*, or 1.25. For catch basin 1 I get:

3.56 × 1.25 = 4.45 CCY

Average Depth Calculations for Storm Sewer Lines

The next step is finding the average excavated depth for each storm sewer line. Figures 14-45 through 14-47 record my calculations for each line. Look at these calculations for storm sewer line 1, which runs from catch basin 1 to catch basin 2. (See Figure 14-40.)

1) Subtract the outlet depth at CB 1 (98.01) from the finish grade at CB 1 (100). The result is 1.99.

2) Subtract the inlet depth at CB 2 (97.01) from the finish grade at CB 1 (100). The result is 2.99.

3) Add 0.5 feet to each for the 6 inches of crushed rock backfill shown in the standard drawing, Figure 14-39.

4) Total the two elevation differences (2.49 + 3.49), and divide by 2. The result (2.99 feet) is the average depth for storm sewer line 1.

Figures 14-46 and 14-47 show the same set of calculations for storm sewer lines 2 and 3 respectively.

Storm Sewer Line Excavation Volumes

Now let's find excavation volumes for each of the three storm sewer lines. My work is summarized in Figure 14-48. Using storm sewer line 1 as our example, let's plug in the numbers:

Length of line 1 (CB 1 to CB 2) is 42 feet as shown in Figure 14-40.

Width is 2 feet.

Average depth is 2.99 feet as computed in Figure 14-45.

Volume is Length times Width times Average depth divided by 27.

(42 × 2 × 2.99) ÷ 27 = 9.3 CY

Actual vol. is *Volume* times the *swell (+) factor* + 1.31.

9.3 CY × 1.31 = 12.19 LCY

Take a look at Figure 14-48, and you'll see the same number in the *Actual vol.* column for Line 1.

Storm Sewer Line Backfill Volumes

Before we can compute the dirt backfill required, we need to know how much of the excavated depth is filled by the pipe and the rock backfill. Figure 14-39, you recall, is the standard drawing showing a section view of the storm sewer after pipe placement. Notice that 6 inches of rock are required over and under the pipe (6 + 6 = 12). Now add the 12-inch diameter of the pipe (12 + 12 = 24). So rock and pipe together fill 24 inches (2 feet) of the total excavated depth.

Dirt Backfill

Figure 14-49 shows my calculations for dirt backfill. The values for *Length*, *Width* and *Average depth* haven't changed so I reuse the data from Figure 14-48. We also know that rock backfill and the pipe reduce the excavation depth by 2 feet. Here are my calculations for Line 1, as an example:

2.99' − 2' = 0.99'
42' × 2' × 0.99' = 83.16 CF
83.16 CF ÷ 27 = 3.08 CY
3.08 CY × 1.25 = 3.85 CCY

The result, 3.85 CCY, matches the number you'll see under *Actual vol.* for Line 1 in Figure 14-49.

Rock Backfill

Figure 14-50 shows how I calculated rock backfill quantities. Numbers in the *Length* and *Width* columns are re-used from Figure 14-46. *Depth* is 24 inches or 2 feet. Of course, pipe fills a portion of this volume. How much of the 24-inch by 24-inch cross section area does the pipe fill? We know the pipe diameter is 12 inches, so its radius is 6 inches. The area of a circle is 3.14 times the radius squared. Substituting the numbers, I get:

6" × 6" × 3.14 = 113.04 square inches
113.04 ÷ 144 = 0.78 SF

In Figure 14-50 this constant appears in the column headed *Area of pipe*.

To find the volume of rock needed for fill, multiply *Width* by *Depth*. The result's the excavated area in square feet. Next, subtract the *Area of pipe* from the excavated area, multiply the answer by the *Length*, and then divide the result by 27. Here's what my rock fill volume calculations for Line 1 look like:

2' × 2' = 4 SF
4 SF - 0.78 SF = 3.22 SF
3.22 SF × 42' = 135.24 CF
135.24 CF ÷ 27 = 5.01 CY

Now look at the number that appears in the *Rock fill vol.* column for Line 1 in Figure 14-50. Repeat these calculations for Lines 2 and 3. Then total the *Rock fill vol.* column and round the result to full cubic yards. As Figure 14-50 shows, the total volume I found for the rock backfill is 30 CY.

A Sample Take-off **255**

Shop Building Footing

For the shop footing, we need to find three volumes: total excavated volume, rock backfill volume, and dirt backfill volume. Most of the data we'll use in finding these volumes comes from two standard drawings, Figures 14-51 and 14-52. We'll refer to Figure 14-51 for excavation details, and to Figure 14-52 for backfill details. For the shop building's dimensions, refer to Figure 14-2. Remember, all measurements are made to the outside of the footing. If you feel lost at any point, try reviewing Chapter 10.

Footing Excavation, V-out

We'll begin with the V-out calculations. V-out for the shop building equals excavation volume plus the total excavated corner volume. To find excavation volume, you use the standard formula:

$$l \times w \times d$$

Or, using the shop building's dimensions (from Figures 14-2 and 14-51) and dividing by 27, so your answer's in cubic yards, the excavation volume comes to:

$$240 \times 2 \times 1 \div 27 = 17.8 \text{ CY}$$

To find the total excavated corner volume in cubic yards use this formula:

$$(1/3 \, \pi \, r^2) \times d \times n \div 27$$

Using pi (π) = 3.1416, corner radius (r) = 1', depth (d) = 1', and number of corners (n) = 4, the total corner volume is:

$$0.3333 \times 3.1416 \times 1^2 \times 1 \times 4 \div 27 = 0.16 \text{ CY}$$

The last step is to add the two volumes together:

V-out = 17.8 CY + 0.16 CY = 17.96 CY

Footing Excavation, V-in

The next volume we'll find is for V-in. This is a simple volume calculation, where volume equals length times width times depth. Your data comes from Figures 14-2 and 14-51:

$$80 \times 40 \times 1 \div 27 = 118.52 \text{ CY}$$

Rock Backfill

Next we'll find the backfill volumes. Figure 14-52 shows rock backfill on the V-in side, and dirt backfill on the V-out side. You calculate the two volumes separately, so let's begin with the rock backfill.

Volume for the rock backfill equals V-in minus the volume displaced by two masses of concrete. Let's start with the volume displaced by the 6-inch-thick concrete slab floor. The depth for V-in (Figure 14-51) was 1 foot.

Subtract the slab's depth, 0.5 feet, and the result is the depth of the rock backfill. Here are the revised dimensions: depth 0.5 foot, length 80 feet, and width 40 feet. The second lump of concrete is the footing, but only the portion on the V-in side. (See Figure 14-52.) Calculate this volume using these dimensions from Figure 14-52: depth 0.5 foot, width 1 foot, and length 240 feet. After finding these two volumes in cubic yards, subtract the footing volume from the depth adjusted volume. The result is the rock backfill volume for the shop building footing.

Dirt Backfill

You find dirt backfill volume by calculating the volume displaced by the footing on the V-out side. Its dimensions (from Figure 14-52) are: width 0.5 foot, depth 0.5 foot, and length, 240 feet. The volume for dirt backfill equals excavation volume for V-out minus the volume of the footing.

Spoil Volume

We need one more volume to finish out the shop building footing calculations: the spoil volume. Spoil is the difference between the volume excavated and the volume replaced as dirt backfill. We already calculated both, so spoil volume is:

23.53 CY – 19.68 CY = 3.85 CY of spoil

See Figure 14-53 for my shop building footing calculations: V-out, V-in, rock backfill, and dirt backfill.

If you noticed the lack of shrink and swell factors in these calculations, good job! The factors apply to both the excavated and the backfill volumes, except for the rock backfill. These actual volumes, the volume multiplied by the applicable factor, are very important. I don't want to lose track of these figures. I also don't want to waste time searching for them in the forest of numbers and calculations. (See Figure 14-53.) Here's my solution; I transfer my totals to a worksheet. My worksheet for the shop building footing is Figure 14-54. This is where I add the shrink or swell factors and round off my results to full cubic yards.

Office Building

The office building has a basement. We need to calculate volumes for V-out, V-in, rock backfill, and dirt backfill. The formulas and sequence of steps are the same as those for the shop building footing. This time we have one standard drawing. Figure 14-55 covers the details for excavation and backfill. Figure 14-2 gives the office building dimensions: length 55 feet and width 50 feet. Try completing the basement volume calculations on your own first. When you're done, or if you hit a snag, read through the step-by-step that follows. Use it to check your work, or jog your memory.

Basement Excavation, V-out

V-out volume is the sum of the excavated volume and the total excavated corner volume. To find the excavated volume you multiply length by width by depth. Here's another way of describing this calculation:

Exc. vol. = perimeter × average slope line area × depth ÷ 27

Perimeter is the same as length or side plus side plus side plus side. Using the dimensions from Figure 14-2 for the office building, we find:

50' + 50' + 55' + 55' = 210'

We saw how to find average slope line areas in Chapter 10. As you recall, this is the total width of the V-out times depth. Total width is the sum of footing width plus work space plus one-half total rise times run divided by rise. Using the data from Figure 14-55 we find:

(0.5 + 4 +1.75) × 7 = 43.75 CF

Put the two together and we find that excavated volume comes to:

210 × 43.75 ÷ 27 = 340.28 CY

Find the total excavated corner volume in cubic yards using this formula:

Corner volume = $^1/_3$ πr^2 × depth × number of corners ÷27

For pi (π) use 3.1416, corner radius is 3.5 feet, depth is 7 feet, and number of corners is 4.

0.3333 × 3.1416 × 3.5^2 × 7 × 4 ÷ 27 = 13.3 CY

Finish your V-out calculations by adding excavated volume and total corner volume together:

9340.28 CY + 13.3 CY = 353.58 CY

Basement Excavation, V-in

Next we calculate the basement V-in. It's a simple volume calculation because there's no slope or ramp to include. The basement dimensions are: length 55 feet, width 50 feet, and depth 7 feet.

955' × 50' × 7' ÷ 27 = 712.96 CY

When you actually excavate a basement, it's not possible to cut perfect corners. That's not a problem in estimating because an allowance is built into the corner volume calculations. Here's how it works. One of the variables in the formula for finding corner volume is corner radius. The dimension used is the horizontal distance between the toe of the slope to the top of the slope. But, in reality, only half of that horizontal distance is additional yardage. The rest is the built-in allowance.

Basement Backfill Volumes

Now let's turn to finding the backfill volumes. Figure 14-55 calls for rock backfill on the V-in side, and dirt backfill on the V-out side. Let's find the rock backfill volume first. Here's your chance to practice what you learned earlier about deducting for a displaced volume. Figure 14-55 includes two pieces of data you'll need:

1) The depth of the rock backfill is the same as the height of the basement footing, 6 inches, or 0.5 feet.

2) The concrete footing fills a 2-foot-wide strip all around the basement perimeter. That's the same as subtracting 2 feet each from the basement length and from the width.

That means you can calculate the rock backfill volume as follows:

$(V_{in}\ length - 2') \times (V_{in}\ width - 2') \times V_{in}\ depth \div 27$
$(55' - 2') \times (50' - 2') \times 0.5' \div 27 = 47.11\ CY$

The basement's dirt backfill volume in cubic yards equals the V-out, after you subtract the volume displaced by concrete. Here are the dimensions, from Figure 14-55, of concrete footing on the V-out side: length 210 feet, width 0.5 foot, and depth 0.5 foot. So, total displaced volume, in cubic yards, equals:

$210' \times 0.5' \times 0.5' \div 27 = 1.94\ CY$

The dirt backfill volume in cubic yards equals V-out volume minus displaced volume:

$353.58\ CY - 1.94\ CY = 351.64\ CY$

See Figure 14-56 for my calculations for the office building basement excavation, and backfill. Then turn to Figure 14-57. This is my worksheet for the basement showing actual volumes, rounded totals (to full cubic yards), and total spoil.

Summary Sheet

Figure 14-58 is the project summary sheet. It brings all the totals together on one page. We estimate a total of 67 CCY of fill, 19,695 LCY of excavation, 1,005 CCY of backfill, and 291 CY of rock fill. In addition, we'll move 2,614 CY of topsoil. We'll have 18,621 LCY of excess material, or spoil, to remove from the site. Note that dirt backfill quantities are given in compacted cubic yards, CCY, and excavation quantities are in loose cubic yards, LCY.

Calculating volumes for any excavation job is easy if you break the project into small tasks and solve each in sequence. Take it a step at a time, work systematically and show all your work. That's the key to accurate excavation estimates.

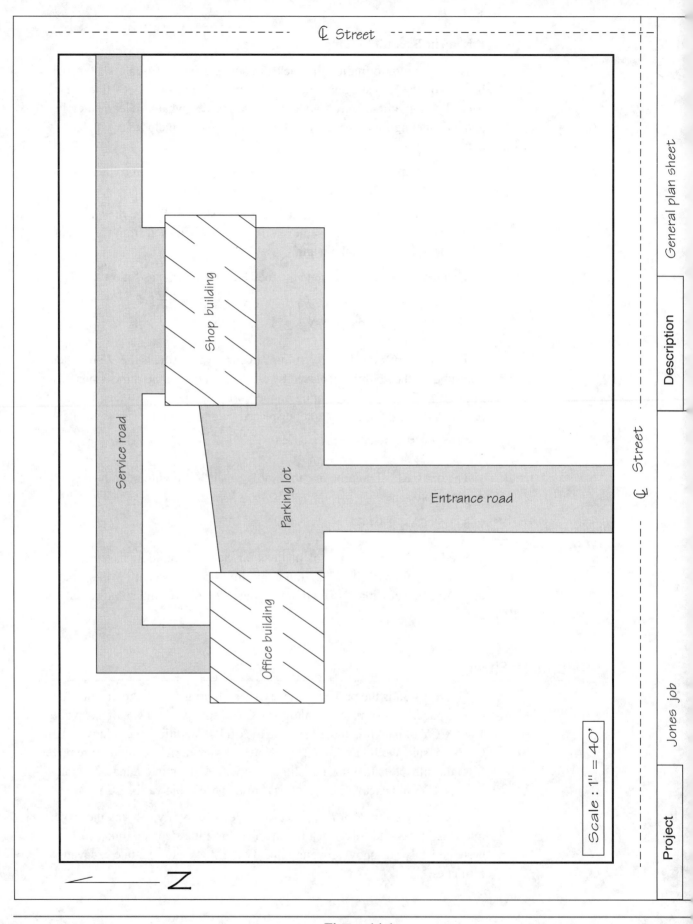

Figure 14-1
General plan sheet

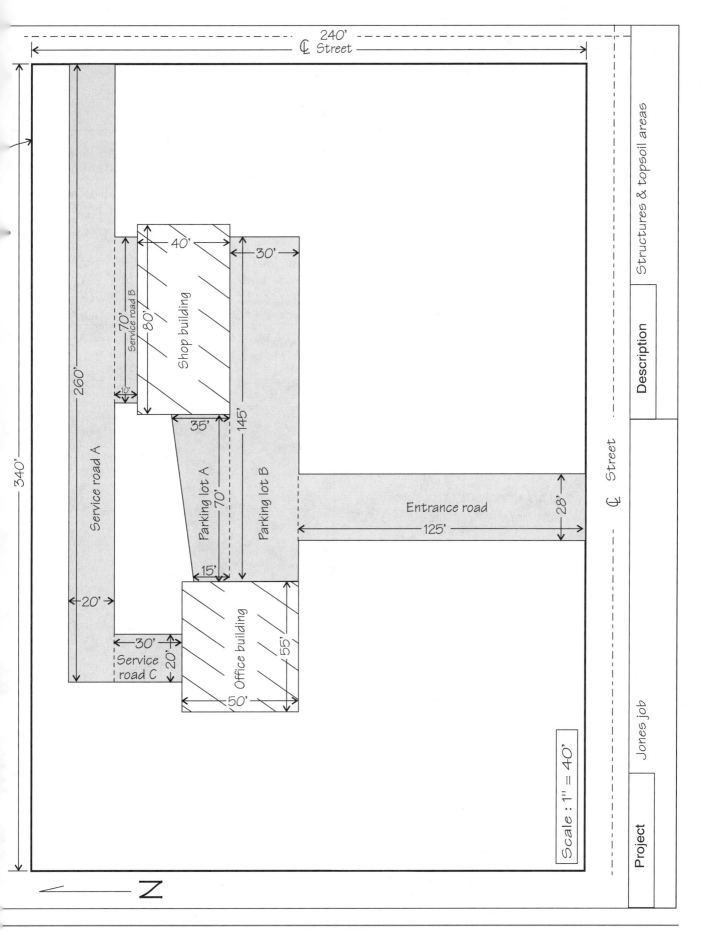

Figure 14-2
Site plan

Calculation Sheet

Project: Jones job **Date:** 4/20

Site dimensions: 340' x 240'

Strip 6", entire site

$$\text{volume (CY)} = \frac{340 \times 240 \times 0.5}{27}$$
$$= 1{,}511.11 \text{ CY}$$

Dispose of top 2" offsite

$$\text{volume (CY)} = \frac{340 \times 240 \times 0.1666}{27}$$
$$= 503.7 \text{ CY}$$

Stockpile remaining 4" for later replacement

$$\text{volume (CY)} = \frac{340 \times 240 \times 0.3333}{27}$$
$$= 1{,}007.41 \text{ CY}$$

Replace 6" on all areas without structures

$$\text{volume (CY)} = \frac{(\text{site area} - \text{structures area}) \times 0.5'}{27}$$

Site area = 340 x 240
= 81,600 SF

Structures area
Entrance road	28 x 125	= 3,500
Office building	50 x 55	= 2,750
Shop building	80 x 40	= 3,200
Parking lot A	$\frac{(15 + 35)}{2} \times 70$	= 1,750
Parking lot B	145 x 30	= 4,350
Service road A	260 x 20	= 5,200
Service road B	10 x 70	= 700
Service road C	20 x 30	= 600
Total structures area		= 22,050 SF

$$\text{Replace volume (CY)} = \frac{(81{,}600 - 22{,}050) \times 0.5}{27}$$
$$= \frac{59{,}550 \times 0.5}{27}$$
$$= \frac{29{,}775}{27}$$
$$= 1{,}102.77 \text{ CY}$$

Conclusion

Need 1,102.77 CY of replacement topsoil. Stockpile contains 1,007.41 CY, leaving shortfall of 95.36 CY to purchase offsite. (1,102.77 − 1,007.41 = 95.36)

Figure 14-3
Topsoil quantities calculations

Quantities Take-off Sheet

Project: Jones job **Date:** 4/20

Quantities for: Topsoil **Sheet** 1 of 1

By: DB   **Checked:** LL   **Misc:**

Item	Length (ft)	Width (ft)	Depth (ft)	Volume (CY)	Notes
Strip 6", entire site	340	240	0.5	1,511.11	Strip and distribute as follows:
Top 2", non-reuseable	340	240	0.1666	503.7	Remove for disposal off site
Lower 4", reuseable	340	240	0.3333	1,007.41	Stockpile on site for reuse
Replace 6" (area without strutures)				1,102.77	Shortfall, 96 CY, to be brought on site

Note:

Figure 14-4
Topsoil calculations summary

A Sample Take-off

Calculation Sheet

Project: Jones job   **Date:** 4/20

Total area service road (SF) = area A* + area B* + area C*
= 5,200 + 700 + 600
= 6,500 SF
* (see Figure 14-3)

6" rock
volume (CY) = $\dfrac{6,500 \times 0.5}{27}$
= 120.37 CY

Conclusion

120 CY of rock needed.

Figure 14-5
Rock on service roads calculations

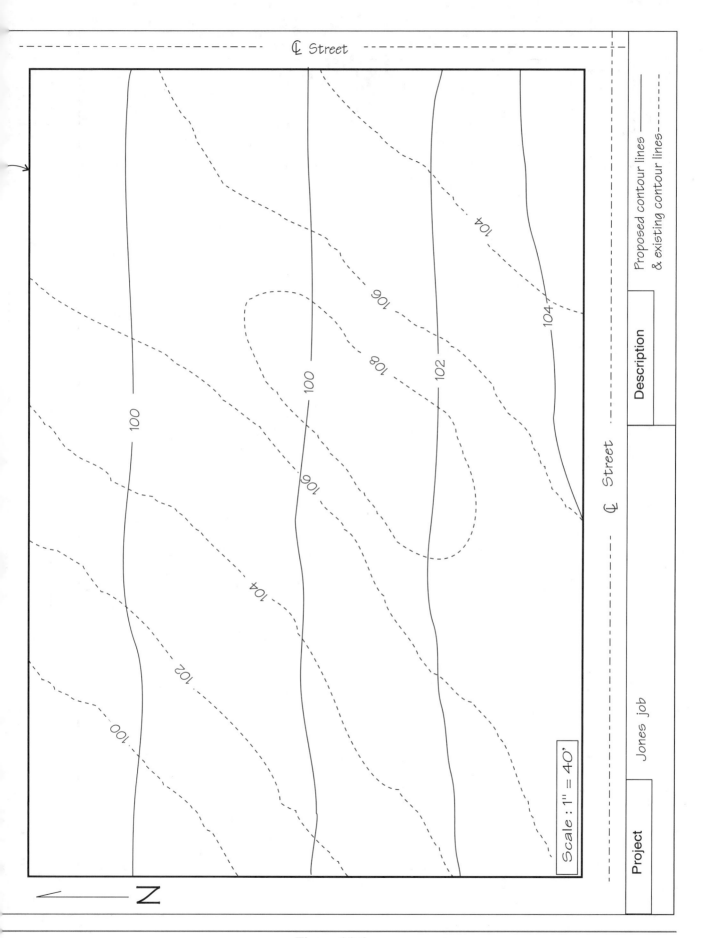

Figure 14-6
Plan showing proposed and existing contour lines

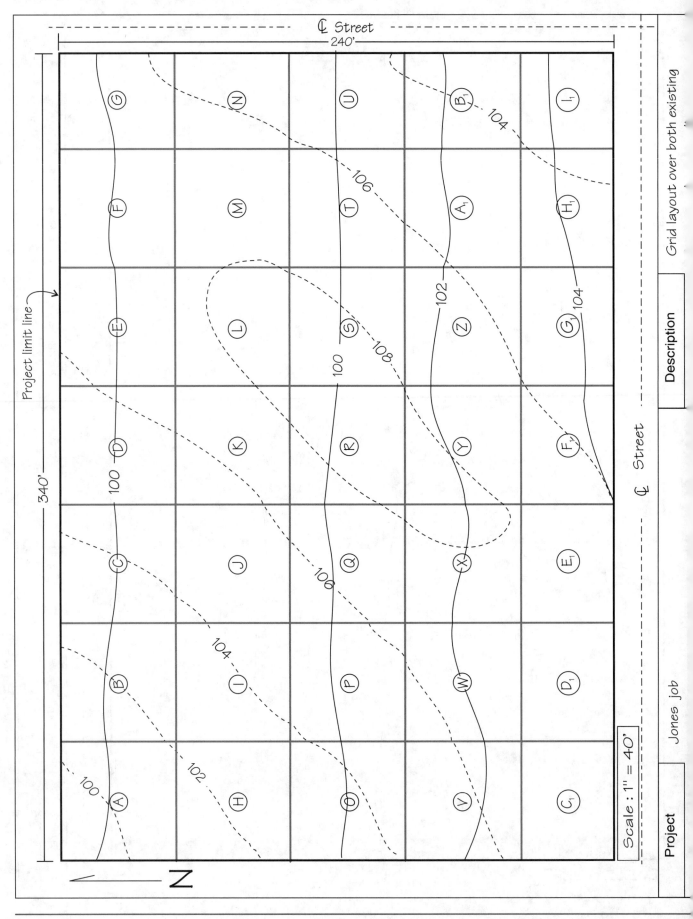

Figure 14-7
Grid layout over existing and proposed contour lines

Grid Take-off, Existing Contour Only

Sheet: __1__ of __6__

Prepared by (initials): __DB__ Date: __4/20__
Approved by (initials): __LL__ Date: __4/22__

Location	Low elevation	High elevation	Scale distance	Contour interval	Add elevation	Point elevation
A-1	100.0	102.0	7/40 (est.)	2.0	0.35	100.35
A-2	102.0	104.0	3/43	2.0	0.14	102.14
A-3	100.0	102.0	25/55 (est.)	2.0	0.91	100.91
A-4	100.0	100.0		2.0		100.0
B-1	102.0	104.0	11/48 (est.)	2.0	0.46	102.46
B-2	102.0	104.0	36/46	2.0	1.57	103.57
B-3	(A-2)			2.0		102.14
B-4	(A-1)			2.0		100.35
C-1	104.0	106.0	9/61 (est.)	2.0	0.15	104.3
C-2	104.0	106.0	32/59	2.0	1.10	105.08
C-3	(B-2)			2.0		103.57
C-4	(B-1)			2.0		102.46
D-1	104.0	106.0	45/55 (est.)	2.0	1.64	105.64
D-2	106.0	108.0	19/45	2.0	0.84	106.84
D-3	(C-2)			2.0		105.08
D-4	(C-1)			2.0		104.3
E-1	106.0	106.0		2.0		106.0
E-2	106.0	106.0		2.0		106.0
E-3	(D-2)			2.0		106.84
E-4	(D-1)			2.0		105.64
F-1	106.0	106.0		2.0		106.0
F-2	106.0	106.0		2.0		106.0
F-3	(E-2)			2.0		106.0
F-4	(E-1)			2.0		106.0

Figure 14-8
Grid square calculations for existing contours

Grid Take-off, Existing Contour Only

Sheet: __2__ of __6__

Prepared by (initials): __DB__ Date: __4/20__
Approved by (initials): __LL__ Date: __4/22__

Location	Low elevation	High elevation	Scale distance	Contour interval	Add elevation	Point elevation
G-1	106.0	106.0		2.0		106.0
G-2	106.0	104.0	11/50 (est.)	2.0	-0.44	105.56
G-3	(F-2)			2.0		106.0
G-4	(F-1)			2.0		106.0
H-1	(A-2)			2.0		102.14
H-2	102.0	104.0	42/48	2.0	1.75	103.75
H-3	102.0	104.0	12/55 (est.)	2.0	0.44	102.44
H-4	(A-3)			2.0		100.91
I-1	(B-2)			2.0		103.57
I-2	104.0	106.0	31/55	2.0	1.13	105.13
I-3	(H-2)			2.0		103.75
I-4	(H-3)			2.0		102.44
J-1	(C-2)			2.0		105.08
J-2	106.0	108.0	10/34	2.0	0.59	106.59
J-3	(I-2)			2.0		105.13
J-4	(I-1)			2.0		103.57
K-1	(D-2)			2.0		106.84
K-2	108.0	108.0		2.0		108.0
K-3	(J-2)			2.0		106.59
K-4	(J-1)			2.0		105.08

Note:

The values in the "Add Elevation" column for locations G-2, and U-2 are negative.

You may find it helpful to recall that: 106.0 + (-0.44) and 106.0 − 0.44 give the same result, 105.56.

Figure 14-8 (continued)
Grid square calculations for existing contours

Grid Take-off, Existing Contour Only

Sheet: __3__ of __6__

Prepared by (initials): __DB__   Date: __4/20__
Approved by (initials): __LL__   Date: __4/22__

Location	Low elevation	High elevation	Scale distance	Contour interval	Add elevation	Point elevation
L-1	(E-2)			2.0		106.0
L-2	106.0	108.0	2/50	2.0	0.08	106.08
L-3	(K-2)			2.0		108.0
L-4	(K-1)			2.0		106.84
M-1	(F-2)			2.0		106.0
M-2	106.0	106.0		2.0		106.0
M-3	(L-2)			2.0		106.08
M-4	(L-1)			2.0		106.0
N-1	(G-2)			2.0		105.56
N-2	104.0	106.0	35/50 (est.)	2.0	1.4	105.4
N-3	(M-2)			2.0		106.0
N-4	(M-1)			2.0		106.0
O-1	(H-2)			2.0		103.75
O-2	104.0	106.0	28/47	2.0	1.19	105.19
O-3	104.0	106.0	2/43	2.0	0.09	104.09
O-4	(H-3)			2.0		102.44
P-1	(I-2)			2.0		105.13
P-2	106.0	108.0	11/59	2.0	0.37	106.37
P-3	(O-2)			2.0		105.19
P-4	(O-1)			2.0		103.75
Q-1	(J-2)			2.0		106.59
Q-2	108.0	108.0		2.0		108.0
Q-3	(P-2)			2.0		106.37
Q-4	(P-1)			2.0		105.13

Figure 14-8 (continued)
Grid square calculations for existing contours

Grid Take-off, Existing Contour Only

Sheet: __4__ of __6__

Prepared by (initials): __DB__ Date: __4/20__
Approved by (initials): __LL__ Date: __4/22__

Location	Low elevation	High elevation	Scale distance	Contour interval	Add elevation	Point elevation
R-1	(K-2)			2.0		108.0
R-2	108.0	108.0		2.0		108.0
R-3	(Q-2)			2.0		108.0
R-4	(Q-1)			2.0		106.59
S-1	(L-2)			2.0		106.08
S-2	106.0	108.0	31/45	2.0	1.38	107.38
S-3	(R-2)			2.0		108.0
S-4	(R-1)			2.0		108.0
T-1	(M-2)			2.0		106.0
T-2	104.0	106.0	22/54	2.0	0.81	104.81
T-3	(S-2)			2.0		107.38
T-4	(S-1)			2.0		106.08
U-1	(N-2)			2.0		105.4
U-2	104.0	102.0	8/51 (est.)	2.0	-0.31	103.69
U-3	(T-2)			2.0		104.81
U-4	(T-1)			2.0		106.0
V-1	(O-2)			2.0		105.19
V-2	106.0	106.0		2.0		106.0
V-3	106.0	106.0		2.0		106.0
V-4	(O-3)			2.0		104.09
W-1	(P-2)			2.0		106.37
W-2	106.0	106.0		2.0		106.0
W-3	(V-2)			2.0		106.0
W-4	(V-1)			2.0		105.19

Figure 14-8 (continued)
Grid square calculations for existing contours

Grid Take-off, Existing Contour Only

Sheet: 5 of 6

Prepared by (initials): DB Date: 4/20
Approved by (initials): LL Date: 4/22

Location	Low elevation	High elevation	Scale distance	Contour interval	Add elevation	Point elevation
X-1	(Q-2)			2.0		108.0
X-2	106.0	108.0	7/42	2.0	0.33	106.33
X-3	(W-2)			2.0		106.0
X-4	(W-1)			2.0		106.37
Y-1	(R-2)			2.0		108.0
Y-2	106.0	108.0	42/45	2.0	1.87	107.87
Y-3	(X-2)			2.0		106.33
Y-4	(X-1)			2.0		108.0
Z-1	(S-2)			2.0		107.38
Z-2	104.0	106.0	23/71	2.0	0.65	104.65
Z-3	(Y-2)			2.0		107.87
Z-4	(Y-1)			2.0		108.0
A_1-1	(T-2)			2.0		104.81
A_1-2	104.0	106.0	56/59	2.0	1.9	105.9
A_1-3	(Z-2)			2.0		104.65
A_1-4	(Z-1)			2.0		107.38
B_1-1	(U-2)			2.0		103.69
B_1-2	102.0	104.0	30/71 (est.)	2.0	0.85	102.85
B_1-3	(A_1-2)			2.0		105.9
B_1-4	(A_1-1)			2.0		104.81
C_1-1	(V-2)			2.0		106.0
C_1-2	106.0	106.0		2.0		106.0
C_1-3	106.0	106.0		2.0		106.0
C_1-4	(V-3)			2.0		106.0

Figure 14-8 (continued)
Grid square calculations for existing contours

Grid Take-off, Existing Contour Only

Prepared by (initials): __DB__ Date: __4/20__
Sheet: __6__ of __6__ Approved by (initials): __LL__ Date: __4/22__

Location	Low elevation	High elevation	Scale distance	Contour interval	Add elevation	Point elevation
D₁-1	(W-2)			2.0		106.0
D₁-2	106.0	106.0		2.0		106.0
D₁-3	(C₁-2)			2.0		106.0
D₁-4	(C₁-1)			2.0		106.0
E₁-1	(X-2)			2.0		106.33
E₁-2	106.0	106.0		2.0		106.0
E₁-3	(D₁-2)			2.0		106.0
E₁-4	(D₁-1)			2.0		106.0
F₁-1	(Y-2)			2.0		107.87
F₁-2	104.0	106.0	29/87 (est.)	2.0	0.67	104.67
F₁-3	(E₁-2)			2.0		106.0
F₁-4	(E₁-1)			2.0		106.33
G₁-1	(Z-2)			2.0		104.65
G₁-2	104.0	106.0	63/79 (est.)	2.0	1.6	105.6
G₁-3	(F₁-2)			2.0		104.67
G₁-4	(F₁-1)			2.0		107.87
H₁-1	(A₁-2)			2.0		105.9
H₁-2	102.0	104.0	17/90 (est.)	2.0	0.38	102.38
H₁-3	(G₁-2)			2.0		105.6
H₁-4	(G₁-3)			2.0		104.65
I₁-1	(B₁-2)			2.0		102.85
I₁-2	102.0	104.0	55/85 (est.)	2.0	1.29	103.29
I₁-3	(H₁-2)			2.0		102.38
I₁-4	(H₁-1)			2.0		105.9

Figure 14-8 (continued)
Grid square calculations for existing contours

Grid Take-off, Proposed Contour Only

Sheet: __1__ of __4__

Prepared by (initials): __DB__ Date: __4/21__
Approved by (initials): __LL__ Date: __4/22__

Location	Low elevation	High elevation	Scale distance	Contour interval	Add elevation	Point elevation
O-1	100.0	100.0		2.0		100.0
O-2	100.0	102.0	37/70	2.0	1.06	101.06
O-3	100.0	102.0	35/83	2.0	0.84	100.84
O-4				2.0		100.0
P-1	100.0	100.0		2.0		100.0
P-2	100.0	102.0	44/78	2.0	1.13	101.13
P-3	(O-2)			2.0		101.06
P-4	(O-1)			2.0		100.0
Q-1	100.0	100.0		2.0		100.0
Q-2	100.0	102.0	44/72	2.0	1.22	101.22
Q-3	(P-2)			2.0		101.13
Q-4	(P-1)			2.0		100.0
R-1	100.0	100.0		2.0		100.0
R-2	100.0	102.0	47/64	2.0	1.47	101.47
R-3	(Q-2)			2.0		101.22
R-4	(Q-1)			2.0		100.0
S-1	100.0	100.0		2.0		100.0
S-2	100.0	102.0	40/67	2.0	1.19	101.19
S-3	(R-2)			2.0		101.47
S-4	(R-1)			2.0		100.0

Note:

Grid squares A to N are not included here because they are all in the limits of proposed elevation (100.0).

Therefore, all these elevations are 100.0.

Figure 14-9
Grid square calculations for proposed contours

Grid Take-off, Proposed Contour Only

Sheet: 2 of 4

Prepared by (initials): DB Date: 4/21
Approved by (initials): LL Date: 4/22

Location	Low elevation	High elevation	Scale distance	Contour interval	Add elevation	Point elevation
T-1	100.0	100.0		2.0		100.0
T-2	100.0	102.0	40/62	2.0	1.29	101.29
T-3	(S-2)			2.0		101.19
T-4	(S-1)			2.0		100.0
U-1	100.0	100.0		2.0		100.0
U-2	100.0	102.0	40/60	2.0	1.33	101.33
U-3	(T-2)			2.0		101.29
U-4	(T-1)			2.0		100.0
V-1	(O-2)			2.0		101.06
V-2	102.0	104.0	21/94 (est.)	2.0	0.45	102.45
V-3	102.0	104.0	22/99 (est.)	2.0	0.44	102.44
V-4	(O-3)			2.0		100.84
W-1	(P-2)			2.0		101.13
W-2	102.0	104.0	23/97 (est.)	2.0	0.47	102.47
W-3	(V-2)			2.0		102.45
W-4	(V-1)			2.0		101.06
X-1	(Q-2)			2.0		101.22
X-2	102.0	104.0	39/81	2.0	0.96	102.96
X-3	(W-2)			2.0		102.47
X-4	(W-1)			2.0		101.13
Y-1	(R-2)			2.0		101.47
Y-2	102.0	104.0	43/80	2.0	1.08	103.08
Y-3	(X-2)			2.0		102.96
Y-4	(X-1)			2.0		101.22

Figure 14-9 (continued)
Grid square calculations for proposed contours

Grid Take-off, Proposed Contour Only

Sheet: __3__ of __4__

Prepared by (initials): __DB__   Date: __4/21__
Approved by (initials): __LL__   Date: __4/22__

Location	Low elevation	High elevation	Scale distance	Contour interval	Add elevation	Point elevation
Z-1	(S-2)			2.0		101.19
Z-2	102.0	104.0	55/80	2.0	1.38	103.38
Z-3	(Y-2)			2.0		103.08
Z-4	(Y-1)			2.0		101.47
A_1-1	(T-2)			2.0		101.29
A_1-2	102.0	104.0	43/60	2.0	1.43	103.43
A_1-3	(Z-2)			2.0		103.38
A_1-4	(Z-1)			2.0		101.19
B_1-1	(U-2)			2.0		101.33
B_1-2	102.0	104.0	43/60	2.0	1.43	103.43
B_1-3	(A_1-2)			2.0		103.43
B_1-4	(A_1-1)			2.0		101.29
C_1-1	(V-2)			2.0		102.45
C_1-2	102.0	102.0		2.0		102.0
C_1-3	102.0	102.0		2.0		102.0
C_1-4	(V-3)			2.0		102.44
D_1-1	(W-2)			2.0		102.47
D_1-2	102.0	102.0		2.0		102.0
D_1-3	(C_1-2)			2.0		102.0
D_1-4	(C_1-1)			2.0		102.45
E_1-1	(X-2)			2.0		102.96
E_1-2	102.0	102.0		2.0		102.0
E_1-3	(D_1-2)			2.0		102.0
E_1-4	(D_1-1)			2.0		102.47

Figure 14-9 (continued)
Grid square calculations for proposed contours

Grid Take-off, Proposed Contour Only

Sheet: __4__ of __4__

Prepared by (initials): __DB__   Date: __4/21__
Approved by (initials): __LL__   Date: __4/22__

Location	Low elevation	High elevation	Scale distance	Contour interval	Add elevation	Point elevation
F₁-1	(Y-2)			2.0		103.08
F₁-2	104.0	106.0	19/49 (est.)	2.0	0.78	104.78
F₁-3	(E₁-2)			2.0		102.0
F₁-4	(E₁-1)			2.0		102.96
G₁-1	(Z-2)			2.0		103.38
G₁-2	104.0	106.0	24/44 (est.)	2.0	1.09	105.09
G₁-3	(F₁-2)			2.0		104.78
G₁-4	(F₁-1)			2.0		103.08
H₁-1	(A₁-2)			2.0		103.43
H₁-2	104.0	106.0	36/44 (est.)	2.0	1.64	105.64
H₁-3	(G₁-2)			2.0		105.09
H₁-4	(G₁-1)			2.0		103.38
I₁-1	(B₁-2)			2.0		103.43
I₁-2	104.0	106.0	40/51 (est.)	2.0	1.57	105.57
I₁-3	(H₁-2)			2.0		105.64
I₁-4	(H₁-1)			2.0		103.43

Figure 14-9 (continued)
Grid square calculations for proposed contours

Individual Grid Square Calculation Sheet

Job number: 498   Project: Jones job   Prepared by (initials): DB   Date: 4/21
Sheet: 1 of 7   Approved by (initials): LL   Date: 4/22

Grid A
Average depth: 0.85

Element	1	2	3	4
Proposed	100.0	100.0	100.0	100.0
Existing	100.35	102.14	100.91	100.0
Depth	0.35	2.14	0.91	0

Grid B
Average depth: 2.13

Element	1	2	3	4
Proposed	100.0	100.0	100.0	100.0
Existing	102.46	103.57	102.14	100.35
Depth	2.46	3.57	2.14	0.35

Grid C
Average depth: 3.85

Element	1	2	3	4
Proposed	100.0	100.0	100.0	100.0
Existing	104.3	105.08	103.57	102.46
Depth	4.3	5.08	3.57	2.46

Grid D
Average depth: 5.47

Element	1	2	3	4
Proposed	100.0	100.0	100.0	100.0
Existing	105.64	106.84	105.08	104.3
Depth	5.64	6.84	5.08	4.3

Grid E
Average depth: 6.12

Element	1	2	3	4
Proposed	100.0	100.0	100.0	100.0
Existing	106.0	106.0	106.84	105.64
Depth	6.0	6.0	6.84	5.64

Figure 14-10
Grid square calculations to find average depth of cut or fill

A Sample Take-off

Individual Grid Square Calculation Sheet

Job number: __498__   Project: __Jones job__   Prepared by (initials): __DB__   Date: __4/21__

Sheet: __2__ of __7__   Approved by (initials): __LL__   Date: __4/22__

Grid F
Average depth: 6.0

Element	1	2	3	4
Proposed	100.0	100.0	100.0	100.0
Existing	106.0	106.0	106.0	106.0
Depth	6.0	6.0	6.0	6.0

Grid G
Average depth: 5.89

Element	1	2	3	4
Proposed	100.0	100.0	100.0	100.0
Existing	106.0	105.56	106.0	106.0
Depth	6.0	5.56	6.0	6.0

Grid H
Average depth: 2.31

Element	1	2	3	4
Proposed	100.0	100.0	100.0	100.0
Existing	102.14	103.75	102.44	100.91
Depth	2.14	3.75	2.44	0.91

Grid I
Average depth: 3.72

Element	1	2	3	4
Proposed	100.0	100.0	100.0	100.0
Existing	103.57	105.13	103.75	102.44
Depth	3.57	5.13	3.75	2.44

Grid J
Average depth: 5.09

Element	1	2	3	4
Proposed	100.0	100.0	100.0	100.0
Existing	105.08	106.59	105.13	103.57
Depth	5.08	6.59	5.13	3.57

Figure 14-10 (continued)
Grid square calculations to find average depth of cut or fill

Individual Grid Square Calculation Sheet

Job number: 498   Project: Jones job   Prepared by (initials): DB   Date: 4/21
Sheet: 3 of 7   Approved by (initials): LL   Date: 4/22

Grid K
Average depth: 6.63

Element	1	2	3	4
Proposed	100.0	100.0	100.0	100.0
Existing	106.84	108.0	106.59	105.08
Depth	6.84	8.0	6.59	5.08

Grid L
Average depth: 6.73

Element	1	2	3	4
Proposed	100.0	100.0	100.0	100.0
Existing	106.0	106.08	108.0	106.84
Depth	6.0	6.08	8.0	6.84

Grid M
Average depth: 6.73

Element	1	2	3	4
Proposed	100.0	100.0	100.0	100.0
Existing	106.0	106.0	106.08	106.0
Depth	6.0	6.0	6.08	6.0

Grid N
Average depth: 5.74

Element	1	2	3	4
Proposed	100.0	100.0	100.0	100.0
Existing	105.56	105.4	106.0	106.0
Depth	5.56	5.4	6.0	6.0

Grid O
Average depth: 3.39

Element	1	2	3	4
Proposed	100.0	101.06	100.84	100.0
Existing	103.75	105.19	104.09	102.44
Depth	3.75	4.13	3.25	2.44

Figure 14-10 (continued)
Grid square calculations to find average depth of cut or fill

A Sample Take-off

Individual Grid Square Calculation Sheet

Job number: __498__   Project: __Jones job__   Prepared by (initials): __DB__   Date: __4/21__

Sheet: __4__ of __7__   Approved by (initials): __LL__   Date: __4/22__

Grid P
Average depth: 4.56

Element	1	2	3	4
Proposed	100.0	101.13	101.06	100.0
Existing	105.13	106.37	105.19	103.75
Depth	5.13	5.24	4.13	3.75

Grid Q
Average depth: 5.94

Element	1	2	3	4
Proposed	100.0	101.22	101.13	100.0
Existing	106.59	108.0	106.37	105.13
Depth	6.59	6.78	5.24	5.13

Grid R
Average depth: 6.98

Element	1	2	3	4
Proposed	100.0	101.47	101.22	100.0
Existing	108.0	108.0	108.0	106.59
Depth	8.0	6.53	6.78	6.59

Grid S
Average depth: 6.7

Element	1	2	3	4
Proposed	100.0	101.19	101.47	100.0
Existing	106.08	107.38	108.0	108.0
Depth	6.08	6.19	6.53	8.0

Grid T
Average depth: 5.45

Element	1	2	3	4
Proposed	100.0	101.29	101.19	100.0
Existing	106.0	104.81	107.38	106.08
Depth	6.0	3.52	6.19	6.08

Figure 14-10 (continued)
Grid square calculations to find average depth of cut or fill

Individual Grid Square Calculation Sheet

Job number: 498   Project: Jones job   Prepared by (initials): DB   Date: 4/21
Sheet: 5 of 7   Approved by (initials): LL   Date: 4/22

Grid U
Average depth: 4.32

Element	1	2	3	4
Proposed	100.0	101.33	101.29	100.0
Existing	105.4	103.69	104.81	106.0
Depth	5.4	2.36	3.52	6.0

Grid V
Average depth: 3.62

Element	1	2	3	4
Proposed	101.06	102.45	102.44	100.84
Existing	105.19	106.0	106.0	104.09
Depth	4.13	3.55	3.56	3.25

Grid W
Average depth: 4.11

Element	1	2	3	4
Proposed	101.13	102.47	102.45	101.06
Existing	106.37	106.0	106.0	105.19
Depth	5.24	3.53	3.55	4.13

Grid X
Average depth: 4.73

Element	1	2	3	4
Proposed	101.22	102.96	102.47	101.13
Existing	108.0	106.33	106.0	106.37
Depth	6.78	3.37	3.53	5.24

Grid Y
Average depth: 5.37

Element	1	2	3	4
Proposed	101.47	103.08	102.96	101.22
Existing	108.0	107.87	106.33	108.0
Depth	6.53	4.79	3.37	6.78

Figure 14-10 (continued)
Grid square calculations to find average depth of cut or fill

Individual Grid Square Calculation Sheet

Job number: __498__   Project: __Jones job__   Prepared by (initials): __DB__   Date: __4/21__
Sheet: __6__ of __7__   Approved by (initials): __LL__   Date: __4/22__

Grid Z
Average depth: 4.7

Element	1	2	3	4
Proposed	101.19	103.38	103.08	101.47
Existing	107.38	104.65	107.87	108.0
Depth	6.19	1.27	4.79	6.53

Grid A$_1$
Average depth: 3.36

Element	1	2	3	4
Proposed	101.29	103.43	103.38	101.19
Existing	104.81	105.9	104.65	107.38
Depth	3.52	2.47	1.27	6.19

Grid B$_1$
Average depth: 1.94

Element	1	2	3	4
Proposed	101.33	103.43	103.43	101.29
Existing	103.69	102.85	105.9	104.81
Depth	2.36	-0.58	2.47	3.52

Grid C$_1$
Average depth: 3.78

Element	1	2	3	4
Proposed	102.45	102.0	102.0	102.44
Existing	106.0	106.0	106.0	106.0
Depth	3.55	4.0	4.0	3.56

Grid D$_1$
Average depth: 3.77

Element	1	2	3	4
Proposed	102.47	102.0	102.0	102.45
Existing	106.0	106.0	106.0	106.0
Depth	3.53	4.0	4.0	3.55

Figure 14-10 (continued)
Grid square calculations to find average depth of cut or fill

Individual Grid Square Calculation Sheet

Job number: __498__   Project: __Jones job__   Prepared by (initials): __DB__   Date: __4/21__

Sheet: __7__ of __7__   Approved by (initials): __LL__   Date: __4/22__

Grid E$_1$
Average depth: 3.73

Element	1	2	3	4
Proposed	102.96	102.0	102.0	102.47
Existing	106.33	106.0	106.0	106.0
Depth	3.37	4.0	4.0	3.53

Grid F$_1$
Average depth: 3.01

Element	1	2	3	4
Proposed	103.08	104.78	102.0	102.96
Existing	107.87	104.67	106.0	106.33
Depth	4.79	-0.11	4.0	3.37

Grid G$_1$
Average depth: 1.62

Element	1	2	3	4
Proposed	103.38	105.09	104.78	103.08
Existing	104.65	105.6	104.67	107.87
Depth	1.27	0.51	-0.11	4.79

Grid H$_1$
Average depth: 0.25

Element	1	2	3	4
Proposed	103.43	105.64	105.09	103.38
Existing	105.9	102.38	105.6	104.65
Depth	2.47	-3.26	0.51	1.27

Grid I$_1$
Average depth: -0.91

Element	1	2	3	4
Proposed	103.43	105.57	105.64	103.43
Existing	102.85	103.29	102.38	105.9
Depth	-0.58	-2.28	-3.26	2.47

Figure 14-10 (continued)
Grid square calculations to find average depth of cut or fill

Quantities Take-off Sheet

Project: Jones job　　　　　　　　　　　　　　　　　　　　**Date:** 4/22

Quantities for: Grid square take-off　　　　　　　　　　　**Sheet** 1 **of** 2

By: DB　　　　　　　　　**Checked:** LL　　　　　　　　　**Misc:**

Location		Length (ft)	Width (ft)	Ave. depth (ft)	Vol. cut (CY)	Vol. fill (CY)
Grid A		50	50	0.85	78.70	
Grid B		50	50	2.13	197.22	
Grid C		50	50	3.85	356.48	
Grid D		50	50	5.47	506.48	
Grid E		50	50	6.12	566.67	
Grid F		50	50	6.0	555.56	
Grid G		50	40	5.89	436.30	
Grid H		50	50	2.31	213.89	
Grid I		50	50	3.72	344.44	
Grid J		50	50	5.09	471.30	
Grid K		50	50	6.63	613.89	
Grid L		50	50	6.73	623.15	
Grid M		50	50	6.02	557.41	
Grid N		50	40	5.74	425.19	
Grid O		50	50	3.39	313.89	
Grid P		50	50	4.56	422.22	
Grid Q		50	50	5.94	550.00	
Grid R		50	50	6.98	646.30	
Grid S		50	50	6.70	620.37	
Grid T		50	50	5.45	504.63	
Subtotal					9,004.09	

Note:

Figure 14-11
Grid square take-off, cut and fill calculations summary

Quantities Take-off Sheet

Project: Jones job **Date:** 4/22

Quantities for: Grid square take-off **Sheet** 2 **of** 2

By: DB **Checked:** LL **Misc:**

Location		Length (ft)	Width (ft)	Ave. depth (ft)	Vol. cut (CY)	Vol. fill (CY)
Grid U		50	40	4.32	320.00	
Grid V		50	50	3.62	335.19	
Grid W		50	50	4.11	380.56	
Grid X		50	50	4.73	437.96	
Grid Y		50	50	5.37	497.22	
Grid Z		50	50	4.70	435.19	
Grid A_1		50	50	3.36	311.11	
Grid B_1		50	40	1.94	143.70	
Grid C_1		40	50	3.78	280.00	
Grid D_1		40	50	3.77	279.26	
Grid E_1		40	50	3.73	276.30	
Grid F_1		40	50	3.01	222.96	
Grid G_1		40	50	1.62	120.00	
Grid H_1		40	50	0.25	18.52	
Grid I_1		40	40	-0.91		53.93
Subtotal (this page)					4,057.97	
Subtotal (page 1)					9,004.09	
Total cut					13,062.06	
Total fill						53.93

Note:

Total spoil = 17,040.65 LCY*

Total spoil = (total cut - total fill) x 1.31 [swell factor]

*LCY = loose cubic yards

Figure 14-11 (continued)
Grid square take-off, cut and fill calculations summary

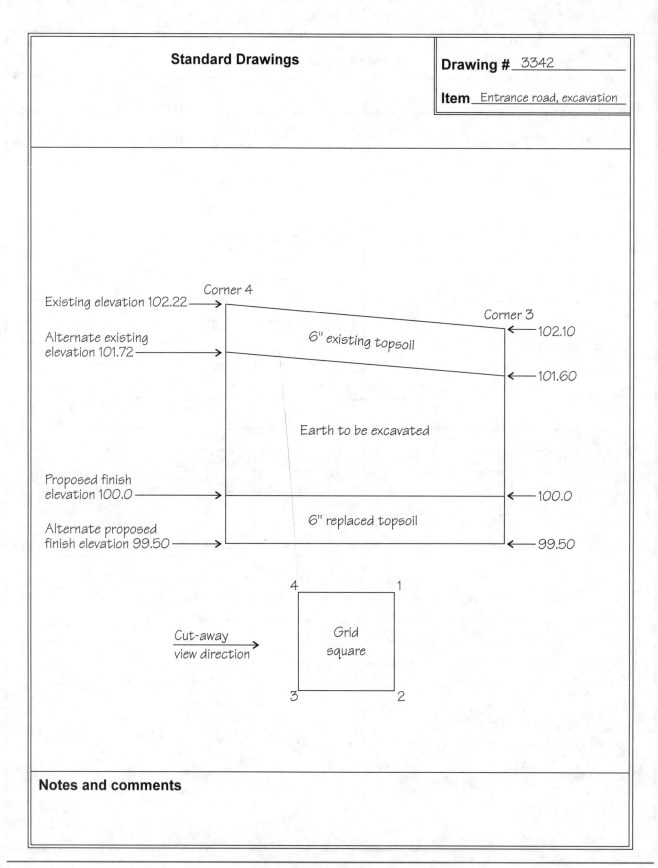

Figure 14-12
Entrance road cross-section view

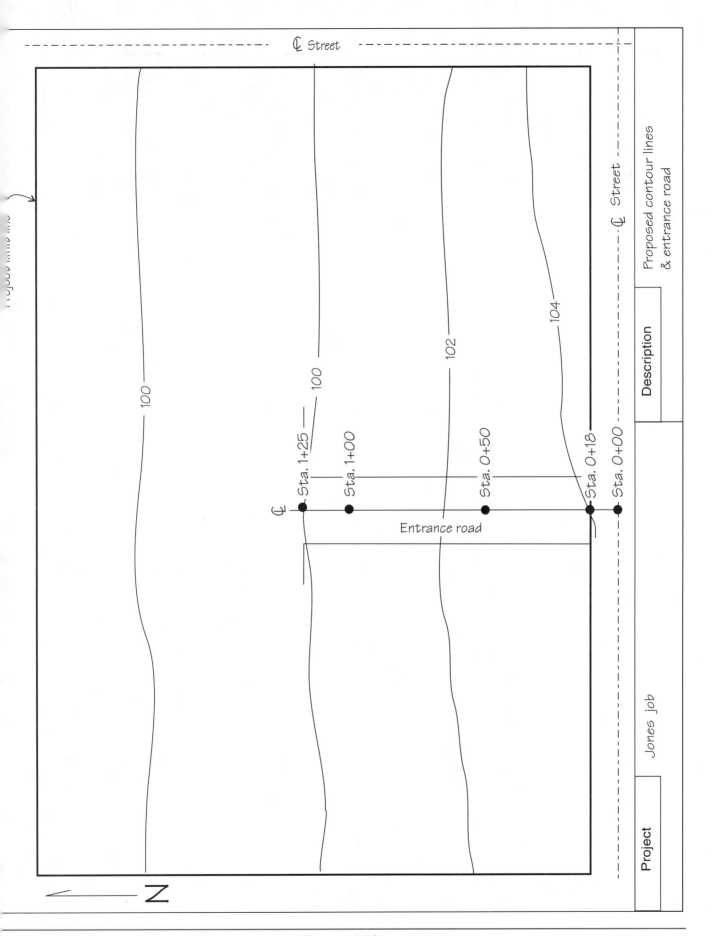

Figure 14-13
Proposed contour lines and entrance road

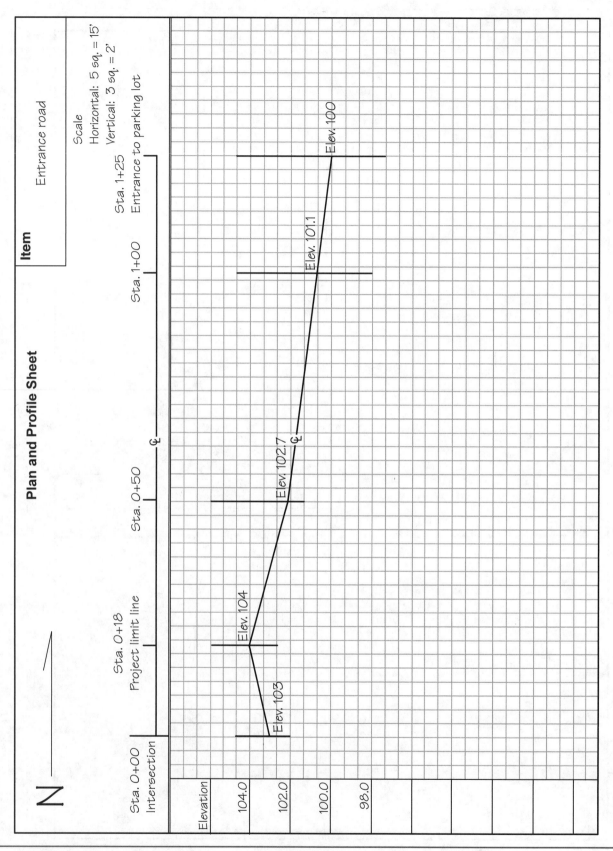

Figure 14-13a
Entrance road plan and profile sheet

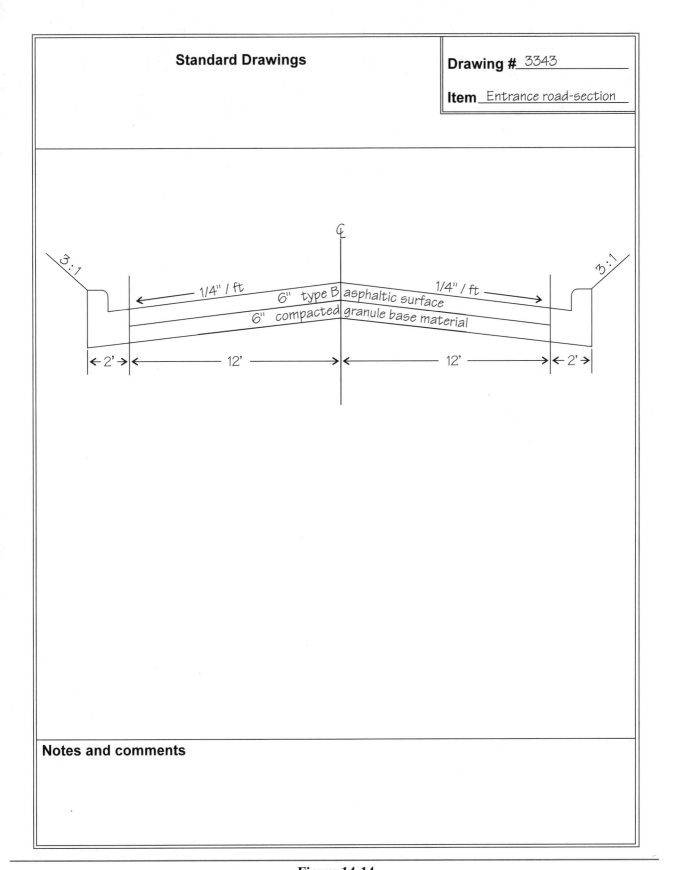

Figure 14-14
Standard drawing, entrance road section

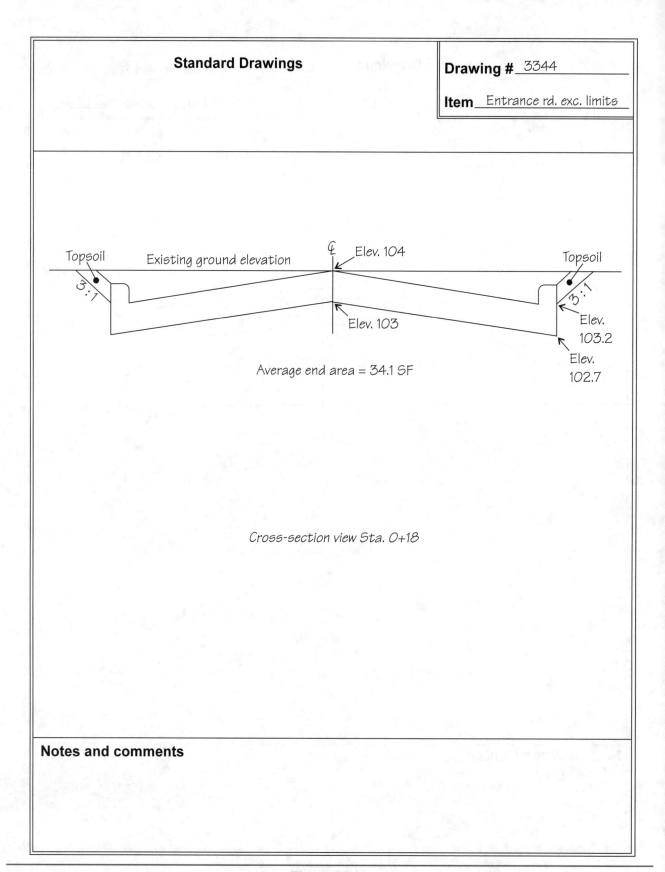

Figure 14-15
Entrance road excavation limits shown in cross section

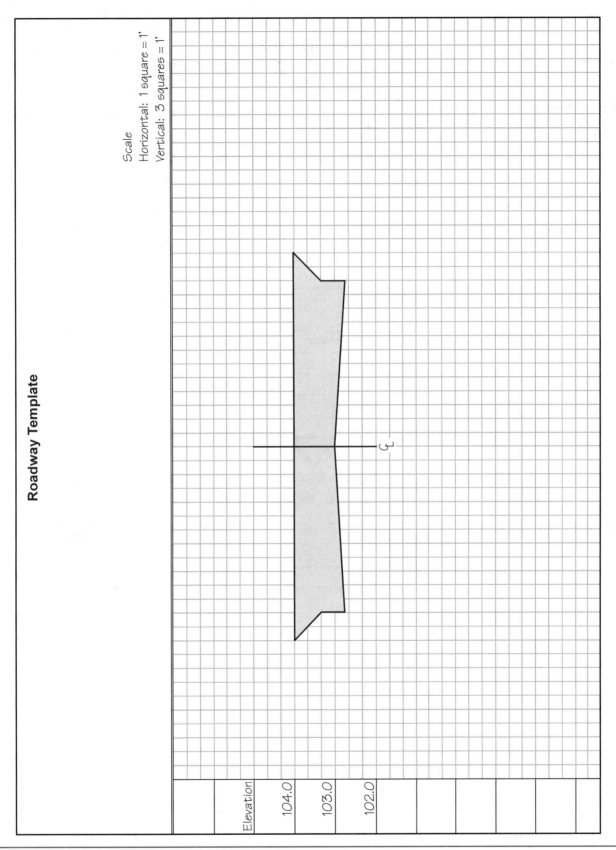

Figure 14-16
Roadway template

A Sample Take-off 291

Quantities Take-off Sheet

Project: Jones job **Date:** 4/22
Quantities for: Entrance drive **Sheet** 1 **of** 1
By: DB   **Checked:** LL **Misc:** _____

Sta.			AEA*	Distance (sta. to sta.)	Vol. (CF)	Vol. (CY)
0 + 00			0	0	0	0
0 + 18			34.1	18	306.9	11.37
0 + 50			34.1	32	1,091.2	40.42
1 + 00			34.1	50	1,705.0	63.15
1 + 25			34.1	25	852.5	31.57
Total volume excavated						146.51
Actual vol. (total vol. x 1.31**)						191.93
Total (rounded to full CY)						192.00

Note:
*AEA = Average end area
**Swell factor

Figure 14-17
Entrance road calculations summary

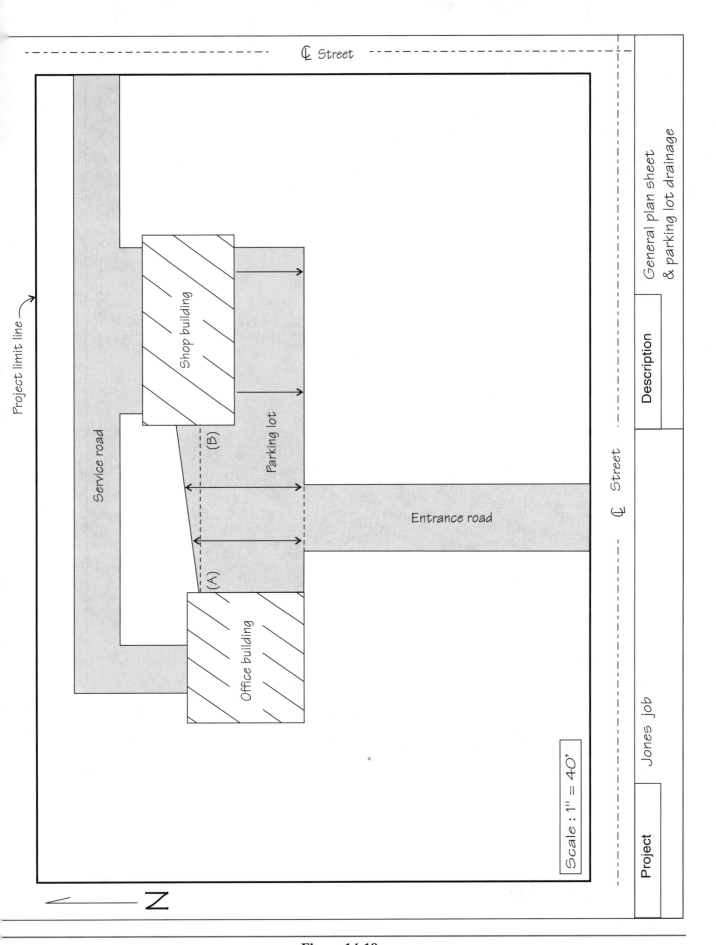

Figure 14-18
Parking lot drainage plan sheet

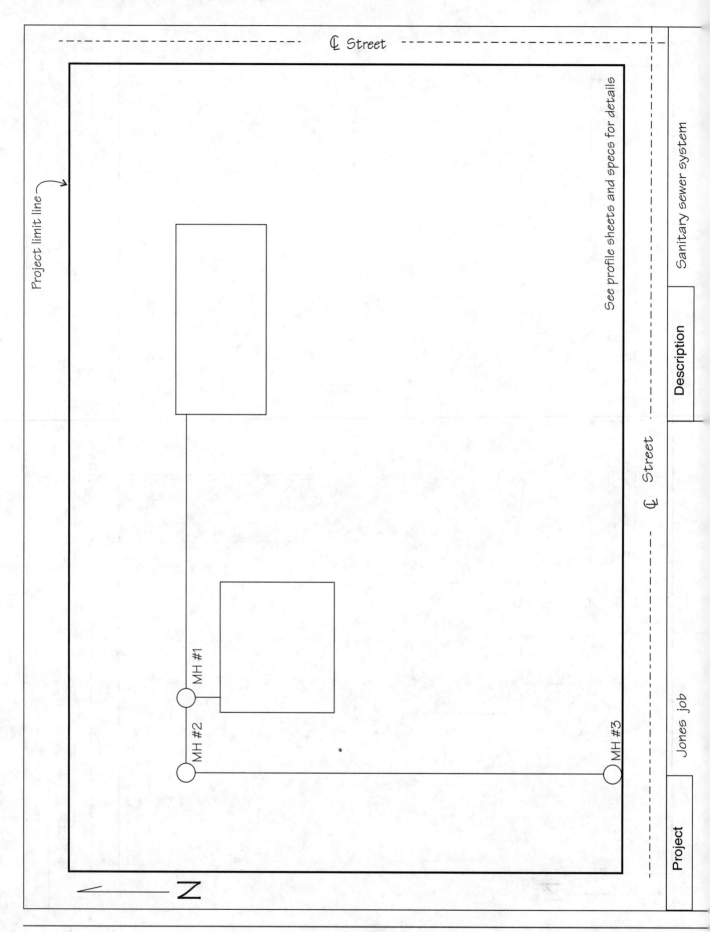

Figure 14-19
Sanitary sewer system plan sheet

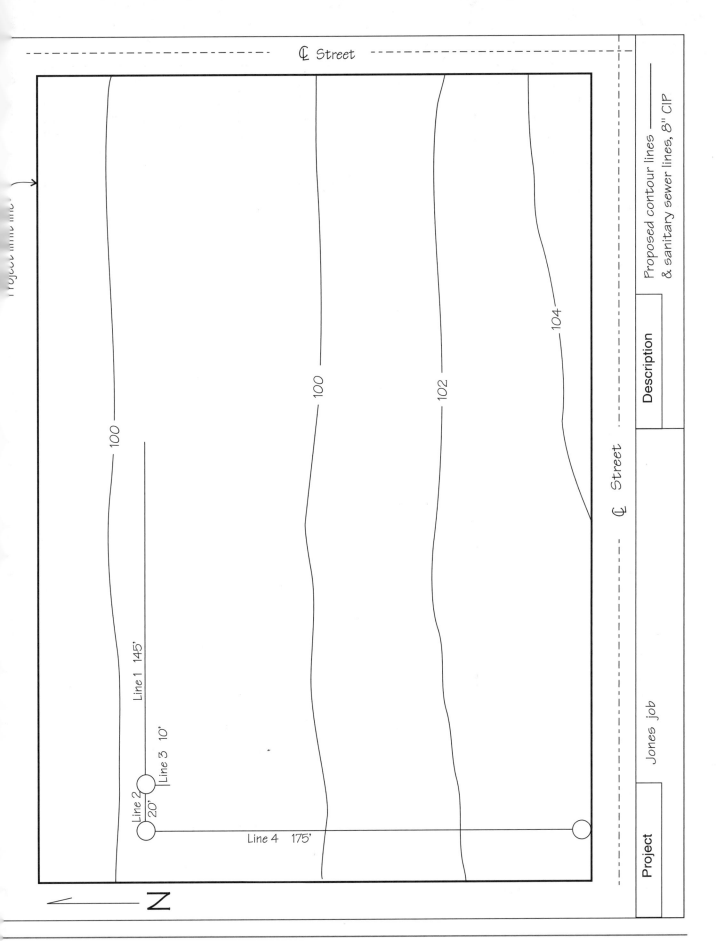

Figure 14-20
Sanitary sewer system plan sheet shown with proposed contours

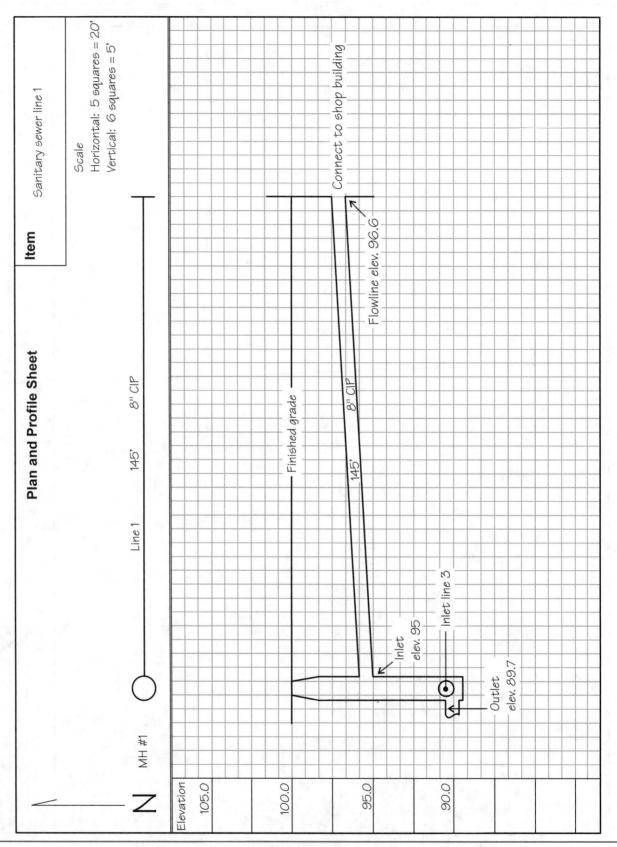

Figure 14-21
Sanitary sewer line 1, plan and profile sheet

296 *Estimating Excavation*

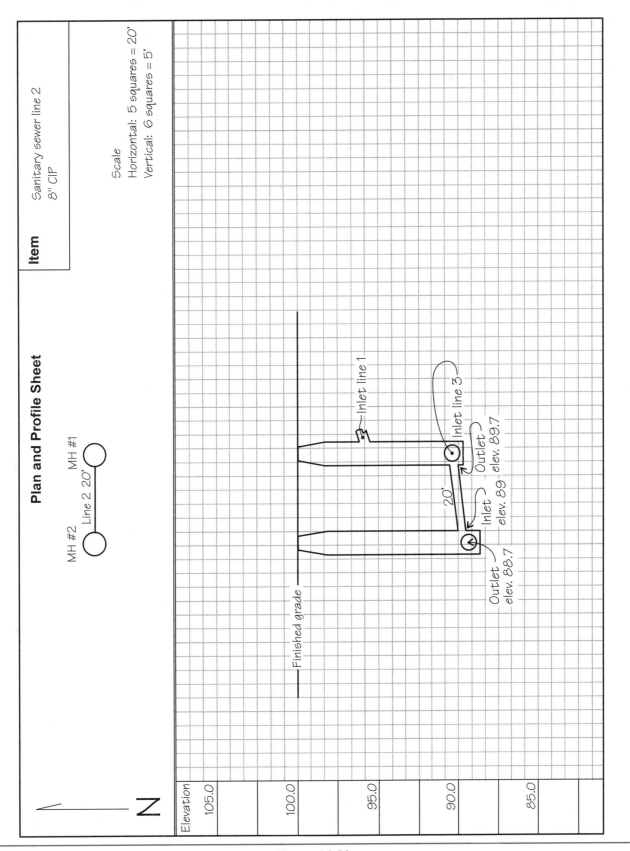

Figure 14-22
Sanitary sewer line 2, plan and profile sheet

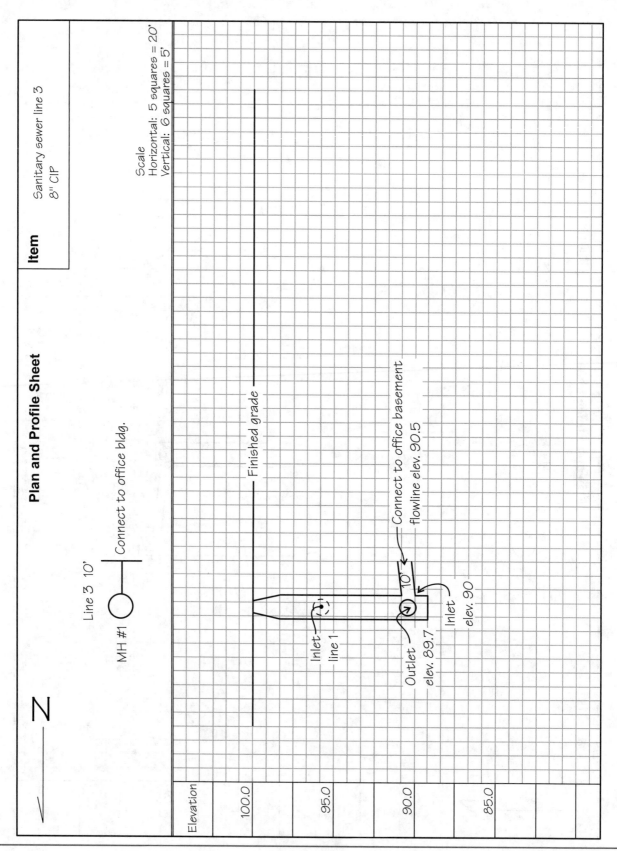

Figure 14-23
Sanitary sewer line 3, plan and profile sheet

298 *Estimating Excavation*

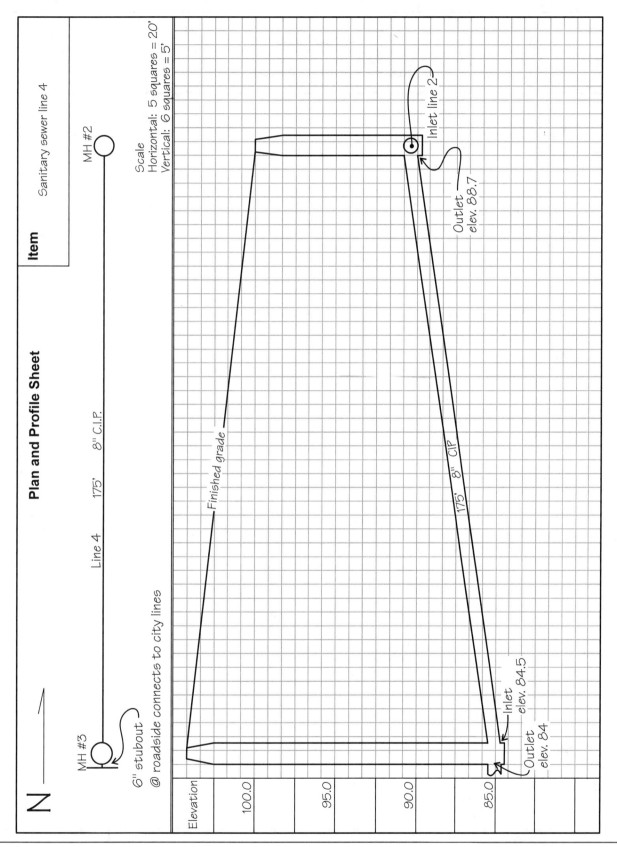

Figure 14-24
Sanitary sewer line 4, plan and profile sheet

Calculation Sheet

Project: <u>Jones job</u> Date: <u>4/22</u>

Line 1

Average depth $= \dfrac{(A) + (B)}{2}$

(A) = (finished grade elev. - flowline elev.) + 0.5' (6" undercut, see standard drawing. Figure 14-30)
  = (100' - 96.6') + 0.5'
  = 3.4' + 0.5'
(A) = 3.9'

(B) = (finish grade elev. - inlet MH1 elev.) + 0.5' (6" undercut)
  = (100' - 95') + 0.5'
  = 5' + 0.5'
(B) = 5.5'

Average depth $= \dfrac{3.9' + 5.5'}{2}$

$= \dfrac{9.4'}{2}$

= 4.7'

Conclusion
Average depth = 4.7'
Width = 2' (see standard drawing, Figure 14-30)
Length = 145' (see Figure 14-20)

Figure 14-25
Sanitary sewer line 1, average depth calculations

Estimating Excavation

Calculation Sheet

Project: Jones job   Date: 4/22

Line 2

$$\text{Average depth} = \frac{(A) + (B)}{2}$$

(A) = (finished grade elev. - outlet MH1 elev.) + 0.5' (6" undercut, see standard drawing. Figure 14-30)
   = (100' - 89.7') + 0.5'
   = 10.3' + 0.5'
(A) = 10.8'

(B) = (finish grade elev. - outlet MH2 elev.) + 0.5' (6" undercut)
   = (100' - 88.7') + 0.5'
   = 11.3' + 0.5'
(B) = 11.8'

$$\text{Average depth} = \frac{10.8' + 11.8'}{2}$$
$$= \frac{22.6'}{2}$$
$$= 11.3'$$

Conclusion
Average depth = 11.3'
Width = 2' (see standard drawing, Figure 14-30)
Length = 20' (see Figure 14-20)

Figure 14-26
Sanitary sewer line 2, average depth calculations

Calculation Sheet

Project: Jones job   Date: 4/22

Line 3

Average depth = $\dfrac{(A) + (B)}{2}$

(A) = (finished grade elev. - flowline elev.) + 0.5' (6" undercut, see standard drawing. Figure 14-30)
   = (100' - 90.5') + 0.5'
   = 9.5' + 0.5'
(A) = 10'

(B) = (finish grade elev. - outlet MH1 elev.) + 0.5' (6" undercut, see standard drawing. Figure 14-30)
   = (100' - 89.7') + 0.5'
   = 10.3' + 0.5'
(B) = 10.8'

Average depth = $\dfrac{10' + 10.8'}{2}$
   = $\dfrac{20.8'}{2}$
   = 10.4'

Conclusion
Average depth = 10.4'
Width = 2' (see standard drawing, Figure 14-30)
Length = 10' (see Figure 14-20)

Figure 14-27
Sanitary sewer line 3, average depth calculations

Calculation Sheet

Project: Jones job   Date: 4/22

Line 4

Average depth $= \dfrac{(A) + (B)}{2}$

(A) = (finished grade MH2 elev. - outlet MH2 elev.) + 0.5' (6" undercut, see Figure 14-30)
   = (100' - 88.7') + 0.5'
   = 11.3' + 0.5'
(A) = 11.8'

(B) = (finish grade MH3 elev. - outlet MH3 elev.) + 0.5' (6" undercut, see Figure 14-30)
   = (103.7 - 84') + 0.5'
   = 19.7' + 0.5'
(B) = 20.2'

Average depth $= \dfrac{11.8' + 20.2'}{2}$

$= \dfrac{32'}{2}$

$= 16'$

Conclusion

Average depth = 16'
Width = 2' (see standard drawing, Figure 14-30)
Length = 175' (see Figure 14-20)

Figure 14-28
Sanitary sewer line 4, average depth calculations

Quantities Take-off Sheet

Project: Jones job **Date:** 4/22

Quantities for: Sanitary sewer lines **Sheet** 1 **of** 1

By: DB **Checked:** LL **Misc:** Excavation only dry clay - 100% Proctor

Location	Length (ft)	Width (ft)	Ave. depth (ft)	Volume (CY)	Shrink (−) or swell (+) factor	Actual vol. (CY)
Line 1	145	2	4.7	50.48	+1.31	66.13
Line 2	20	2	11.3	16.74	+1.31	21.93
Line 3	10	2	10.4	7.70	+1.31	10.09
Line 4	175	2	16.0	207.41	+1.31	271.71
Total volume excavated				282.33	+1.31	369.86
Total (rounded to full CY)						370.00

Note:

Figure 14-29
Sanitary sewer lines excavation calculations summary

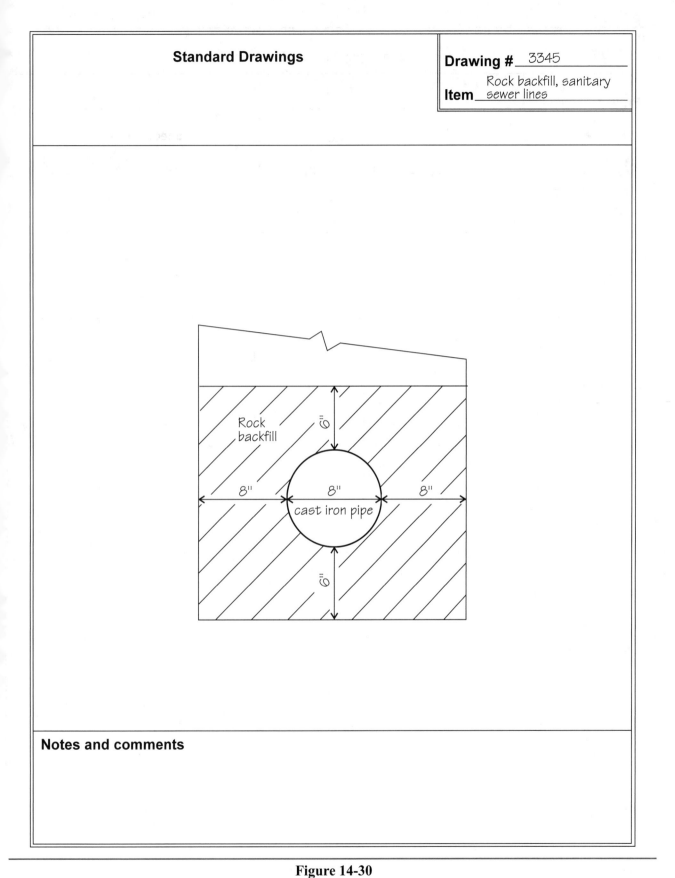

Figure 14-30
Standard drawing, sanitary sewer lines with rock backfill

Quantities Take-off Sheet

Project: Jones job **Date:** 4/23

Quantities for: Sanitary sewer lines **Sheet** 1 **of** 1

By: DB **Checked:** LL **Misc:** Dirt backfill only dry clay - 90% Proctor

Location	Length (ft)	Width (ft)	Ave. depth (ft)	Rock depth (ft)	Shrink (-) or swell (+) factor	Actual vol. (CY)
Line 1	145	2	4.7	1.67	-1.25	40.68
Line 2	20	2	11.3	1.67	-1.25	17.84
Line 3	10	2	10.4	1.67	-1.25	8.08
Line 4	175	2	16.0	1.67	-1.25	232.20
Total vol. dirt backfill						298.79
Total (rounded to full CY)						299.00

Note:
370 exc -299 backfill = 71 CY spoil

Figure 14-31
Sanitary sewer lines, dirt backfill calculations summary

Quantities Take-off Sheet

Project: Jones job **Date:** 4/23

Quantities for: Sanitary sewer lines **Sheet** 1 **of** 1

By: DB **Checked:** LL **Misc:** Rock backfill only

Location	Length (ft)	Width (ft)	Depth (ft)	Area of pipe (SF)	Rock fill vol. (CY)	
Line 1	145	2	1.67	0.35	16.06	
Line 2	20	2	1.67	0.35	2.21	
Line 3	10	2	1.67	0.35	1.11	
Line 4	175	2	1.67	0.35	19.38	
Total vol. rock backfill					38.76	
Total (rounded to full CY)					39.00	

Note:

Figure 14-32
Sanitary sewer lines, rock backfill calculations summary

Calculation Sheet

Project: <u>Jones job</u> Date: <u>4/23</u>

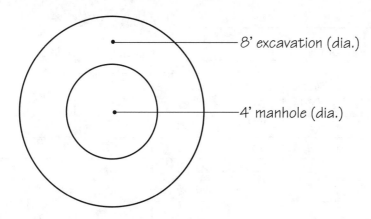

Area of circle = πr^2
$\pi = 3.14$
$r = \dfrac{dia.}{2}$

Manhole area = 3.14×2^2
= 12.6 SF

Excavation area = 3.14×4^2
= 50.24 SF

Areas are for each foot of manhole depth

Total manhole depth = finished elev. - outlet elev. + 1'

Conclusion

Figure 14-33
Manhole area calculations

Quantities Take-off Sheet

Project: Jones job **Date:** 4/23

Quantities for: Manhole volumes **Sheet** 1 **of** 1

By: DB **Checked:** LL **Misc:** Excavation only

Location	Elevation difference* (ft)	Depth** (ft)	Excavation area (SF)	Exc. vol. (CY)	Shrink (−) or swell (+) factor	Actual vol. (CY)
No. 1	100−89.7 = 10.3	11.3	50.24	21.03	+1.31	27.55
No. 2	100−88.7 = 11.3	12.3	50.24	22.89	+1.31	29.99
No. 3	103.7−84 = 19.7	20.7	50.24	38.52	+1.31	50.46
Total volume excavated						108.01
Total (rounded to full CY)						108.00

Note:

*Elevation difference = finished elevation − outlet elevation

**Depth = elevation difference + 1'

Figure 14-34
Manhole excavation volume calculations

Quantities Take-off Sheet

Project: Jones job **Date:** 4/23

Quantities for: Manhole volumes **Sheet** 1 **of** 1

By: DB **Checked:** LL **Misc:** Backfill only 90% Proctor

Location	Depth (ft)	Excavation (SF)	Manhole (MH) area (SF)	Exc. area - MH area (SF)	Backfill vol (CY)	Shrink (-) or swell (+) factor	Actual vol. (CY)
No. 1	11.3	50.24	12.6	37.64	425.33	-1.25	19.69
No. 2	12.3	50.24	12.6	37.64	462.97	-1.25	21.43
No. 3	20.7	50.24	12.6	37.64	779.15	-1.25	36.07
Total vol. dirt backfill							77.19
Total (rounded to full CY)							77.00

Note:
108 exc - 77 backfill = 31 CY spoil

Figure 14-35
Manhole backfill calculations summary

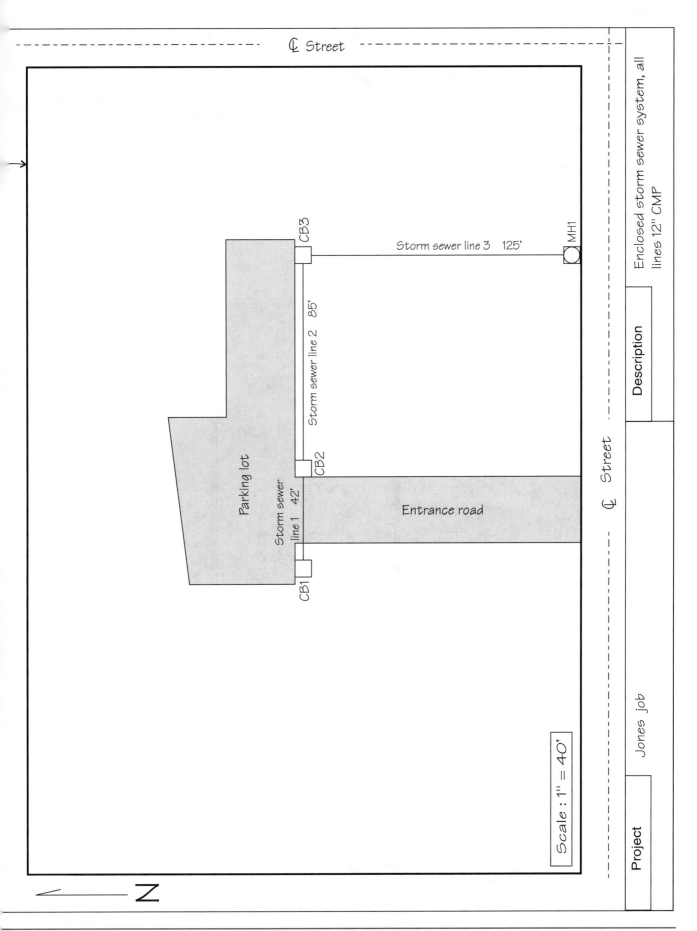

Figure 14-36
Enclosed storm sewer system, general plan sheet

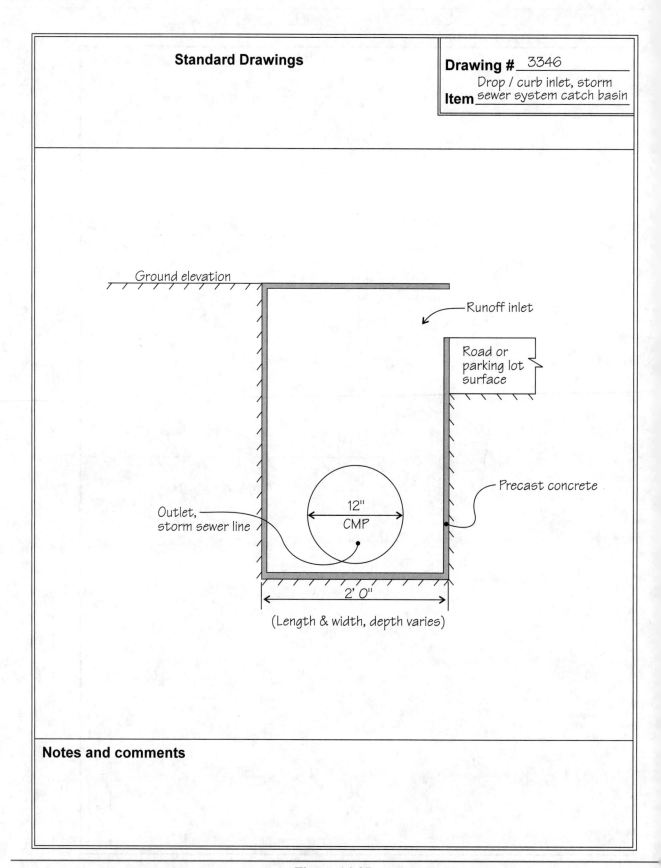

Figure 14-37
Standard drawing, section view, catch basin drop / curb inlet

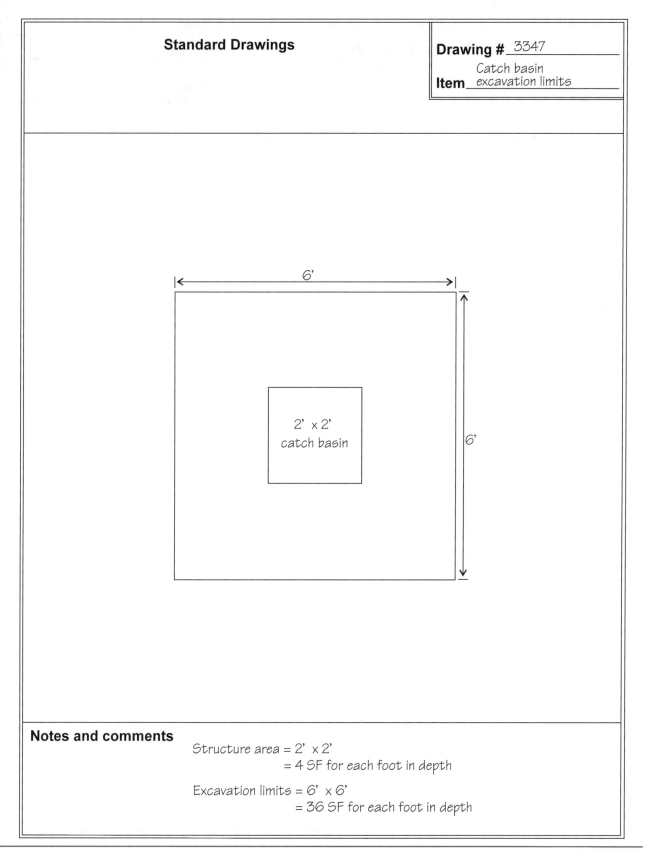

Figure 14-38
Catch basin excavation limits

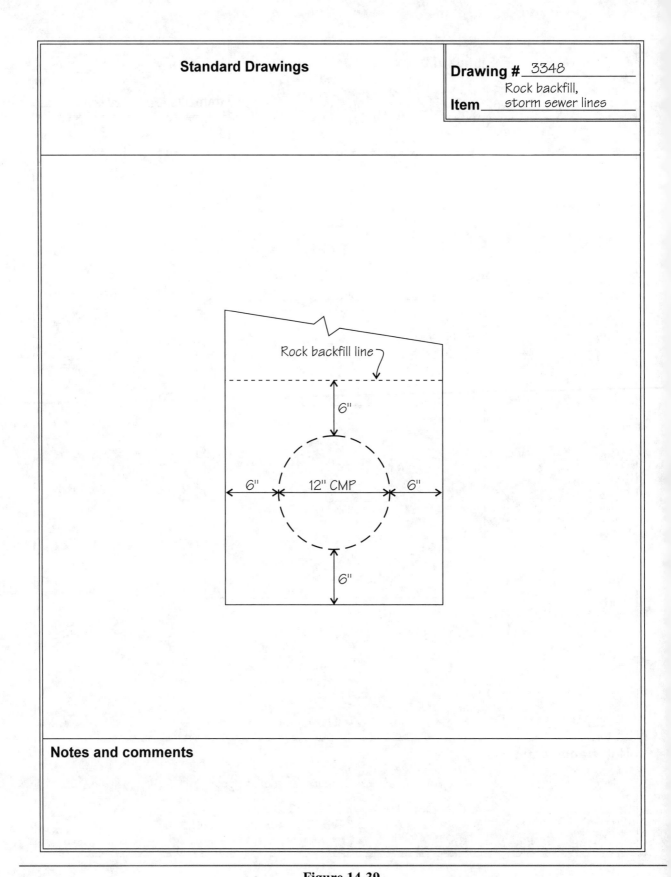

Figure 14-39
Standard drawing, storm sewer lines with rock backfill

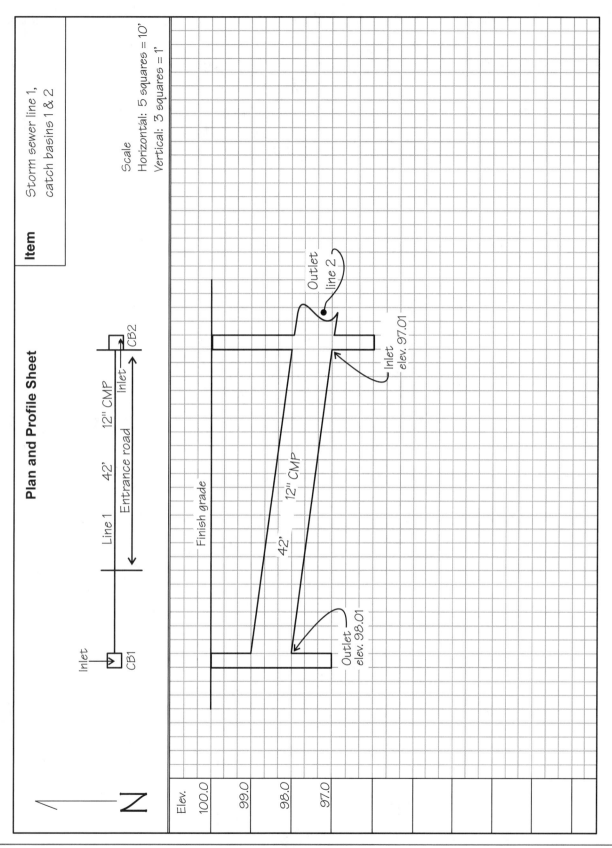

Figure 14-40
Storm sewer line 1, plan and profile sheet

A Sample Take-off

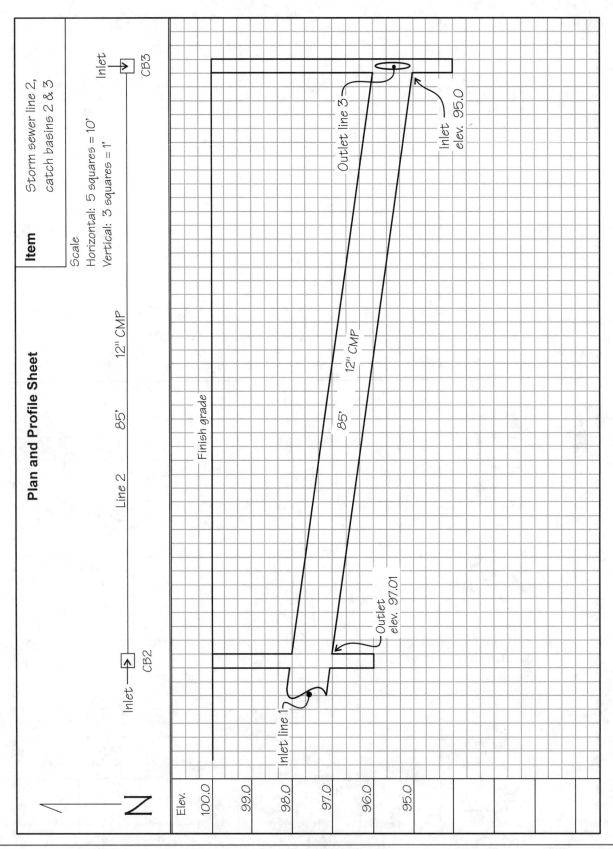

Figure 14-41
Storm sewer line 2, plan and profile sheet

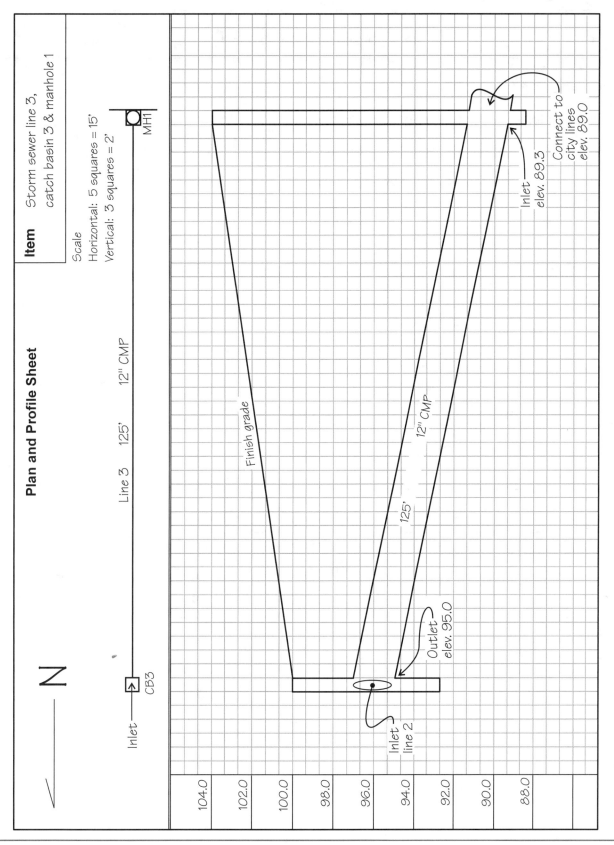

Figure 14-42
Storm sewer line 3, plan and profile sheet

A Sample Take-off

Quantities Take-off Sheet

Project: Jones job **Date:** 4/23

Quantities for: Structure volumes **Sheet** 1 **of** 1

By: DB **Checked:** LL **Misc:** Excavation only

Location	Elev. dif.* (ft)	Depth** (ft)	Struct. exc. area (SF)	Volume (CY)	Shrink (−) or swell (+) factor	Actual vol. (CY)
Catch basin no. 1 (CB1)	100−98 = 2	3	36	4	+1.31	5.24
Catch basin no. 2 (CB2)	100−97 = 3	4	36	5.33	+1.31	6.98
Catch basin no. 3 (CB3)	100−95 = 5	6	36	8	+1.31	10.48
Manhole no. 1 (MH1)	104−89 = 15	16	50.24	29.77	+1.31	39.00
Total volume excavated						61.70
Total (rounded to full CY)						62.00

Note:

*Elev. dif. = finished elev. − outlet elev.

**Depth = elev. dif. + 1'

Figure 14-43
Storm sewer system catch basins, excavation calculations summary

Quantities Take-off Sheet

Project: Jones job   **Date:** 4/23

Quantities for: Structure volume   **Sheet** 1 **of** 1

By: DB   **Checked:** LL   **Misc:** Backfill only 90% Proctor

Location	Depth (ft)	Struct. area (SF)	Exc. area (SF)	Backfill area* (SF)	Vol. (CY)	Shrink (-) or swell (+) factor	Actual vol. (CY)
Catch basin no. 1 (CB1)	3	4	36	32	3.56	-1.25	4.45
Catch basin no. 2 (CB2)	4	4	36	32	4.74	-1.25	5.93
Catch basin no. 3 (CB3)	6	4	36	32	7.11	-1.25	8.89
Manhole no. 1 (MH1)	16	12.6	50.24	37.64	22.31	-1.25	27.89
Total vol. dirt backfill							47.16
Total (rounded to full CY)							47.00

Note:
*Backfill area = exc. area - struct. area
62 exc - 47 backfill = 15 CY spoil

Figure 14-44
Storm sewer system catch basins, backfill calculations summary

Calculation Sheet

Project: Jones job **Date:** 4/23

Storm sewer line 1
  run = CB1 to CB2

$$\text{Average depth} = \frac{(A) + (B)}{2}$$

(A) = (finished grade elev. - outlet CB1 elev.) + 0.5' (6" rock fill see standard drawing, Figure 14-39)
   = (100' - 98.01') + 0.5'
   = 1.99' + 0.5'
(A) = 2.49'

(B) = (finish grade elev. - outlet CB2 elev.) + 0.5' (6" rock fill see standard drawing, Figure 14-39)
   = (100' - 97.01') + 0.5'
   = 2.99' + 0.5'
(B) = 3.49'

$$\text{Average depth} = \frac{2.49' + 3.49'}{2}$$
$$= \frac{5.98'}{2}$$
$$= 2.99'$$

Conclusion
Average depth = 2.99'
Width = 2' (see standard drawing, Figure 14-39)
Length = 42' (see Figure 14-36)

Figure 14-45
Storm sewer line 1, average depth calculations

Calculation Sheet

Project: _Jones job_ Date: _4/24_

Storm sewer line 2
  run = CB2 to CB3

$$\text{Average depth} = \frac{(A) + (B)}{2}$$

(A) = (finished grade elev. - outlet CB2 elev.) + 0.5' (6" rock fill see standard drawing, Figure 14-39)
  = (100' - 97.01') + 0.5'
  = 2.99' + 0.5'
(A) = 3.49'

(B) = (finish grade elev. - outlet CB3 elev.) + 0.5' (6" rock fill see standard drawing, Figure 14-39)
  = (100' - 95.01') + 0.5'
  = 4.99' + 0.5'
(B) = 5.49'

$$\text{Average depth} = \frac{3.49' + 5.49'}{2}$$
$$= \frac{8.98'}{2}$$
$$= 4.49'$$

Conclusion
Average depth = 4.49'
Width = 2' (see standard drawing, Figure 14-39)
Length = 85' (see Figure 14-36)

Figure 14-46
Storm sewer line 2, average depth calculations

Calculation Sheet

Project: _Jones job_ Date: _4/24_

Storm sewer line 3
 run = CB3 to MH1

$$\text{Average depth} = \frac{(A) + (B)}{2}$$

(A) = (finished grade elev. - outlet CB3 elev.) + 0.5' (6" rock fill see standard drawing, Figure 14-39)
 = (100' - 95.01') + 0.5'
 = 4.99' + 0.5'
(A) = 5.49'

(B) = (finish grade elev. - outlet MH1 elev.) + 0.5' (6" rock fill see standard drawing, Figure 14-39)
 = (100' - 89') + 0.5'
 = 15' + 0.5'
(B) = 15.5'

$$\text{Average depth} = \frac{5.49' + 15.5'}{2}$$
$$= \frac{20.99'}{2}$$
$$= 10.5'$$

Conclusion
Average depth = 10.5'
Width = 2' (see standard drawing, Figure 14-39)
Length = 125' (see Figure 14-36)

Figure 14-47
Storm sewer line 3, average depth calculations

Quantities Take-off Sheet

Project: Jones job   **Date:** 4/24

Quantities for: Storm sewer lines   **Sheet** 1 **of** 1

By: DB   **Checked:** LL   **Misc:** Excavation only dry clay

Location	Length (ft)	Width (ft)	Average depth (ft)	Volume (CY)	Shrink (-) or swell (+) factor	Actual vol. (CY)
Line 1	42	2	2.99	9.3	+1.31	12.19
Line 2	85	2	4.49	28.27	+1.31	37.03
Line 3	125	2	10.5	97.22	+1.31	127.36
Total volume excavated						176.58
Total (rounded to full CY)						177.00

Note:

Figure 14-48
Storm sewer lines, excavation calculations summary

A Sample Take-off **323**

Quantities Take-off Sheet

Project: Jones job   **Date:** 4/24

Quantities for: Storm sewer lines   **Sheet** 1 **of** 1

By: DB   **Checked:** LL   **Misc:** Dirt backfill only dry clay 90% Proctor

Location	Length (ft)	Width (ft)	Ave. depth (ft)	Rock depth (ft)	Shrink (-) or swell (+) factor	Actual vol. (CY)
Line 1	42	2	2.99	2	-1.25	3.85
Line 2	85	2	4.49	2	-1.25	19.6
Line 3	125	2	10.50	2	-1.25	98.38
Total vol. dirt backfill						121.83
Total (rounded to full CY)						122.00

Note:
177 exc - 122 backfill = 55 CY spoil

Figure 14-49
Storm sewer lines, dirt backfill calculations summary

Quantities Take-off Sheet

Project: Jones job **Date:** 4/24

Quantities for: Storm sewer lines **Sheet** 1 **of** 1

By: DB **Checked:** LL **Misc:** Rock backfill

Location	Length (ft)	Width (ft)	Depth (ft)	Area of pipe (SF)	Rock fill vol. (CY)	
Line 1	42	2	2	0.78	5.01	
Line 2	85	2	2	0.78	10.14	
Line 3	125	2	2	0.78	14.91	
Total vol. rock backfill					30.06	
Total (rounded to full CY)					30.00	

Note:

Rock backfill volume (CF) = (width × depth − area of pipe) × length.
CY = CF divided by 27

Figure 14-50
Storm sewer lines, rock backfill calculations summary

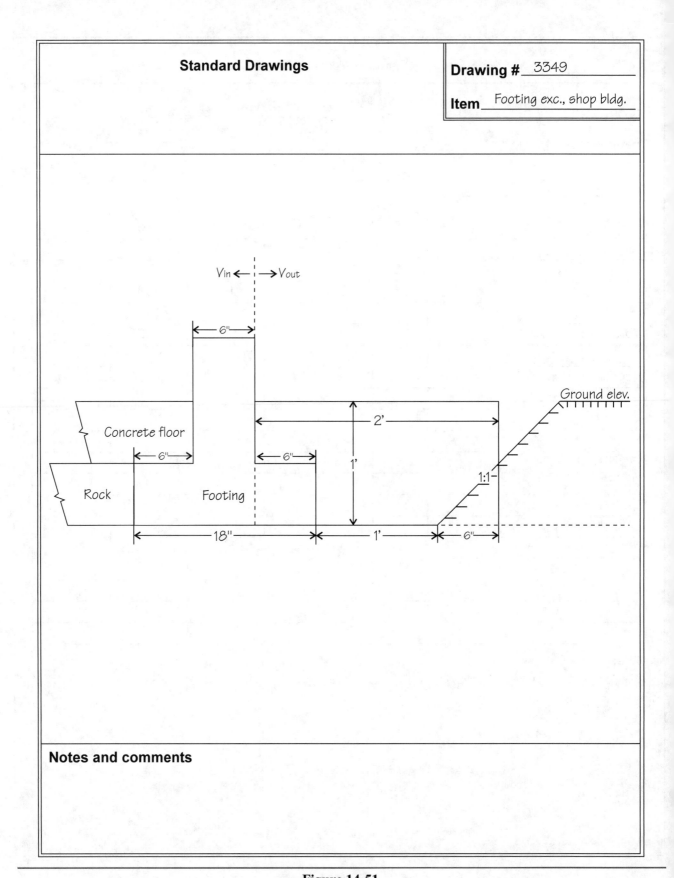

Figure 14-51
Shop building, footing excavation detail

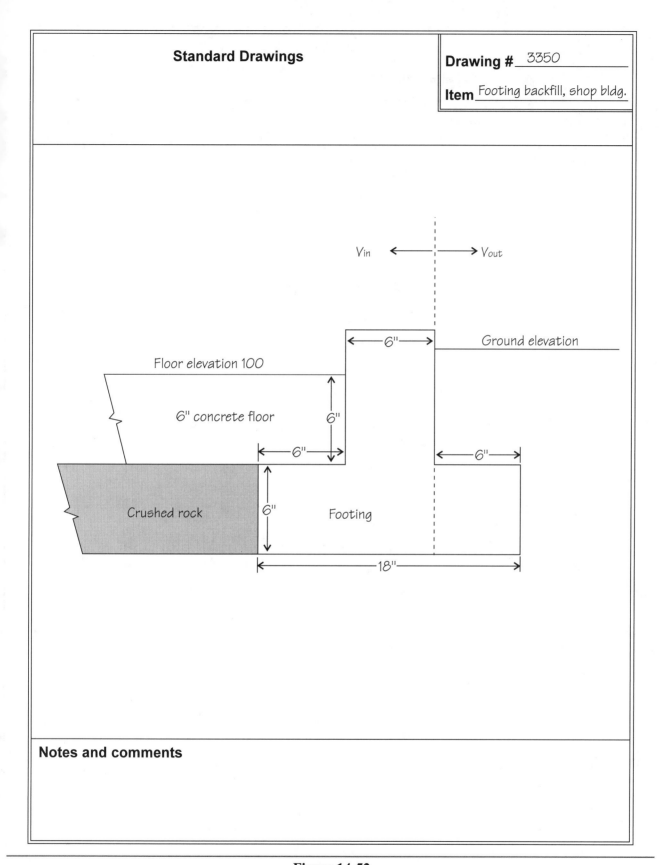

Figure 14-52
Shop building, footing backfill detail

Calculation Sheet

Project: __Jones job__ Date: __4/24__

Shop building dimensions: 80' x 40' (see Figure 14-2)

$$V_{out}\ (CY) = \frac{\text{excavation vol. (CF)}}{27} + \frac{\text{total exc. corner volume (CF)}}{27}$$

excavation vol. (CF) = length x width x depth

total exc. corner volume (CF) = ($^1/_3\ \pi r^2$) x depth x # of corners

$$V_{out}\ (CY) = \frac{(80 + 80 + 40 + 40) \times 2 \times 1}{27} + \frac{0.3333 \times 3.1416 \times 1^2 \times 1 \times 4}{27}$$

$$= 17.8 + 0.16$$
$$= 17.96\ CY$$

$$V_{in}\ (CY) = \frac{\text{excavation vol. (length x width x depth)}}{27}$$

$$= \frac{80 \times 40 \times 1}{27}$$

$$= 118.52\ CY$$

$$\text{Rock backfill (CY)} = \frac{V_{in}\ \text{vol. (CF)}}{27} - \frac{\text{interior concrete footing vol. (CF)}}{27}$$

V_{in} vol. = exc. length x exc. width x (exc. depth – depth concrete floor) (6", see Figure 14-52)

interior concrete footing vol. (see Figure 14-52) = length x width x depth

$$\text{Rock backfill (CY)} = \frac{80 \times 40 \times (1 - 0.5)}{27} - \frac{240 \times 1 \times 0.5}{27}$$

$$= \frac{1{,}600}{27} - \frac{120}{27}$$

$$= 59.26 - 4.44$$
$$= 54.82\ CY$$

$$\text{Dirt backfill (CY)} = V_{out}\ (CY) - \frac{\text{exterior concrete footing vol. (CF)}}{27}$$

V_{out} = 17.96 CY (see above)

exterior concrete footing vol. (see Figure 14-52) = length x width x depth

$$\text{Dirt backfill (CY)} = 17.96 - \frac{240 \times 0.5 \times 0.5}{27}$$

$$= 17.96 - 2.22$$
$$= 15.74\ CY$$

Conclusion

V_{out} = 17.96 CY
V_{in} = 118.52 CY
Rock backfill = 54.82 CY
Dirt backfill = 15.74 CY

Figure 14-53
Shop building footing excavation, and backfill calculations

Quantities Take-off Sheet

Project: Jones job **Date:** 4/24

Quantities for: Shop footing **Sheet** 1 **of** 1

By: DB **Checked:** LL **Misc:**

Item		Exc volume (CY)	Backfill, dirt (d) or rock (r) vol. (CY)	Shrink (−) or swell (+) factor	Actual vol. (CY)	Spoil vol. (CY)
V_{out}		17.96		+1.31	23.53	
V_{out}			15.74 (d)	−1.25	19.68	3.85
V_{in}		118.52		+1.31	155.26*	155.26*
V_{in}			54.82 (r)		54.82	
Total volume excavated		136.48		+1.31	178.79	
(rounded to full CY)					179.00	
Total volume dirt backfill			15.74 (d)	−1.25	19.68	
(rounded to full CY)					20.00	
Total volume rock backfill			54.82 (r)		54.82	
(rounded to full CY)					55.00	
Total volume spoil						159.11
(rounded to full CY)						159.00

Note:
*All V_{in} excavated material is spoil because only rock backfill is used for the V_{in}.

Figure 14-54
Shop building, footing calculations summary

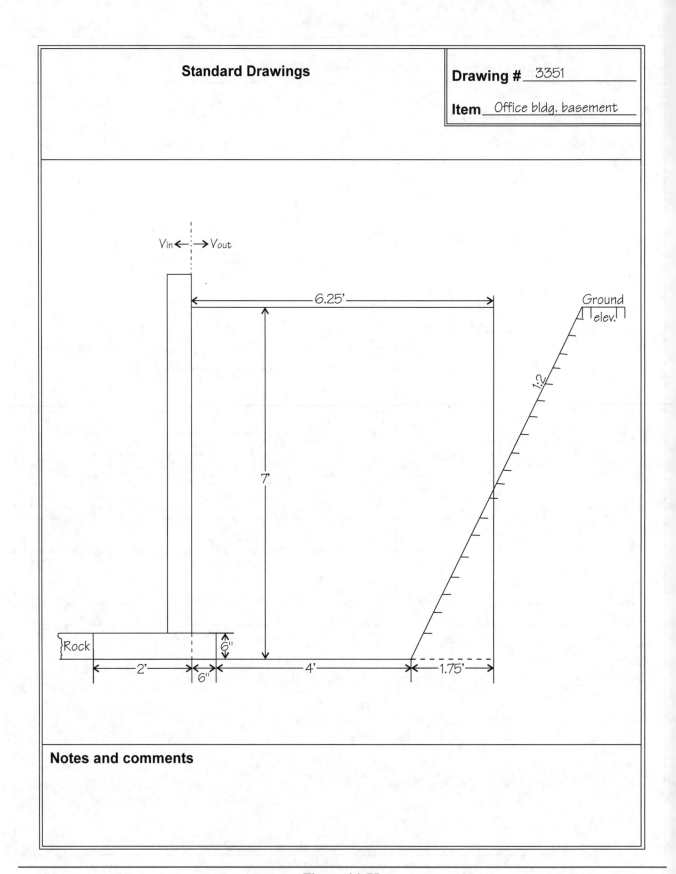

Figure 14-55
Office building basement, excavation and backfill details

330 *Estimating Excavation*

Calculation Sheet

Project: Jones job **Date:** 4/24

Office building dimensions: 55' x 50' (see Figure 14-2)

$$V_{out} = \frac{\text{perimeter} \times \text{average slope line area (SF)}}{27} + \frac{\text{total exc. corner volume (CF)}}{27}$$

average slope line area (SF) = (total width) x depth

total exc. corner volume (CF) = ($\frac{1}{3}\pi r^2$) x depth x # of corners

$$V_{out}\ (CY) = \frac{(50 + 50 + 55 + 55) \times (0.5 + 4 + 1.75) \times 7}{27} + \frac{0.3333 \times 3.1416 \times 3.5^2 \times 7 \times 4}{27}$$

$$= \frac{9{,}187.5 + 359.1536}{27}$$

$$= 340.28 + 13.3$$

$$= 353.58\ CY$$

$$V_{in}\ (CY) = \frac{\text{excavation vol. (length x width x depth)}}{27}$$

$$= \frac{55 \times 50 \times 7}{27}$$

$$= \frac{19{,}250}{27}$$

$$= 712.96\ CY$$

$$\text{Rock backfill (CY)} = \frac{V_{in}\ \text{backfill vol. (CF)}}{27}$$

V_{in} backfill vol. = (length - footing width) x (width - footing width) x depth

$$\text{Rock backfill (CY)} = \frac{(55 - 2) \times (50 - 2) \times 0.5}{27}$$

$$= \frac{1{,}272}{27}$$

$$= 47.11\ CY$$

$$\text{Dirt backfill (CY)} = V_{out}\ (CY) - \frac{\text{concrete footing vol. (CF)}}{27}$$

V_{out} = 353.58 CY (see above)

concrete footing vol. = length x width x depth

$$\text{Dirt backfill (CY)} = 353.58 - \frac{210 \times 0.5 \times 0.5}{27}$$

$$= 353.58 - 1.94$$

$$= 351.64\ CY$$

Conclusion

V_{out} = 353.58 CY
V_{in} = 712.96 CY
Rock backfill = 47.11 CY
Dirt backfill = 351.64 CY

Figure 14-56
Office building, basement excavation and backfill calculations

Quantities Take-off Sheet

Project: Jones job   **Date:** 4/24

Quantities for: Shop footing   **Sheet** 1 **of** 1

By: DB   **Checked:** LL   **Misc:**

Item	Exc. volume (CY)	Backfill, dirt (d) or rock (r) vol. (CY)	Shrink (-) or swell (+) factor	Actual vol. (CY)	Spoil vol. (CY)
V_{out}	353.58		+1.31	463.19	
V_{out}		351.64 (d)	-1.25	439.55	23.64
V_{in}	712.96		+1.31	933.97*	933.97*
V_{in}		47.11 (r)		47.11	
Total volume excavated	1,066.54		+1.31	1,397.17	
(rounded to full CY)				1,397.00	
Total volume dirt backfill		351.64 (d)	-1.25	439.55	
(rounded to full CY)				440.00	
Total volume rock backfill		47.11 (r)		47.11	
(rounded to full CY)				47.00	
Total volume spoil					957.61
(rounded to full CY)					958.00

Note:
*All V_{in} excavated material is spoil because only rock backfill is used for the V_{in}.

Figure 14-57
Office building, basement excavation and backfill calculations summary

Quantities Take-off Sheet

Project: Jones job **Date:** 4/24

Quantities for: Summary sheet **Sheet** 1 **of** 1

By: DB **Checked:** LL **Misc:** All quantities rounded to full CY

Item	Fill vol. (CCY†)	Exc. vol. (LCY‡)	Dirt bkfill (CCY)	Spoil vol. (LCY)	Rock fill vol. (CY)	Topsoil vol.* (CY)	Notes
Topsoil (top 2")						504	Non-usable material
Topsoil (usable)						1,007	Stockpile & replace
Topsoil (total to replace)						1,103	Need additional 96 CY
Grid square site take-off	67	17,111		17,041			
Service road					120		
Entrance road		192		192			
Parking lot		99		99			
Sanitary sewer all lines		370	299	71	39		
Sanitary sewer all mnhls.		108	77	31			
Storm sewer all struct.		62	47	15			
Storm sewer all lines		177	122	55	30		
Shop footing/floor area		179	20	159	55		
Office bldg. basement		1,397	440	958	47		
Totals	67	19,695	1,005	18,621	291	2,614	

Note:

†CCY = compact cubic yards, shrink factor (x 1.25) applied
‡LCY = loose cubic yards, swell factor (x 1.31) applied
*No shrink or swell factor applied to topsoil quantities

Figure 14-58
Project summary sheet

15
Costs and Final Bid for the Sample Estimate

In the last chapter we did a quantity take-off for a sample project. Now we'll take it one step further. In this chapter, we'll price the work. I'll show how to use the estimated quantities and costs to create an accurate bid and win a profitable contract.

As an excavation subcontractor, most of your bids will be submitted to general contractors. On every bid, the general contractor has to decide what work will be subcontracted and what work will be done with the contractor's crews and equipment. That's an important decision — the choice of subcontractors can make or break the contractor. When the general decides not to do the excavation with his company's crews, he'll request bids from subcontractors who specialize in that work.

The general contractor signs the contract with the owner to complete the project according to the plans and specifications. In doing so, he's accepting responsibility for the entire project, even though subcontractors are doing much of the work. The owner pays the general contractor as work is completed and the general contractor disburses payments to subcontractors and suppliers.

Some general contractors are no more than "paper contractors." They own no construction equipment and have no work crews. Instead, they subcontract it all, just administering and coordinating the job with subcontractors who actually do the work. Some states now restrict that type of contracting. They require that the general contractor's crews handle most of the work on a project.

The trend in the construction industry for at least the last 20 years has been toward specialization. More and more contractors and subcontractors are specializing in certain types of work. Fewer and fewer general contractors routinely handle the whole job, from excavation, concrete work and masonry to carpentry. When I began my career in construction, most larger general contracting companies owned at least some earthmoving equipment. Today, because of the high cost of equipment (and the high wages paid to experienced operators), even the larger construction companies leave excavation work to specialists.

That's good news for excavation contractors. You'll have a chance to bid on most projects that require excavation equipment. But with the good news comes extra responsibility. You're bidding because the general contractor expects you'll do the work faster, better and at a lower cost. If you can do that, keeping your equipment busy should be no problem.

The Bid Preparation Process

The best way to explain how to price an estimate is to work through an example. Most of this chapter is that example, based on the take-off in Chapter 14. Study the details in this sample estimate, use the information I've presented earlier in the book, and you'll have no trouble mastering the skills required for excavation estimating.

We're preparing this sample estimate for J. Q. Corporation, a land development company. J. Q. Corp. bought some property last year. Now they're developing the northwest corner as a manufacturing plant and office complex. A major regional manufacturer has agreed to buy the building under a turnkey contract once construction is completed.

J. Q. Corp. has asked our company, Quality Construction, to submit a bid for the sitework. J. Q. Corp. has told all bidders that they can dispose of excess material in an adjoining area. (You can look ahead to area "A" on Figure 15-C-8, if you're interested.) Temporary storage space for topsoil is available on the same property. The only condition is that topsoil in the storage area has to be stripped off before stockpiling and then replaced when the job is finished. Any additional topsoil needed can be taken from this area. This is a condition of the bid, not part of it.

Quality Construction is a small excavation and site improvement contracting company that handles work in this size range. Quality does most of its own excavation and has late model equipment. Though the equipment fleet is small, it's all in good condition.

In this sample estimate, we're going to work backwards. First, you can see the bid we prepared, and then study the supporting documents to understand how we arrived at the figures.

The Scope of Work

We've reviewed the specifications and plans and have visited the job site. Figure 15-A-1 is the bid summary we've prepared for submission to J. Q. Corp.

The Bid Sheet

Figure 15-B-1 is our worksheet for compiling the bid. It shows bid prices for the individual line items and who will do the work — either Quality or a subcontractor. The column headed *By* shows either the name of our sub or Self, meaning Quality will do the work. The *Unit Cost* column shows Quality's cost for each line item. Notice the *Profit* column. In this case, Quality decided to add a 7 percent profit to each line item. How much you add depends on competitive conditions and the market for excavation work.

Let's review each of the bid items in Figure 15-B-1 so you can see how these costs were developed.

Mobilization

This is the lump sum cost of moving an office trailer, equipment and materials to the job site and getting ready to work. It may also include getting utilities hooked up. On small projects, this cost may be insignificant. On larger jobs in remote areas, it may be substantial. This amount will be included in the first payment from the general contractor. A unit cost of $500 is appropriate for this job, so the bid total including profit is $535.

Clearing and grubbing

Very little clearing or grubbing is required on this job. Normally clearing and grubbing includes removing trees and brush or existing structures, usually calculated on a square yard basis. On this project, we only need to clear a few small trees and some light brush. I've allowed a lump sum of $535.

Topsoil

This includes all the required topsoil work. Topsoil requires special consideration because Quality has to separate the material and move it twice. Quality will save the good topsoil and dispose of the waste portion off site. On this job there are 504 CY of unusable material to haul off, and 1,007 CY of usable material to store and then return. We'll need an additional 96 CY from off site to total the 1,103 CY we need for total replacement material. That totals 2,614 CY of material to move for the topsoil item. On some jobs there's no available on-site space for topsoil storage. Then you'll have to haul the topsoil to temporary off-site storage. Later, after construction's finished, you haul the topsoil back to the site. As we saw earlier, this job site has a storage area for topsoil. As a result we can summarize all the topsoil costs on just one line.

You can follow along with my calculations for this job on Figures 15-C-1 through C-7. Figure C-1 shows the job site, the off-site stockpile site (A) and the topsoil center of mass (CM). Since the average depth of removal and replacement is the same throughout the project, the center of mass is the center of the project. Figures C-2 and C-3 show the production output for the machines we'll use and establish the haul time of 0.11 minute when loaded. Figure C-4 is the production output for the empty segment. The result is an hourly production rate of 472 CY. But look at Figure C-7. Even though the theoretical rate would be about one day, I'm using two days because of the difficulty of maneuvering in the small area.

My actual cost comes to $.86 per CY. When I adjust that for my 7 percent profit, I'll bid it at $.92 per CY.

Earthwork, cut, general

This is the largest cost on this project. Notice that cut and fill are separated into two categories because the cost and volume of each is different. The general cut volume is 17,111 CY. This includes the 99 CY of additional excavation for the parking lot, which isn't a separate bid item. We'll use 67 CY of the total for on-site fill. That leaves 17,044 to haul off and dispose of. Figures 15-D-1 through D-10 show how I arrived at a unit cost of $.80 per CY, or $.86 with my profit.

Earthwork, fill, general

The procedure for calculating this volume is the same as the procedure for calculating cut. This is 67 CY fill at the $.86 unit cost.

Earthwork, cut, roads

Calculations are the same as for cut and fill. We have 192 CY of cut for roads at the same unit cost.

Utility trenches

This is the usual way of bidding trench work, with a separate line for each range of depths. Here depths are in ranges of 5 feet. The range you'll select depends on terrain, type of project, soil conditions and design data. The range might be 1 to 2 feet for a small job where accuracy is important. On a larger job with high capacity trenching equipment, the range might be 10 or 20 feet. Generally, ranges of 5 feet will be accurate enough. Precision usually isn't necessary. It's not uncommon for grades to be raised or lowered during excavation to avoid obstacles.

To find the specified sections, simply review each section of pipe to locate where the trench depth goes from one depth limit to another. When you know the length of each section, take it off from the plans. We'll only account for excavation time in this part of the estimate. Figures 15-E-1 through E-5 show my calculations for the cost of each segment of trench.

In this section, trenching is kept separate from the other production costs. The excavator time includes trenching only. An equal amount of crew time is shown for the excavator for placing bedding and pipe, moving the trench box, and other miscellaneous work. Estimated cycle times and production amounts are from the charts in the back of this chapter and the general production figures we discussed earlier in this book.

Throughout this section, I've increased times slightly to allow for unanticipated delays and inefficiency. These estimates are based on my experiences but even with no experience you can anticipate some of the potential problems.

All material prices are shown in the worksheets. Crew times on the trench compactor and small crawler tractor are split half-and-half. Use the procedures from Chapter 13 to calculate the time and costs, including the storm sewer lines.

All of the costs to this point will be handled with Quality crews and equipment.

AB-3 rock bedding

The base rock is the material used as bedding under pipes and structures and as a base for the service road. There will also be base rock in parking areas, but this cost is included in the asphalt placement bid.

Asphalt surface

Because Quality doesn't do this type of work, we got a bid from a subcontractor. The bid assumes that the base is prepared to within 0.1 foot, plus or minus, before asphalt is placed.

Cast iron pipe

Quality will install this pipe. Figure 15-B-1 lists the name of the pipe supplier. The bid price must include pipe, rock bedding and backfill including labor and equipment cost. The price in the By column is the material-only price. Figures 15-F and 15-G show my calculations for the utility line structures, and storm and sanitary sewer line work. You'll find the price workup for cast iron pipe, corrugated metal pipe, and the precast manholes and catch basins.

Precast manholes and catch basins

Like the utility trenches these structures are also grouped into ranges based on depth. These bid prices include all the installation costs, equipment, materials and labor. I like to list the material-only price for each precast structure in the *By* column.

Type G curb and gutter

Because Quality doesn't do concrete work, we got a bid from a concrete subcontractor on items 20, 21, and 22. The bid includes concrete materials and labor only. Quality will do the necessary excavation.

Shop building footing

Quality will dig the footing and haul it off as spoil, using the track excavator and one-wheel scraper. Although the job is small, the cycle time is fairly long because of the loading and maneuver times. From the cycle time and bucket capacity, we can calculate the volume as shown in Figure H-1. Since the bid forms request a bid per linear foot, we figure the cost of excavation and translate it into linear feet.

Costs and Final Bid for the Sample Estimate **339**

Sands Construction, operating as our subcontractor, has been awarded a contract for placing of the concrete footing at $11.90 per linear foot.

Office building footing and walls

In this situation, the amount of material to be excavated for the basement and work area, and backfill material, aren't separate bid items. We figure the excavation cost and convert it into a unit cost per linear foot. Quality will do the excavation with the track loader and use a one-wheel scraper to haul off spoil. Cycle time is relatively long because the loader must move into the pit, load, and come back out to fill the scraper. The calculations for cycle time and bucket capacity in Figure 15-I-1 show the time required. I figured a time of 9.5 hours to do this. I've decided that this excess material can be done at the same time as the general excavation, so we'll use the compactor and grader with no additional time charged to this line item. Figure 15-I-2 shows the results.

Sands Construction will do concrete work in the basement at $31.00 per linear foot.

Overhead

Figure J-1 shows Quality's annual overhead cost calculations. Again, pay attention to the method, not the actual costs. They may be entirely different from your costs.

Machine Selection

We've gone through the estimate step-by-step, calculating quantities and costs, then preparing the bid prices. But it took a lot of work to get to the point where we could do that. Before it's possible to figure actual costs, we have to have accurate machine owing and operating costs. Then we can begin planning for equipment. We need to know what machines are needed. Do we have the equipment needed for this job or will we have to rent other equipment for better efficiency?

I've used the forms and procedures from Chapter 13 and the data from the bid list. But I have one disclaimer to make: The costs here may not be accurate for the work you do. I've tried to make this example realistic. But the prices are my prices, not yours. Concentrate on the process, not the bid costs. Labor, fuel and other costs vary widely throughout the country. Your cycle times and production figures may not be even close to the figures I'm using. *In your bids, always rely on your current costs and accurate production figures for your machines.*

Most construction equipment manufacturers provide manuals, guidelines and other publications that contain the exact specifications for their machines. Cycle time, hauling and maneuvering time, bucket capacity, gear ranges and power available are all available from these books. Don't be afraid to ask for all the information they have.

Ownership and Operating Cost

Using the individual machine cost and the total ownership and operating cost of all machines in inventory, we can calculate each machine's hourly cost. That leads to the line item cost, based on production and time used.

My calculations are in Figures 15-K and L. Each machine has been assigned a company number. Figure 15-K contains 40 pages of specific operating and ownership information for each machine used on this project. Be sure you've mastered the information in Chapter 13 before following along on this portion of the bid. It's unlikely that any contractor will have all the needed equipment on hand — or use all the equipment that's available. The total machine cost includes rental on leased machines and the expense for owned machines that aren't used. Figure 15-L-1 shows a summary of the O&O costs for each machine.

Bid Sheet

Project: C-17   **No:** _____

Owner: J.Q. Corp.

Item	Quantity	Unit	Cost per unit	Total cost
Mobilization	1	LS	535.00	535.00
Clearing / grubbing	1	LS	535.00	535.00
Topsoil	2,614	CY	.92	2,404.88
Earthwork, cut, general	17,111	CY	0.86	14,715.46
Earthwork, fill, general	67	CY	0.86	57.62
Earthwork, cut, roads	192	CY	0.86	165.12
Utility trenches, 24" W, 1 - 5' D	272	LF	0.50	136.00
Utility trenches, 24" W, 6' - 10' D	135	LF	1.62	218.70
Utility trenches, 24" W, 11' - 15' D	20	LF	2.71	54.20
Utility trenches, 24" W, over 15' D	175	LF	2.55	446.25
AB-3 rock bedding, roads & parking lots	600	Ton	Included in individual line item costs	
Asphaltic surface in place	276	Ton	34.78	9,599.28
8" C.I.P. in place	350	LF	12.42	4,347.00
12" C.M.P. in place	252	LF	14.86	3,744.72
Precast manhole set 10' to 13' D	2	Ea	4,980.87	9,961.74
Precast manhole set 13' to 21' D	2	Ea	7,121.41	14,242.82
Precast catch basin set 2' to 4' D	2	Ea	1,353.36	2,706.72
Precast catch basin set 4' to 6' D	1	Ea	2,493.61	2,493.61
Type "G" curb & gutter	405	LF	5.53	2,239.65
Shop building footing 1' - 1-1/2'	240	LF	20.43	4,903.20
Office building walls/footings	210	LF	46.04	9,668.40
			JOB TOTAL	83,175.37

Figure 15-A-1
Bid summary

Bid Preparation Form

Item	By	Unit	Quantity	Unit cost	Profit	Bid price
Mobilization	Self	LS	1	500.00	35.00	535.00
Clearing / grubbing	Self	LS	1	500.00	35.00	535.00
Topsoil	Self	CY	2,614	0.86	0.06	0.92
Earthwork, cut, general	Self	CY	17,111	0.80	0.06	.86
Earthwork, fill, general	Self	CY	67	0.80	0.06	.86
Earthwork, cut, roads		CY	192	0.80	0.06	.86
Utility trenches, 24" W, 1 - 5' D	Self	LF	272	0.47	0.03	.50
Utility trenches, 24" W, 6' - 10' D	Self	LF	135	1.51	0.11	1.62
Utility trenches, 24" W, 11' - 15' D	Self	LF	20	2.53	0.18	2.71
Utility trenches, 24" W, over 15' D	Self	LF	175	2.38	0.17	2.55
AB-3 rock bedding, roads & parking lots	Murray Quarry bid received	Ton	600	Included in other line items		
Asphaltic surface in place	Citywide	Ton	276	32.50	2.28	34.78
8" C.I.P. in place	Self/Manns Sup.	LF	350	11.61	0.81	12.42
12" C.M.P. in place	Self/Manns Sup.	LF	252	13.89	0.97	14.86
Precast manhole set 10' to 13' D	Material only $4000 ea on site	Ea	2	4,655.02	325.85	4,980.87
Precast manhole set 13' to 21' D	Material only $6000 ea on site	Ea	2	6,655.52	465.89	7,121.41
Precast catch basin set 2' to 4' D	Material only $1100 ea on site	Ea	2	1,264.82	88.54	1,353.36
Precast catch basin set 4' to 6' D	Material only $1880 on site	Ea	1	2,330.48	163.13	2,493.61
Type "G" curb & gutter	Sands Const.	LF	405	5.17	0.36	5.53
Shop building footing 1' - 1-1/2'	Sands Const.	LF	240	19.09	1.34	20.43
Office building walls/footings	Sands Const.	LF	210	43.03	3.01	46.04

Figure 15-B-1
Bid preparation form

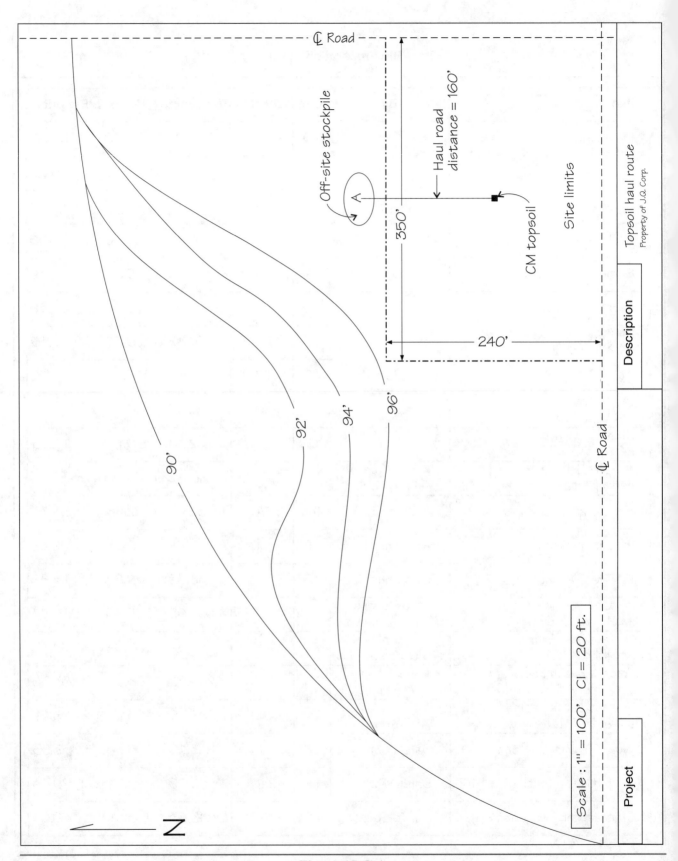

Figure 15-C-1
Topsoil haul route

Haul Road Criteria

Job No.: _C-17_ Date: _____ Checked: _____

Loaded

Road length _160'_

High elevation _108'_   Difference _8'_

Low elevation _100'_

% grade = elevation difference / road length

% grade = = / - _5% (Favorable)_

Surface type _Dirt_   RR factor _100_

Notes:

This % will remain the same

Same

Empty

Road length _____

High elevation _____   Difference _____

Low elevation _____

% grade = elevation difference / road length

% grade = = / - _5% Uphill_

Surface type _____   RR factor _____

Notes:

Figure 15-C-2
Haul road criteria

Production Work Form

Job No. C-17 **Area** Topsoil **Segment** Loaded
Empty wt / lbs 65,000 **Ton** 32.5 **Mach. no.** 501
Capacity / CY 21 **Ton** 24 **Type** Wheel scraper
No. wheels 4 **No. drivers** 2 **Add** ---

Resistance

Rolling = weight on wheels x RR factor

= 56.5 ton x 100 (RRF)

= 5650 lb

Grade = total weight x 20 lb / ton x unit of % grade

= 56.5 x 20 lb / ton x 5

= 5650

Total R = RR +/- GR

= 5650 +/⊖ 5650

= 0 *This is odd, but it does show use of retarder curve is needed

Effective grade = $\dfrac{\text{RR lb / ton}}{20 \text{ lb / ton / \% grade}}$ + % grade

= $\dfrac{100}{20 \text{ lb / ton / \% grade}}$ + 5

= 10%

Power / Gear selection From retarder curve

Power required = Rct 6th gear **In** 17 MPH
Power available = _____ **In gear** 6 **at** 17 MPH

Note: this data from machine inventory form

Usable power = Weight on drivers x coefficient of traction

= 103,000 lb x .55

= 56,650 lb

Note: coefficient of traction from charts

Page 1

Figure 15-C-3
Production work form

Production Work Form

Cycle time

Haul time = $\dfrac{\text{Distance}}{\text{MPH} \times 88}$ = $\dfrac{160}{17 \times 88}$ = $\dfrac{160}{1496}$ = .11

Load time _____ + Maneuver time _____ + Dump time _____
+ Haul time __.11__ + Return time _____ = Total cycle time _____

Actual production   N/A

Trips / hour = $\dfrac{60 \text{ minutes}}{\text{Cycle time}}$ = $\dfrac{60}{}$ = _____

Hr / production = CY / trip × No of trips / hr

= _____ × _____

= _____

Actual hourly production = Hr / production × Eff. factor

= _____ × _____

= _____

Compaction

Compacted CY / Hr = $\dfrac{W \times S \times L \times 16.3}{P}$

= _____ × _____ × _____ × 16.3

= _____

Where:
P = No. of passes
W = Compactor width
S = Average speed
L = Lift thickness
16.3 = Constant

Final production results, notes, units and miscellaneous calculations

Page 2

Figure 15-C-3 (continued)
Production work form

Production Work Form

Job No. _C-17_ Area _Topsoil_ Segment _Empty_
Empty wt / lbs _65,000_ Ton _32.5_ Mach. no. _501_
Capacity / CY _---_ Ton _---_ Type _Wheel scraper_
No. wheels _4_ No. drivers _2_ Add _---_

Resistance

Rolling = weight on wheels x RR factor

= _32.5 ton_ x _100_

= _3250 lb_

Grade = total weight x 20 lb / ton x unit of % grade

= _32.5_ x 20 lb / ton x _5_

= _3250_

Total R = RR +/- GR

= _3250_ ⊕/- _3250_

= _6500 lbs_

Effective grade = $\dfrac{\text{RR lb / ton}}{20 \text{ lb / ton / \% grade}}$ + % grade

= $\dfrac{}{20 \text{ lb / ton / \% grade}}$ + _____

= _10% uphill_

Power / Gear selection

Power required = _6500_ In _lbs rim pull_
Power available = _____ In gear _8th_ at _30 MPH_

Note: this data from machine inventory form

Usable power = Weight on drivers x coefficient of traction

= _65,000 lb_ x _.55_

= _35,750 lbs_

Note: coefficient of traction from charts

Page 1

Figure 15-C-4
Production work form

Production Work Form

Cycle time

Haul time = $\dfrac{\text{Distance}}{\text{MPH} \times 88}$ = $\dfrac{160}{30 \times 88}$ = $\dfrac{160}{2640}$ = .06

Load time __.95__ + Maneuver time __.60__ + Dump time __.25__
+ Haul time __.06__ + Return time __.06__ = Total cycle time __1.92 or 2 min.__

Actual production

Trips / hour = $\dfrac{60 \text{ minutes}}{\text{Cycle time}}$ = $\dfrac{60}{2.00}$ = __30__

Hr / production = CY / trip x No of trips / hr
= __21__ x __30__
= __630__

Actual hourly production = Hr / production x Eff. factor
= __630__ x __.75__
= __472.5 CY/hr__

Compaction

Compacted CY / Hr = $\dfrac{W \times S \times L \times 16.3}{P}$

= $\dfrac{\_\_\_ \times \_\_\_ \times \_\_\_ \times 16.3}{}$

= _____

Where:
P = No. of passes
W = Compactor width
S = Average speed
L = Lift thickness
16.3 = Constant

Final production results, notes, units and miscellaneous calculations

Page 2

Figure 15-C-4 (continued)
Production work form

Calculation Form

Job No. _____ **Date** _____

Unit capacity _21 CY_   **Job Amount** _2614_

Units per day _3780 per 8 hour_   **No. days** _2_

Calculations

Machine & crew
1 - scraper
1 - grader
1 - pickup

Note: Because of small area, moving material at a different times, will use 2 days even though theoretical value would be about 1 day or less.

Figure 15-C-5
Topsoil calculations

Calculation Form

Line item Topsoil   **Segment** All movement

No. units 2614   **Type** CY   **Special** _____

Total cost 2252.96   **Unit cost** .86

Section	Mach. no.	Mach. hrs.	Mach. $/hr	Mach. total/$	Materials used	Material cost	Other	Notes
	101	8	23.17	185.36				
	401	8	48.27	386.16				
	501	8	69.37	554.96				
Subtotals				1126.48				

Calculations and notes

$$\frac{1126.48 \times 2 \text{ days}}{2614 \text{ units}} = \frac{2252.96}{2614} = .86 / CY$$

Figure 15-C-5 (continued)
Topsoil calculations

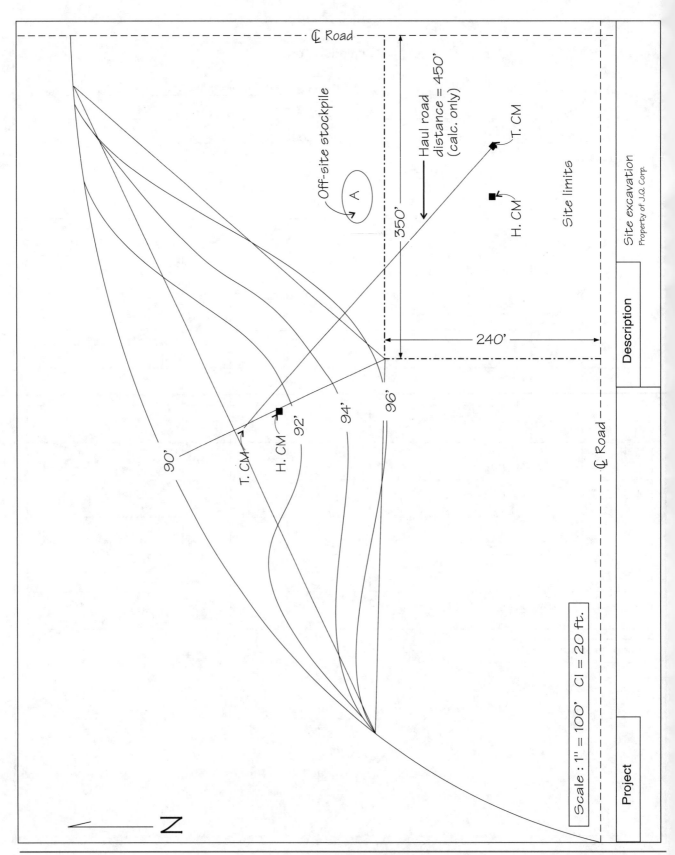

Figure 15-D-1
Main site excavation haul route

352 *Estimating Excavation*

Haul Road Criteria

Job No.: __C-17__ Date: _____ Checked: _____

Loaded

Road length __450'__
High elevation __108'__   Difference __17'__
Low elevation __91'__

% grade = elevation difference / road length

% grade = = / - __4% (2%) (Favorable)__
Surface type __Clay-rutted__   RR factor __150__

Notes:

Will use 2% grade because averages

Same

↓

Empty

Road length _____
High elevation _____   Difference _____
Low elevation _____

% grade = elevation difference / road length

% grade = = / - __2% Uphill__
Surface type _____   RR factor _____

Notes:

Figure 15-D-2
Haul road criteria

Production Work Form

Job No. _C-17_   Area _E & E_   Segment _Loaded_

Empty wt / lbs _65,000_   Ton _32.5_   Mach. no. _501 / 502_

Capacity / CY _21_   Ton _24_   Type _Wheel scraper_

No. wheels _4_   No. drivers _2_   Add _---_

Resistance

Rolling = weight on wheels x RR factor

= _56.5 ton_ x _150_

= _8475_

Grade = total weight x 20 lb / ton x unit of % grade

= _56.5_ x 20 lb / ton x _2_

= _2260_

Total R = RR +/- GR

= _8475_ +/⊖ _2260_

= _6215_

Effective grade = $\dfrac{\text{RR lb / ton}}{20 \text{ lb / ton / \% grade}}$ + % grade

= $\dfrac{150}{20 \text{ lb / ton / \% grade}}$ + _2_

= _9% uphill_

Power / Gear selection

Power required = _6215_   In _Rimpull_

Power available = _7211_   In gear _7_   at _24_

Note: this data from machine inventory form

Usable power = Weight on drivers x coefficient of traction

= _103,000_ x _.40_

= _41,200 lbs_

Note: coefficient of traction from charts

Page 1

Figure 15-D-3
Production work form

Production Work Form

Cycle time

Haul time = $\dfrac{\text{Distance}}{\text{MPH} \times 88}$ = $\dfrac{450}{\underline{24} \times 88}$ = $\dfrac{450}{2112}$ = .21

Load time _____ + Maneuver time _____ + Dump time _____

+ Haul time ___.21___ + Return time _____ = Total cycle time _____

Actual production

Trips / hour = $\dfrac{60 \text{ minutes}}{\text{Cycle time}}$ = $\dfrac{60}{}$ = _____

Hr / production = CY / trip x No of trips / hr

= _____ x _____

= _____

Actual hourly production = Hr / production x Eff. factor

= _____ x _____

= _____

Compaction

Compacted CY / Hr = $\dfrac{\text{W x S x L x 16.3}}{\text{P}}$

= $\dfrac{\underline{} \times \underline{} \times \underline{} \times 16.3}{}$

= _____

Where:
P = No. of passes
W = Compactor width
S = Average speed
L = Lift thickness
16.3 = Constant

Final production results, notes, units and miscellaneous calculations

Haul time = .21

Page 2

Figure 15-D-3 (continued)
Production work form

Production Work Form

Job No. C-17   **Area** E & E   **Segment** Empty
Empty wt / lbs 65,000   **Ton** 32.5   **Mach. no.** 501/502
Capacity / CY ---   **Ton** ---   **Type** ---
No. wheels 4   **No. drivers** 2   **Add** ---

Resistance

Rolling = weight on wheels x RR factor

= 32.5 ton x 150
= 4875

Grade = total weight x 20 lb / ton x unit of % grade

= 32.5 x 20 lb / ton x 2
= 1300

Total R = RR +/- GR

= 4875 ⊕ / - 1300
= 6175

Effective grade = $\dfrac{\text{RR lb / ton}}{20 \text{ lb / ton / \% grade}}$ + % grade

= $\dfrac{150}{20 \text{ lb / ton / \% grade}}$ + 2

= 9%

Power / Gear selection

Power required = 6175   **In** Rimpull
Power available = 7211   **In gear** 7th   **at** 24

Note: this data from machine inventory form

Usable power = Weight on drivers x coefficient of traction

= 65,000 lb x .40
= 26,000 lbs

Note: coefficient of traction from charts

Page 1

Figure 15-D-4
Production work form

Production Work Form

Cycle time

Haul time = $\dfrac{\text{Distance}}{\text{MPH} \times 88}$ = $\dfrac{460}{24 \times 88}$ = $\dfrac{460}{2112}$ = .22

Load time __.75__ + Maneuver time __.80__ + Dump time __.40__
+ Haul time __.22__ + Return time __.21__ = Total cycle time __2.38__

Actual production

Trips / hour = $\dfrac{60 \text{ minutes}}{\text{Cycle time}}$ = $\dfrac{60}{2.38}$ = __25__

Hr / production = CY / trip x No of trips / hr
= __21__ x __25__
= __525 CY__

Actual hourly production = Hr / production x Eff. factor
= __525__ x __.75__
= __394__

Compaction

Compacted CY / Hr = $\dfrac{W \times S \times L \times 16.3}{P}$

= _____ x _____ x _____ x 16.3

= _____

Where:
P = No. of passes
W = Compactor width
S = Average speed
L = Lift thickness
16.3 = Constant

Final production results, notes, units and miscellaneous calculations

2 machines each @ 394 / hr = 3152 / 8 hr

Page 2

Figure 15-D-4 (continued)
Production work form

Total weight = 90,000 lbs

Production Work Form

Job No. C-17   **Area** E & E   **Segment** Compaction
Empty wt / lbs 62,000   **Ton** 31   **Mach. no.** 302
Capacity / CY ---   **Ton** ---   **Type** Crawler tr.
No. wheels _____   **No. drivers** 2   **Add** No 901 sheepsfoot

---

Resistance

Rolling = weight on wheels x RR factor

= _____ x _____

= _____

Grade = total weight x 20 lb / ton x unit of % grade

= _____ x 20 lb / ton x _____

= _____

Total R = RR +/- GR

= _____ +/- _____

= _____

$$\text{Effective grade} = \frac{\text{RR lb / ton}}{20 \text{ lb / ton / \% grade}} + \text{\% grade}$$

$$= \frac{\underline{\qquad}}{20 \text{ lb / ton / \% grade}} + \underline{\qquad}$$

= _____

Power / Gear selection

Power required = _____ **In** _____
Power available = _____ **In gear** 3rd **at** 3.2

Note: this data from machine inventory form

Usable power = Weight on drivers x coefficient of traction

= _____ x _____

= _____

Note: coefficient of traction from charts

Page 1

Figure 15-D-5
Production work form

Production Work Form

Cycle time

Haul time = $\dfrac{\text{Distance}}{\text{MPH} \times 88}$ = $\dfrac{\rule{1cm}{0.4pt}}{\rule{1cm}{0.4pt} \times 88}$ = _____ = _____

Load time _____ + Maneuver time _____ + Dump time _____

+ Haul time _____ + Return time _____ = Total cycle time _____

Actual production

Trips / hour = $\dfrac{60 \text{ minutes}}{\text{Cycle time}}$ = $\dfrac{60}{\rule{1cm}{0.4pt}}$ = _____

Hr / production = CY / trip × No of trips / hr

= _____ · x _____

= _____

Actual hourly production = Hr / production × Eff. factor

= _____ × _____

= _____

Compaction

Compacted CY / Hr = $\dfrac{W \times S \times L \times 16.3}{P}$

= $\dfrac{6 \times 3.2 \times 6 \times 16.3}{5}$

= 376/hr = 3008 / 8 hr

Where:
P = No. of passes
W = Compactor width
S = Average speed
L = Lift thickness
16.3 = Constant

Final production results, notes, units and miscellaneous calculations

Will keep ahead of hauler, but will need to be on job same time.

Page 2

Figure 15-D-5 (continued)
Production work form

Calculation Form

Job No. _____ **Date** _____

Unit capacity _21 CY_____ **Job Amount** _17402_____

Units per day _3152_____ **No. days** _5.4 use 6_____

Calculations

```
Using one scraper      - 501
1 - grader             - 401
1 - trac & compactor   - 301+
1 - trac — push load   - 302
1 - pickup / foreman   - 101
```

Figure 15-D-6
Calculations

Calculation Form

Line item *Excavation **Segment** All

No. units 17402 **Type** CY **Special** _____

Total cost 13,989.12 **Unit cost** .80

Section	Mach. no.	Mach. hrs	Mach $/hr	Mach. total/$	Materials used	Material cost	Other	Notes
	101	48	23.17	1112.16				
	301	48	59.77	2868.96				
	302	48	84.23	4043.04				
	401	48	48.26	2316.48				
	501	48	69.37	3329.76				
	1101	48	6.64	318.72				
Subtotals				13989.12				

Calculations and notes

* Calc. will be for all earthwork including cut, fill, roadway & parking lot

Figure 15-D-7
Final calculations

Calculation Form Trench work only

Job No. _____ **Date** _____

   Unit capacity _____ **Job Amount** _____

   Units per day _____ **No. days** _____

Calculations

Depth	Storm	Sanitary	Total	Exc. Hr.
0'-6'	127'	145'	272'	1.25
6'-11'	125'	10'	135'	2.0
11'-16'		20'	20'	0.5
16'-20'		175'	175'	4.0
0'-6'	45 min.	30 min.	1.25 hr.	
6'-11'	45 min.	30 min.	1.25 hr.	
11'-16'	1.5 hr.	60 min.	2.5 hr.	
16'-20'		30 min.	0.5 hr.	

Figure 15-E-1
Trench work calculations

Calculation Form

Line item _Trenching_ **Segment** _0'-6'_

No. units _272_ **Type** _LF_ **Special** _____

Total cost _126.61_ **Unit cost** _$0.47_

Section	Mach. no.	Mach. hrs.	Mach. $/hr.	Mach. total/$	Materials used	Material cost	Other	Notes
	701	1.25	101.29	126.61				
	Subtotals							

Calculations and notes

Figure 15-E-2
Trenching 0'-6'

Costs and Final Bid for the Sample Estimate **363**

Calculation Form

Line item Trenching   **Segment** 6'-11'

No. units 135   **Type** LF   **Special** _____

Total cost 202.58   **Unit cost** 1.51

Section	Mach. no.	Mach. hrs.	Mach. $/hr	Mach. total/$	Materials used	Material cost	Other	Notes
	701	2.0	101.29	202.58				
Subtotals								

Calculations and notes

Figure 15-E-3
Trenching 6'-11'

Calculation Form

Line item __Trenching__ **Segment** __11'-16'__

No. units __20__ **Type** __LF__ **Special** _____

Total cost __50.61__ **Unit cost** __2.53__

Section	Mach. no.	Mach. hrs.	Mach. $/hr	Mach. total/$	Materials used	Material cost	Other	Notes
	701	0.5	101.29	50.65				
	Subtotal							

Calculations and notes

Figure 15-E-4
Trenching 11'-16'

Calculation Form

Line item Trenching **Segment** 16'-20'

No. units 175 **Type** LF **Special** _____

Total cost 405.16 **Unit cost** 2.38

Section	Mach. no.	Mach. hrs.	Mach. $/hr	Mach total/$	Materials used	Material cost	Other	Notes
	701	4.0	101.29	405.16				
Subtotals								

Calculations and notes

Figure 15-E-5
Trenching 16'-20'

366 *Estimating Excavation*

Calculation Form

Line item __Catch basin__   Segment __2'-4'__

No. units __2__   Type __Precast__   Special __#1 & 2__

Total cost __2529.64__   Unit cost __1264.82__

Section	Mach. no.	Mach. hrs.	Mach. $/hr.	Mach. total/$	Materials used	Material cost	Other	Notes
	701	.5	101.29	50.64	Precast	1104.51	Labor	3 @ 7.00 = 21.00/crew
	601	.5	79.52	39.76				
	801	.5	29.36	14.68				
	101	.5	23.17	11.58				
	1001	.5	2.84	1.42				
	901	.25	42.69	10.67				
	301	.25	84.23	21.06				
Subtotal				149.81				

Calculations and notes

Eq. = 149.81
Mat. = 1104.51
Lab. =   10.50
         1264.82

Figure 15-F-1
Catch basin calculations

Structure Construction Production Cost

Structure type __Catch basin__ **No.** __1 & 2__ **Location** __Storm line 1__

Width __2' x 2'__ **Depth** __2-4 (3)__ **Length** _____

Material __Precast concrete__ **CY volume for structure** __4.5 CY__

Vol. / _____

Work description:

   Excavate, place bedding, install & backfill

Equipment needed:

   1- track excavator #701   1- compactor @ 1/2 time
   1- track loader #601   1- small dozer @ 1/2 time
   1- boom truck #801
   1- pu / foreman #101

Crew:

   3 additional laborers

Struct. material __Precast concrete__ **From** __Royal__ **Cost / LF** __1100.00 lump sum__

Bedding material __AB-3__ **From** __Murray__ **Cost / LF** __4.51 lump sum__

Other _____ **From** _____ **Cost / LF** _____

Calculations:

$$\frac{2.5 \times 2.5}{27} = .23 \text{ CY} = .46 \text{ ton @ } 9.75 = 4.51$$

Total material cost _____ / LF

Production cycle time _____ **Bucket volume** _____ **Job eff.** _____

Hr. production dig time __15 min.__

Other time _____

Total time = __30 min.__ **Ft / Hr.** _____

Job hours _____ **Job days** _____

Figure 15-F-2
Catch basin calculations

Calculation Form

Line item __Catch basin__   Segment __4'-6'__

No. units __1__   Type __Precast__   Special __#3__

Total cost _____   Unit cost __2330.48__

Section	Mach. no.	Mach. hrs.	Mach. $/hr	Mach. total/$	Materials used	Material cost	Other	Notes
	701	1.5	101.29	151.94	Precast	1804.51	Labor	3@7.00 = 21.00/crew hr.
	601	1.5	79.52	119.28				
	801	1.5	29.36	44.04				
	101	1.5	23.17	34.76				
	901	.75	42.69	4.26				
	1001	1.5	2.84	32.02				
	301	.75	84.23	63.17				
Subtotal				494.47				

Calculations and notes

Eq. = 494.47
Mat. = 1804.51
Lab. =   31.50
        2330.48

Figure 15-F-3
Catch basin calculations

Structure Construction Production Cost

Structure type __Catch basin__ No. __3__ Location __Storm line 1-2__
Width __2' x 2'__ Depth __4'-6' 5'__ Length _____
Material __Precast concrete__ CY volume for structure __6.00__
Vol. / _____

Work description:

Same

Equipment needed:

Crew:

Struct. material _____ From _____ Cost / ~~LF~~ __$1800.00 lump sum__
Bedding material _____ From _____ Cost / LF _____
Other _____ From _____ Cost / LF _____
Calculations:

Total material cost __$1804.51__ / LF

Production cycle time _____ Bucket volume _____ Job eff. _____
Hr. production dig time _____
Other time _____

Total time = __1.5 hr all__ Ft / Hr. _____
Job hours _____ Job days _____

Figure 15-F-4
Catch basin calculations

370 *Estimating Excavation*

Calculation Form

Line item __Manhole__ Segment __16'-18'__
No. units __2__   Type _____ Special __Storm tie-in__
Total cost _____   Unit cost __6655.52__

Section	Mach. no.	Mach. hrs.	Mach. $/hr	Mach. total/$	Materials used	Material cost	Other	Notes
	701	2.0	101.29	202.58	Precast	6014.24	Labor	3@7.00 = 21.00 crew hr
	601	2.0	79.52	159.04				
	801	2.0	29.36	58.72				
	101	2.0	23.17	46.34				
	1001	2.0	2.84	5.68				
	901	1.0	42.69	42.69				
	301	1.0	84.23	84.23				
Subtotal				599.28				

Calculations and notes

Eq. = 599.28
Mat. = 6014.24
Lab. =   42.00
       ―――――――
       6655.52

Figure 15-F-5
Manhole calculations

Structure Construction Production Cost

Structure type ___Manhole___ **No.** ___1___ **Location** ___Storm___
Width ___4' dia.___ **Depth** ___16'___ **Length** _____
Material ___Precast concrete___ **CY volume for structure** ___30___
Vol. / _____

Work description:
  Same

Equipment needed:
  Same

Crew:
  Same

Struct. material ___Precast concrete___ **From** ___Royal___ **Cost / LF** ___6000.00 lump sum___
Bedding material ___AB-3___ **From** ___Murray___ **Cost / LF** ___14.20 lump sum___
Other _____ **From** _____ **Cost / LF** _____

Calculations:

Total material cost ___6014.24___ **/ LF**

Production cycle time ___---___ **Bucket volume** _____ **Job eff.** _____
Hr. production dig time ___---___
Other time ___---___

Total time = ___2 hr all___ **Ft / Hr.** _____
Job hours _____ **Job days** _____

Figure 15-F-6
Manhole calculations

372 *Estimating Excavation*

Utility Line Construction Production Cost

Utility line type __Storm sewer__   No. __1__   Length __42'__

Width __3'__   Depth __1-5 2-5__   (Average for both) Size _____

Line type __12" C.M.P.__   CY volume for line __11.7__

Vol. /LF __0.28__

Work description:

Same as sanitary sewer

Equipment needed:

Same

Crew:

Same

Line material __C.M.P__   From __Murray Supply__   Cost / LF __4.70__

Bedding material __Same as sanitary__   From __Murray Supply__   Cost / LF __2.32__

Other _____   From _____   Cost / LF _____

Calculations:

$$R = \frac{(2' \times 2') - 6" = 5^2 \times 3.1416}{27} = .12 \, CY = .24 \, ton \, @ \, 9.75/ton = 2.32$$

Total material cost __$6.02__ / LF

Production cycle time __17 sec.__   Bucket volume __1.31__   Job eff. _____

Hr. production dig time __15 min.__

Other time __30 min. for all except 901/301 = 1 hr., 701 = 15 min.__

Total time = __30 min.__   Ft / Hr. _____

Job hours _____   Job days _____

Figure 15-F-7
Storm sewer calculations

Utility Line Construction Production Cost

Utility line type __Storm sewer__ No. __2__ Length __85'__
Width __3.0'__ Depth __1-5 2-5__ (Average for both) Size _____
Line type __12" C.M.P.__ CY volume for line __23.62__
Vol. /LF __0.28__

Work description:
  Same

Equipment needed:
  Same

Crew:

Same
(arrow pointing down)

Line material _____ From _____ Cost / LF _____
Bedding material _____ From _____ Cost / LF _____
Other _____ From _____ Cost / LF _____

Calculations:

Total material cost __$6.02__ / LF

Production cycle time __Same__ Bucket volume __---__ Job eff. _____
Hr. production dig time __use 1/2 hr__
Other time __Use 1 hr all except 901/301 = 2 hr, 701 = 30 min__

Total time = _____ Ft / Hr. _____
Job hours _____ Job days _____

Figure 15-F-8
Storm sewer calculations

Utility Line Construction Production Cost

Utility line type _Storm sewer_ **No.** _3_ **Length** _60'_
Width _3.0'_ **Depth** _6-10 8'_ (Average for both) **Size** _____
Line type _12" C.M.P._ **CY volume for line** _53_
Vol. /LF _0.89_

Work description:
  Same

Equipment needed:
  Same

Crew:

Same

Line material _____ **From** _____ **Cost / LF** _____
Bedding material _____ **From** _____ **Cost / LF** _____
Other _____ **From** _____ **Cost / LF** _____

Calculations:

Total material cost _$6.02_ / LF
Production cycle time _30 sec._ **Bucket volume** _1.31_ **Job eff.** _____
Hr. production dig time _use 3/4 hr_
Other time _1.5 all, 3 compactor_

Total time = _____ **Ft / Hr.** _____
Job hours _____ **Job days** _____

Figure 15-F-9
Storm sewer calculations

Utility Line Construction Production Cost

Utility line type _Storm sewer_ **No.** _3_ **Length** _65'_

Width _3.0'_ **Depth** _11-15 13'_ **(Average for both) Size** _____

Line type _12" C.M.P._ **CY volume for line** _93.9_

Vol. /LF _1.44_

Work description:
  Same

Equipment needed:
  Same

Crew:

Same

↓

Line material _____ **From** _____ **Cost / LF** _____
Bedding material _____ **From** _____ **Cost / LF** _____
Other _____ **From** _____ **Cost / LF** _____

Calculations:

Total material cost _$6.02_ **/ LF**

Production cycle time _42 sec._ **Bucket volume** _1.31_ **Job eff.** _____

Hr. production dig time _use 1-1/2 hr_

Other time _3 all, 6 hr compactor_

Total time = _____ **Ft / Hr.** _____

Job hours _____ **Job days** _____

Figure 15-F-10
Storm sewer calculations

Estimating Excavation

Calculation Form

Job No. _____   Date _____

    Unit capacity _____   Job amount __252_____

    Units per day _____   No. days _____

Calculations

In minutes

Line	701	601	801	101	1001	901	301	Extra labor
1	15	30	30	30	30	60	60	30
2	30	60	60	60	60	120	120	60
3A	45	90	90	90	90	180	180	90
3B	90	180	180	180	180	360	360	180
	180M / 3 hr	360M / 6 hr	6 hr.	6 hr.	6 hr.	720M / 12 hr	720M / 12 hr	6 hr.

252 feet total

Figure 15-F-11
Equipment calculations

Calculation Form

Line item __12" C.M.P.__ Segment _____

No. units __252__ Type __LF__ Special _____

Total cost __4512.09__ Unit cost __13.89__

Section	Mach. no.	Mach. hrs	Mach. $/hr	Mach. total/$	Materials used	Material cost	Other	Notes
	701	3	101.29	303.87			Labor	3@7.00 = 21 x 6 hr = 126
	601	6	79.52	477.12				
	801	6	14.68	88.08				
	101	6	23.17	139.02				
	1001	6	2.84	17.04				
	901	12	42.67	512.04				
	301	12	59.77	717.24				
Subtotal				2869.05		1517.04	126	

Calculations and notes

Eq. = 2869.05
Mat. = 1517.04
Lab. = 126.00
     4512.09

Figure 15-F-12
Pipe calculation summary

Utility Line Construction Production Cost

Utility line type Sanitary sewer   **No.** 1   **Length** 145'

Width 3'   **Depth** 2.5'   (Average for both) **Size** _____

Line type 8" C.I.P   **CY volume for line** 40

Vol. /LF 0.28 CY

Work description:
Evacuate trench, place bedding, install pipe, cover with bedding material, backfill and compact to 90% Proctor. Trench box will be used. Hoe will dig trench, move box, place bottom bedding and pipe. Hoe will have 36" bucket to allow for work room and trench box. Extra bedding material will not be charged to job. Pipe is in 20' sections. Small dozer and ditch compactor will be run by one man with a 50-50 time and cost split. Track loader will place top bedding and help backfill. Boom truck will place pipe along trench.

Equipment needed:

1 - Track excavator no. 701   1 - Laser Gun no. 1001
1 - Track loader no. 601   1 - Trench compactor no. 901
1 - Boom truck no. 801   1 - Small dozer no. 301
1 - Pickup / foreman no. 101   1 - operator / both

Crew:
3 laborers in addition to above operator and foreman

Line material 8" C.I.P.   **From** Manns Supply   **Cost / LF** 3.77

Bedding material AB-3   **From** Murray Quarries   **Cost / LF** 2.38

Other _____   **From** _____   **Cost / LF** _____

Calculations:

$$\frac{\text{Rock/LF} = (3 \times 1) - 8" \text{ pipe converted. to dec.} = (3.14)(.33)^2}{27} = .12 \text{ CY} = .24 \text{ ton @ } 9.75/\text{ton} = 2.38 \text{ LF}$$

Total material cost $6.15 / LF

Production cycle time 17 sec.   **Bucket volume** 1.31 CY   **Job eff.** _____

Hr. production dig time 1 hr = 110 CY /// use 30 min.

Other time lay pipe. 2 min. to place, 2 min. to align, 2 min. to seal, 1 min. to move and 3 min. to hand shade pipe = 10 min. / 20 ft. joint

Total time = _____   **Ft / Hr.** 120   Same as others
                                               701 - 30 min
Job hours n/a   **Job days** n/a   main crew - 60 min.
                                               901 / 301 - 120 min.

Figure 15-G-1
Utility line production cost

Utility Line Construction Production Cost

Utility line type ___Sanitary sewer___ **No.** __2__ **Length** __20'__
Width __3'__ **Depth** __5-10 7.5__ (Average for both) **Size** _____
Line type ___8" C.I.P.___ **CY volume for line** __17__
Vol. /Line foot __0.83__

Work description:
Same

Equipment needed:
Same

Crew:

Same
↓

Line material _____ **From** _____ **Cost / LF** _____
Bedding material _____ **From** _____ **Cost / LF** _____
Other _____ **From** _____ **Cost / LF** _____

Calculations:

Total material cost __$6.15__ / LF

Production cycle time __17__ **Bucket volume** __131__ **Job eff.** _____
Hr. production dig time __1/4 (15 min.) = 55 CY__
Other time __Crew 30__

Total time = _____ **Ft / Hr.** _____ 701 = 15 min.
Job hours _____ **Job days** _____ main crew = 30 min.
901/301 = 60 min.

Figure 15-G-2
Utility line production cost

380 *Estimating Excavation*

Utility Line Construction Production Cost

Utility line type _Sanitary sewer_ **No.** _4_ **Length** _45'_
Width _3'_ **Depth** _16-20 18'_ (Average for both) **Size** _____
Line type _8" C.I.P._ **CY volume for line** _90_
Vol. /Line foot _2_

Work description:
Same

Equipment needed:
Same

Crew:

Same

Line material _____ **From** _____ **Cost / LF** _____
Bedding material _____ **From** _____ **Cost / LF** _____
Other _____ **From** _____ **Cost / LF** _____

Calculations:

Total material cost _$6.15_ / LF
Production cycle time _17 sec._ **Bucket volume** _1.31_ **Job eff.** _____
Hr. production dig time _1/2 hr = 110 CY_
Other time _1/2 hr for 2+ joints_

Total time = _____ **Ft / Hr.** _120_   701 = 30 min
Job hours _____ **Job days** _____   main crew = 60 min
                                           901/301 = 120 min

Figure 15-G-3
Utility line production cost

Utility Line Construction Production Cost

Utility line type __Sanitary sewer__   No. __4__   Length __130'__
Width __3'__   Depth __11-15 13__   (Average for both) Size _____
Line type __8" C.I.P.__   CY volume for line __188__
Vol. /Line foot __1.44__

Work description:
  Same

Equipment needed:
  Same

Crew:

                                    Same
                                      ↓

Line material _____ From _____ Cost / LF _____
Bedding material _____ From _____ Cost / LF _____
Other _____ From _____ Cost / LF _____

Calculations:

Total material cost __$6.15__ / LF

Production cycle time __17 sec.__   Bucket volume __1.31__   Job eff. _____
Hr. production dig time __1 hr = 220 CY__
Other time __6+ joints @ 10 min.__

Total time = _____   Ft / Hr. _____   701 = 60 min.
Job hours _____ Job days _____   main crew = 120 min.
                                             901/301 = 240 min.

Figure 15-G-4
Utility line production cost

382 *Estimating Excavation*

Calculation Form

Job No. _____   Date _____

    Unit capacity _____   Job amount _____

    Units per day _____   No. days _____

Calculations

In minutes

Line	701	601	801	101	1001	901	301	Extra labor
1	30	60	60	60	60	120	120	60
2	15	30	30	30	30	60	60	30
4	30	60	60	60	60	120	120	60
11	60	120	120	120	120	240	240	120
	2.25	4.5	4.5	4.5	4.5	9 hr.	9 hr.	4.5 hr.

Figure 15-G-5
Utility line summary

Calculation Form

Line item __8" C.I.P.__ Segment _____

No. units __350__ Type __LF__ Special _____

Total cost __3945.75__ Unit cost __11.27__

Section	Mach. no.	Mach. hrs.	Mach. $/hr	Mach. total/$	Materials used	Material cost	Other	Notes
	701	2.25	101.29	227.90	8" C.I.P.		Labor	3@7.00 = 21.00 x 4.5 = 94.50
	601	4.5	79.52	357.84				
	801	4.5	29.36	132.12				
	101	4.5	23.17	104.27				
	1001	4.5	2.84	12.78				
	901	9.0	42.67	384.02				
	301	9.0	59.77	537.92				
Subtotals				1756.85		2094.40	94.50	

Calculations and notes

Eq. = 1756.85
Mat. = 2094.40
Lab. =   94.50
       3945.75

Figure 15-G-6
Cast iron pipe

Calculation Form

Line item __Manhole__   Segment __16'-18'__

No. units __1__   Type __LF__   Special _____

Total cost _____   Unit cost __$6,641.28__

Section	Mach. no.	Mach. hrs.	Mach. $/hr	Machine total/$	Materials used	Material cost	Other	Notes
3	701	2.0	101.29	202.58	Precast concrete	6000.00	Labor	3@7.00/hr = 21.00 crew
	601	2.0	79.52	159.04				
	801	2.0	29.36	58.72				
	101	2.0	23.17	46.34				
	1001	2.0	2.84	5.68				
	901	1.0	42.69	42.69				
	301	1.0	84.23	84.23				
Subtotal				599.28		6000.00	42.00	

Calculations and notes

Eq. = 6000.00
Mat. = 599.28
Lab. = 42.00
6641.28

Figure 15-G-7
Manhole

Calculation Form

Line item: **Manhole** Segment: **10'-12'**
No. units: **2** Type: **---** Special: _____
Total cost: **---** Unit cost: **4540.21**

Section	Mach. no.	Mach. hrs.	Mach. $/hr	Mach. total/$	Materials used	Material cost	Other	Notes
1 & 2	701	1.5	101.29	151.94	Precast conc. 10'-12	4014.24	Labor	3@7.00/hr. = 21.00 crew hr
	601	1.5	79.52	119.28				
	801	1.5	29.36	44.04				
	101	1.5	23.17	34.76				
	1001	1.5	2.84	4.26				
	901	.75	42.69	32.02				
	301	.75	84.23	63.17				
Subtotal				494.47		4014.24	31.50	Subtotals are for each MH

Calculations and notes

```
 4014.24
  494.47
 4508.71
   31.50
 4540.21
```

Figure 15-G-8
Manholes

Structure Construction Production Cost

Structure type _Manhole_ **No.** _1_ **Location** _Sanitary Line 1-2_

Width _MH = 4' Dia. Exc = 8' Dia._ **Depth** _11.30_ **Length** _____

Material _Precast concrete_ **CY volume for structure** _28_

Vol. / _____

Work description:
Excavate, install 6" AB-3 under structure with 6" overhang. Backfill & compact 20% Proctor. Manholes will remain above grade 1 - 2.5 ft. All else is same as sanitary lines.

Equipment needed:
Same

Crew:
Same

Struct. material _Precast concrete_ **From** _Royal Conc. (1-10'-12')_ **Cost / LF** _4000.00 lump sum_

Bedding material _AB-3_ **From** _Murray_ **Cost / LF** _14.24 lump sum_

Other _____ **From** _____ **Cost / LF** _____

Calculations:

Bedding 4' MH dia + 6" overhang = 5 ft   R = 2.5 $A = \dfrac{(2.5)^2(3.1416)}{27} = .73$ CY

1.5 ton @ 9.75/ton = 1.5 ton

Total material cost _4014.24_ / ~~LF~~ LS

Production cycle time _.17_ **Bucket volume** _1.31_ **Job eff.** _____

Hr. production dig time _---_

Other time _Use 1.5 hr for all to allow for all work as this is tight working_

Total time = _1.5 hr all_ **Ft / Hr.** _____

Job hours _____ **Job days** _____

Figure 15-G-9
Structure construction cost

Structure Construction Production Cost

Structure type _Manhole_ **No.** _1_ **Location** _Sanitary Line 2-3_
Width _MH = 4' dia. Exc. = 8' dia._ **Depth** _12.3_ **Length** _____
Material _Precast concrete_ **CY volume for structure** _30_
Vol. / _____

Work description:
  Same

Equipment needed:
  Same

Crew:

Same
↓

Struct. material _____ **From** _1-10'-12'_ **Cost / LF** _____
Bedding material _____ **From** _____ **Cost / LF** _____
Other _____ **From** _____ **Cost / LF** _____

Calculations:

Total material cost _4014.24_ ~~LF~~ LS

Production cycle time _____ **Bucket volume** _____ **Job eff.** _____
Hr. production dig time _____
Other time _Same_ _____

Total time = _1.5 hr all_   **Ft / Hr.** _____
Job hours _____ **Job days** _____

Figure 15-G-10
Structure construction cost

Structure Construction Production Cost

Structure type _Manhole_ **No.** _3_ **Location** _Sanitary line - 3 end_
Width _MH = 4' dia. Exc. = 8' dia._ **Depth** _20.70_ **Length** _____
Material _Precast concrete_ **CY volume for structure** _51_
Vol. / _____

Work description:
Same

Equipment needed:
Same

Crew:
Same

Struct. material _Precast concrete_ **From** _Royal Conc. 1 -MH 16-19'_ **Cost / LF** _6000.00 lump sum_
Bedding material _Same_ **From** _____ **Cost / LF** _____
Other _____ **From** _____ **Cost / LF** _____

Calculations:

Total material cost _6014.24_ / LF LS

Production cycle time _---_ **Bucket volume** _____ **Job eff.** _____
Hr. production dig time _---_
Other time _Use 2 hr._

Total time = _2.0 / all_ **Ft / Hr.** _____
Job hours _____ **Job days** _____

Figure 15-G-11
Structure construction cost

Structure Construction Production Cost

Structure type _Concrete_ **No.** _---_ **Location** _Shop footing_

Width _____ **Depth** _____ **Length** _____

Material _____ **CY volume for structure** _155.23_

Vol. / _____

Work description:

   701 to load
   501 to haul   Because of short work areas - cycle time will be about 30 sec. considering maneuver time.

Equipment needed:

Crew:

Struct. material _Concrete_ **From** _Sands_ **Cost / LF** _11.90_

Bedding material _AB-3_ **From** _Murray_ **Cost / LF** _4.35_

Other _____ **From** _____ **Cost / LF** _____

Calculations:

$$\text{AB-3 - 35 CY} = \frac{110 \text{ ton @ } 9.50}{240}$$

Total material cost _____ **/ LF**

Production cycle time _30"_ **Bucket volume** _2_ **Job eff.** _____

Hr. production dig time _4 hr. Exc CY = 136± Backfill = 16_

Other time _____

Total time = _____ **Ft / Hr.** _____

Job hours _____ **Job days** _____

Figure 15-H-1
Structure production cost

Calculation Form

Line item __Shop building footing__   Segment _____

No. units __240__   Type __LF__   Special _____

Total cost __4110.60__   Unit cost __17.09__

Section	Mach. no.	Mach. hrs.	Mach. $/hr	Mach. total/$	Materials used	Material cost	Other	Notes
	701	4	101.29	405.16	Concrete	11.90		
	501	4	69.37	277.48	AB-3 bedding	4.35		
								Conc. in place by Sands Const. Co. @ 11.90/LF
Subtotal				682.64		16.25		

Calculations and notes

Eq. = .72   Material   = 16.25 / LF
Conc.= 11.90   Exc. = 682.64 = 2.84 / LF
         12.62   240
                                = 19.09 / LF

Figure 15-H-2
Shop building footing

Structure Construction Production Cost

Structure type ___Office basement___ **No.** ___---___ **Location** ___---___

Width _____ **Depth** _____ **Length** _____

Material _____ **CY volume for structure** ___934___

Vol. / _____

Work description:

   Excavate for basement. Concrete to be placed by Sands Construction for 31.00 / LF of wall

Equipment needed:

   601 to load
   502 to haul
   101 supervise

Crew:

Struct. material ___Concrete___ **From** ___Sands___ **Cost / LF** ___31.00___
Bedding material ___AB-3___ **From** ___Murray___ **Cost / LF** ___4.25___
Other _____ **From** _____ **Cost / LF** _____

Calculations:

   Prior experience says it will take about 1 min.   Bedding = 47 CY = 94 ton
   cycle time to go into & out of the hole.

Total material cost _____ / LF

Production cycle time ___1 min.___ **Bucket volume** ___2___ **Job eff.** ___.83___
Hr. production dig time ___9.5 hr.___   120 CY/hr   100
Other time _____

Total time = _____ **Ft / Hr.** _____
Job hours _____ **Job days** _____

Figure 15-I-1
Office basement

Calculation Form

Line item **Office basement** Segment _____

No. units **210** Type **LF** Special _____

Total cost **9036.30** Unit cost **43.03**

Section	Mach. no.	Mach. hrs.	Mach. $/hr	Mach. total/$	Materials used	Material cost	Other	Notes
	601	9.5	79.52	755.44	Concrete	31.00		Concrete in place by Sands Const. @ 31.00
	502	9.5	69.37	659.02	AB-3	4.25		
	101	9.5	23.17	220.12				
Subtotal				1634.58		35.25		

Calculations and notes

1634.58/210 = 7.78 / LF Exc.
              = 35.25 / Material
                43.03

Figure 15-I-2
Office basement

Overhead Inventory List

Item	Quantity	Cost Each	Yearly cost
Office and shop grounds	1	n/a	25,000.00
Shop Inventory	1	n/a	7,000.00
Service truck	1	n/a	3,000.00
Low boy / trailer	1	n/a	2,800.00
Pickups	2	2,500.00	5,000.00
Office help and supplies	n/a	n/a	19,000.00
Insurance, taxes, etc. other than mach's	n/a	n/a	4,000.00
Dues, subscriptions	n/a	n/a	750.00
Advertising	n/a	n/a	1,500.00
Computer rental	n/a	n/a	2,000.00
Utilities/all	n/a	n/a	2,900.00
Field supervision	1	n/a	25,000.00
Miscellaneous	n/a	n/a	5,000.00
		Page totals	
		Item totals	102,950.00

Figure 15-J-1
Overhead inventory

Operating Costs No. 101

Fuel			
Unit price	Used / hr		
.72	1.9 gal/hr		1.75

Lubricants / filters			
Item	Unit price	Used / hr	
Engine	4.50 / gal	.06 gal / hr	0.27
Trans	6.00 / gal	.02 gal / hr	0.12
Finals			
Hyd			
Grease	1.25 / lb	104 lb / hr	0.05
Filters	4 sets @ 20.00		.04
Other _____			
Total lubricants			2.23

Tires

$$\frac{\text{Replacement cost}}{\text{Estimated hours}} = \frac{500}{10,000} \qquad 0.50$$

Repairs

$$\frac{\text{Factor} \times \text{del price - tires}}{1000} = \frac{.09 \times 13.500}{1000} \qquad .01$$

Other	
Total operation cost	2.74
Operator wages Supervisor	15.00
Ownership cost	3.09
Total operation and ownership cost	20.83

Figure 15-K-1
Machine no. 101

Hourly Ownership Cost Estimate

Machine type Pickup No. 101

Purchase date _____ **Purchase price** $14,000

Depreciation value

Delivered price (total cost)				14,000
Minus tire replacement cost				
Loc	Size	Qt	Amount	
Front			250	
Rear			250	
Drive				
Total tires				500
Delivered price minus tire cost				13,500
Minus resale or trade-in value				1,500
Net depreciation value				12,000

Ownership cost

Depreciation value

$$= \frac{\text{Net depreciation value (from above)}}{\text{Depreciation period in hours}}$$

$$= \frac{12{,}000}{5{,}000}$$

= 2.40

Interest, insurance, taxes

Rate Int. _9%_ Insc _4%_ Taxes _5%_

Estimated yearly use in hours _2,200_

$$\frac{\text{Factor x delivered price}}{1000} = \frac{.049 \times 14{,}000}{1000}$$

Owning cost = .69

Total ownership cost = Depreciation cost + owning cost = 3.09

Figure 15-K-2
Machine no. 101

| Machine inventory |

Mach. no. __101__ Type __Pickup__ Brand __Ace__

Purchase date _____ Purchase price __$14,000__

Average hours per year use __2,200__

HP _____ Operating weight _____ / ton _____

Capacity full _____ CY Scraped _____ CY

Rated load _____ Rated RPM _____

Weight distribution

Empty drive _____ No. of drivers _____ Loaded drive _____

Rear _____ Rear _____

Maximum height _____ Maximum reach _____ Maximum depth _____

Dig unit width _____

| Gear / power / weight chart |

Gear	Speed	Pounds of rim pull		Drawbar pull		RPM
		Rated	Maximum	Rated	Maximum	
1						
2						
3						
4						
5						
6						
7						
8						

| Comments |

Figure 15-K-3
Machine no. 101

Operating Costs No. 301

Fuel

Unit price	Used / hr		
.72	20 gal/hr		14.40

Lubricants / filters

Item	Unit price	Used / hr	
Engine	4.50 / gal	.06 gal / hr	0.27
Trans	6.00 / gal	.05 gal / hr	0.30
Finals			
Hyd	6.00 / gal	.05 gal / hr	0.30
Grease	1.25 / lb	.03 lb / hr	0.04
Filters	6 sets @ 60.00 / 1700		0.21
Other No. 101			
Total lubricants			15.52

Tires

$$\frac{\text{Replacement cost}}{\text{Estimated hours}} = \frac{-0-}{-0-}$$

			-0-

Repairs

$$\frac{\text{Factor} \times \text{del price - tires}}{1000} = \frac{.09 \times 110,000}{1000}$$

			9.99
Other			
Total operation cost			25.51
Operator wages No. II			12.00
Ownership cost			16.20
Total operation and ownership cost			53.71

Figure 15-K-4
Machine no. 301

Hourly Ownership Cost Estimate

Machine type __Crawler tractor   No. 301__

Purchase date _____   Purchase price __$110,000__

Depreciation value

Delivered price (total cost)				110,000.00

Minus tire replacement cost

Loc	Size	Qt	Amount	
Front				
Rear				
Drive				

Total tires	-0-
Delivered price minus tire cost	110,000.00
Minus resale or trade in value	10,000.00
Net depreciation value	100,000.00

Ownership cost

Depreciation value

$$= \frac{\text{Net depreciation value (from above)}}{\text{Depreciation period in hours}}$$

$$= \frac{100,000}{10,000}$$

= 10.00

Interest, insurance, taxes

Rate   Int. __9%__   Insc __4%__   Taxes __5%__

Estimated yearly use in hours __1,700__

$$\frac{\text{Factor x delivered price}}{1000} = \frac{.062 \times 100,000}{1000}$$

Owning cost	6.20
Total ownership cost = depreciation cost + owning cost	16.20

Figure 15-K-5
Machine no. 301

Machine inventory

Mach. no. __301__ Type __Crawler tractor__ Brand __Ace - A10__

Purchase date _____ Purchase price __$110,000__

Average hours per year use __1,700__

HP __140__   Operating weight __39,000__ / ton __19.5__

Capacity full _____ CY   Scraped _____ CY

Rated load _____   Rated RPM _____

Weight distribution

Empty drive _____ No. of drivers _____ Loaded drive _____

Rear _____   Rear _____

Maximum height _____ Maximum reach _____ Maximum depth _____

Dig unit width _____

Gear / power / weight chart

| Gear | Speed | Pounds of rim pull | | Drawbar pull | | RPM |
		Rated	Maximum	Rated	Maximum	
1	1.6			36,000	47,000	
2	2.4			25,000	32,000	
3	3.5			17,000	21,000	
4	4.7			11,000	14,000	
5	7.1			8,000	10,000	
6						
7						
8						

Comments

Figure 15-K-6
Machine no. 301

Operating Costs No. 302

Fuel

Unit price	Used / hr		
.72	28 Gal/hr		20.16

Lubricants / filters

Item	Unit price	Used / hr	
Engine	4.50 / gal	.07 gal / hr	0.32
Trans	6.00 / gal	.07 gal / hr	0.42
Finals			
Hyd	6.00 / gal	.06 gal / hr	0.36
Grease	1.25 / lb	.03 lb / hr	0.04
Filters	6 sets @ 55.00 / 1400		0.24
Other No. 101			
Total lubricants			21.54

Tires

$$\frac{\text{Replacement cost}}{\text{Estimated hours}} = \frac{-0-}{-0-}$$

-0-

Repairs

$$\frac{\text{Factor x del price - tires}}{1000} = \frac{.09 \times 165{,}000}{1000}$$

14.85

Other			
Total operation cost			36.39
Operator wages No II			12.00
Ownership cost			27.38
Total operation and ownership cost			75.77

Figure 15-K-7
Machine no. 302

Hourly Ownership Cost Estimate

Machine type Crawler tractor   **No.** 302

Purchase date _____   **Purchase price** $165,000

Depreciation value

Delivered price (total cost)				165,000

Minus tire replacement cost

Loc	Size	Qt	Amount
Front			
Rear			
Drive			

Total tires	-0-
Delivered price minus tire cost	165,000
Minus resale or trade in value	15,000
Net depreciation value	150,000

Ownership cost

Depreciation value

$$= \frac{\text{Net depreciation value (from above)}}{\text{Depreciation period in hours}}$$

$$= \frac{150,000}{10,000}$$

= 15.00

Interest, insurance, taxes

Rate **Int.** 9%   **Insc** 4%   **Taxes** 5%

Estimated yearly use in hours 1,400

$$\frac{\text{Factor} \times \text{delivered price}}{1000} = \frac{.075 \times 165,000}{1000}$$

Owning cost = 12.38

Total ownership cost = depreciation cost + owning cost = 27.38

Figure 15-K-8
Machine no. 302

Machine inventory			

Mach. no. __302__   Type __Crawler tractor__   Brand __Ace - A15__

Purchase date _____   Purchase price __$165,000__

Average hours per year use __1,400__

HP __300__   Operating weight __62,000__   / ton __31__

Capacity full _____ CY   Scraped _____ CY

Rated load _____   Rated RPM _____

Weight distribution

Empty drive _____   No. of drivers _____   Loaded drive _____

Rear _____   Rear _____

Maximum height _____   Maximum reach _____   Maximum depth _____

Dig unit width _____

Gear / power / weight chart						
		Pounds of rim pull		Drawbar pull		
Gear	Speed	Rated	Maximum	Rated	Maximum	RPm
1	1.8			58,000	68,000	
2	2.5			49,000	59,000	
3	3.2			38,000	49,000	
4	4.4			27,000	37,000	
5	5.7			19,000	26,500	
6	7.1			12,000	18,000	
7						
8						

Comments

Figure 15-K-9
Machine no. 302

Operating Costs No. 401

Fuel

Unit price	Used / hr		
.72	18 gal/hr		12.96

Lubricants / filters

Item	Unit price	Used / hr	
Engine	4.50 / gal	.06 gal / hr	0.27
Trans	6.00 / gal	.04 gal / hr	0.24
Finals			
Hyd	6.00 / gal	.06 gal / hr	0.36
Grease	1.25 / lb	.04 lb / hr	0.05
Filters	5 sets @ 75.00 / 1800		0.21
Other Bits	7 sets @ 125.00 / 1800		0.49
Total lubricants			14.58

Tires

$$\frac{\text{Replacement cost}}{\text{Estimated hours}} = \frac{3{,}000}{5{,}000}$$

0.60

Repairs

$$\frac{\text{Factor} \times \text{del price} - \text{tires}}{1000} = \frac{.05 \times 76.000}{1000}$$

3.80

Other

Total operation cost	18.98
Operator wages No II	12.00
Ownership cost	12.14
Total operation and ownership cost	43.12

Figure 15-K-10
Machine no. 401

Hourly Ownership Cost Estimate

Machine type __Motor grader__　　　No. __401__

Purchase date _____　Purchase price __$79,000__

Depreciation value

Delivered price (total cost)				79,000

Minus tire replacement cost

Loc	Size	Qt	Amount	
Front			1000	
Rear			1000	
Drive			1000	
Total tires				3,000
Delivered price minus tire cost				76,000
Minus resale or trade in value				10,000
Net depreciation value				66,000

Ownership cost

Depreciation value

$$= \frac{\text{Net depreciation value (from above)}}{\text{Depreciation period in hours}}$$

$$= \frac{66,000}{8000} = 8.25$$

Interest, insurance, taxes

Rate　Int. __9%__　Insc __4%__　Taxes __5%__

Estimated yearly use in hours __1,800__

$$\frac{\text{Factor x delivered price}}{1000} = \frac{.049 \times 79,000}{1000}$$

Owning cost　　　　　　　　　　　　　　3.89

Total ownership cost = depreciation cost + owning cost　　12.14

Figure 15-K-11
Machine no. 401

Machine inventory		
Mach. no. __401__ Type __Motor grader__		Brand __Ace - A21__
Purchase date _____		Purchase price __$79,000__
Average hours per year use __1,800__		

HP __150__   Operating weight __30,000__   / ton __15.0__

Capacity full _____ CY   Scraped _____ CY

Rated load _____   Rated RPM _____

Weight distribution

Empty drive _____   No. of drivers _____   Loaded drive _____

Rear _____   Rear _____

Maximum height _____   Maximum reach _____   Maximum depth _____

Dig unit width _____

Gear / power / weight chart

Gear	Speed	Pounds of rim pull		Drawbar pull		RPM
		Rated	Maximum	Rated	Maximum	
1						
2						
3						
4						
5						
6						
7						
8						

Comments

Figure 15-K-12
Machine no. 401

Operating Costs No. 501

Fuel

Unit price	Used / hr		
.72	27 gal/hr		19.44

Lubricants / filters

Item	Unit price	Used / hr	
Engine	4.50 / gal	.07 gal / hr	0.31
Trans	6.00 / gal	.06 gal / hr	0.36
Finals			
Hyd	6.00 / gal	.09 gal / hr	0.54
Grease	1.25 / lb	.08 lb / hr	0.81
Filters	5 sets @ 300.00 / 1450		1.03
Other Cutting bits	3 sets @ 400.00 / 1450		0.83
Total lubricants			

Tires

$$\frac{\text{Replacement cost}}{\text{Estimated hours}} = \frac{10,000}{10,000}$$

1.00

23.78

Repairs

$$\frac{\text{Factor} \times \text{del price - tires}}{1000} = \frac{.09 \times 95,000}{1000}$$

8.55

Other			
Total operation cost			33.33
Operator wages No. 1			15.00
Ownership cost			13.94
Total operation and ownership cost			62.27/hr

Figure 15-K-13
Maching no. 501

Hourly Ownership Cost Estimate

Machine type Wheel scraper   **No.** 501

Purchase date _____   **Purchase price** $95,000

Depreciation value

Delivered price (total cost)				95,000
Minus tire replacement cost				
Loc	Size	Qt	Amount	
Front				
Rear			5000	
Drive			5000	
Total tires				10,000
Delivered price minus tire cost				85,000
Minus resale or trade in value				15,000
Net depreciation value				70,000

Ownership cost

Depreciation value

$$= \frac{\text{Net depreciation value (from above)}}{\text{Depreciation period in hours}}$$

$$= \frac{70{,}000}{10{,}000}$$

= 7.00

Interest, insurance, taxes

Rate Int. 9%   Insc 4%   Taxes 5%

Estimated yearly use in hours 1,450

$$\frac{\text{Factor x delivered price}}{1000} = \frac{.073 \times 95{,}000}{1000}$$

Owning cost = 6.94

Total ownership cost = depreciation cost + owning cost = 13.94

Figure 15-K-14
Machine no. 501

Machine inventory

Mach. no. __501__   Type __Wheel scraper__   Brand __Ace - 250__

Purchase date _____   Purchase price __$95,000.00__

Average hours per year use __1,450__

HP __1,900__   Operating weight __65,000__   / ton __32.5__

Capacity full __21__ CY   Scraped __15__ CY

Rated load __42,000 lb__   Rated RPM __1,900__

Weight distribution

Empty drive __65%__   No. of drivers __2__   Loaded drive __57%__

Rear __35%__   Rear __43%__

Maximum height _____   Maximum reach _____   Maximum depth _____

Dig unit width _____

Gear / power / weight chart

| Gear | Speed | Pounds of rim pull | | Drawbar pull | | RPM |
		Rated	Maximum	Rated	Maximum	
1	2.5	52,105	67,910			
2	5.0	48,000	61,235			
3	7.5	40,005	51,063			
4	10.0	32,000	46,729			
5	14.0	21,979	37,031			
6	19.5	12,163	26,019			
7	24.0	7,211	19,306			
8	30.0	4,107	15,601			

Comments

Figure 15-K-15
Machine no. 501

Operating Costs No. 502

Fuel

Unit price	Used / hr		
.72	29 gal/hr		20.88

Lubricants / filters

Item	Unit price	Used / hr	
Engine	4.50 / gal	.06 gal / hr	0.27
Trans	6.00 / gal	.08 gal / hr	0.48
Finals			
Hyd	6.00 / gal	.10 gal / hr	0.60
Grease	1.25 / lb	.09 lb / hr	0.11
Filters	5 sets @ 300.00 / 1450		1.03
Other Cutting bits	3 sets @ 400.00 / 1450		0.83
Total lubricants			25.25

Tires

$$\frac{\text{Replacement cost}}{\text{Estimated hours}} = \frac{10{,}000}{10{,}000}$$

1.00

Repairs

$$\frac{\text{Factor} \times \text{del price - tires}}{1000} = \frac{.09 \times 85.000}{1000}$$

7.65

Other			
Total operation cost			33.90
Operator wages No. 1			15.00
Ownership cost			13.94
Total operation and ownership cost			62.84

Figure 15-K-16
Machine no. 502

Hourly Ownership Cost Estimate

Machine type Wheel scraper No. 502
Purchase date _____ Purchase price $95,000

Depreciation value

Delivered price (total cost)				95,000
Minus tire replacement cost				
Loc	Size	Qt	Amount	
Front				
Rear			5000	
Drive			5000	
Total tires				10,000
Delivered price minus tire cost				85,000
Minus resale or trade in value				15,000
Net depreciation value				70,000

Ownership cost

Depreciation value

$$= \frac{\text{Net depreciation value (from above)}}{\text{Depreciation period in hours}}$$

$$= \frac{70{,}000}{10{,}000}$$

7.00

Interest, insurance, taxes

Rate Int. 9% Insc 4% Taxes 5%

Estimated yearly use in hours 1,450

$$\frac{\text{Factor x delivered price}}{1000} = \frac{.073 \times 95{,}000}{1000}$$

Owning cost — 6.94

Total ownership cost = depreciation cost + owning cost — 13.94

Figure 15-K-17
Machine no. 502

Machine inventory

Mach. no. __502__   Type __Wheel scraper__   Brand __Ace - 250__

Purchase date _____   Purchase price __$95,000.00__

Average hours per year use __1,450__

HP __330__   Operating weight __65,000__   / ton __32.5__

Capacity full __21__ CY   Scraped __15__ CY

Rated load __42,000 Lb__   Rated RPM __1,900__

Weight distribution

Empty drive __65%__   No. of drivers __2__   Loaded drive __57%__

Rear __35%__   Rear __43%__

Maximum height _____   Maximum reach _____   Maximum depth _____

Dig unit width _____

Gear / power / weight chart

Gear	Speed	Pounds of rim pull		Drawbar pull		RPM
		Rated	Maximum	Rated	Maximum	
1	2.5	52,105	67,910			
2	5.0	48,000	61,235			
3	7.5	40,005	51,063			
4	10.0	32,000	46,729			
5	14.0	21,979	37,031			
6	19.5	12,163	26,019			
7	24.0	7,211	19,306			
8	30.0	4,107	15,601			

Comments

Figure 15-K-18
Machine no. 502

Operating Costs No. 601

Fuel

Unit price	Used / hr		
.72	18 gal/hr		12.96

Lubricants / filters

Item	Unit price	Used / hr	
Engine	4.50 / gal	.08 gal / hr	0.36
Trans	6.00 / gal	.05 gal / hr	0.30
Finals			
Hyd	6.00 / gal	.08 gal / hr	0.48
Grease	1.25 / lb	.07 lb / hr	0.09
Filters	6 sets @ 150.00 / 2000		0.45
Other Teeth bucket	5 sets @ 375.00 / 2000		0.94
Total Lubricants			

Tires

$$\frac{\text{Replacement cost}}{\text{Estimated hours}} = \frac{-0-}{-0-}$$

-0-
15.58

Repairs

$$\frac{\text{Factor x del price - tires}}{1000} = \frac{.09 \times 175{,}000}{1000}$$

15.75

Other			
Total operation cost			31.33
Operator wages No. 1			15.00
Ownership cost			24.95
Total operation and ownership cost			71.28

Figure 15-K-19
Machine no. 601

Hourly Ownership Cost Estimate

Machine type _Tractor loader   No. 601_

Purchase date _____   Purchase price _$175,000_

Depreciation value

Delivered price (total cost)				175,000
Minus tire replacement cost				
Loc	Size	Qt	Amount	
Front				
Rear				
Drive				
Total tires				-0-
Delivered price minus tire cost				175,000
Minus resale or trade in value				20,000
Net depreciation value				155,000

Ownership cost

Depreciation value

$$= \frac{\text{Net depreciation value (from above)}}{\text{Depreciation period in hours}}$$

$$= \frac{155{,}000}{10{,}000}$$

15.50

Interest, insurance, taxes

Rate   Int. _9%_   Insc _4%_   Taxes _5%_

Estimated yearly use in hours _2,000_

$$\frac{\text{Factor} \times \text{delivered price}}{1000} = \frac{.054 \times 175{,}000}{1000}$$

Owning cost — 9.45

Total ownership cost = depreciation cost + owning cost — 24.95

Figure 15-K-20
Machine no. 601

Machine inventory		

Mach. no. _601_   Type _Track loader_   Brand _Ace - H5_

Purchase date _____   Purchase price _$175,000.00_

Average hours per year use _2,000_

HP _110_   Operating weight _40,000_ / ton _____

Capacity full _____ CY   Scraped _____ CY

Rated load _____   Rated RPM _____

Weight distribution

Empty drive _____   No. of drivers _____   Loaded drive _____

Rear _____   Rear _____

Maximum height _____   Maximum reach _____   Maximum depth _____

Dig unit width _____

Gear / power / weight chart						
		Pounds of rim pull		**Drawbar pull**		
Gear	Speed	Rated	Maximum	Rated	Maximum	RPM
1	1.5					
2	3.1					
3	7.0					
4						
5						
6						
7						
8						

Comments

Figure 15-K-21
Machine no. 601

Operating Costs No. 701

Fuel			
Unit price	Used/Hr		
.72	19 gal/hr		13.68

Lubricants / filters			
Item	Unit price	Used / hr	
Engine	4.50 / gal	.07 gal / hr	0.32
Trans	6.00 / gal	.03 gal / hr	0.18
Finals			
Hyd	6.00 / gal	.09 gal / hr	0.54
Grease	1.25 / lb	.08 lb / hr	0.10
Filters	9 sets @ 95.00 / 1800		.48
Other Bucket teeth	12 sets @ 180 per set / 1800		1.20
Total lubricants			16.50

Tires

$$\frac{\text{Replacement cost}}{\text{Estimated hours}} = \frac{-0-}{-0-}$$

-0-

Repairs

$$\frac{\text{Factor x del price - tires}}{1000} = \frac{.09 \times 250{,}000}{1000}$$

22.50

Other	
Total operation cost	39.00
Operator wages No. 1	15.00
Ownership cost	37.00
Total operation and ownership cost	91.00

Figure 15-K-22
Machine no. 701

Hourly Ownership Cost Estimate

Machine type Track excavator No. 701

Purchase date _____   Purchase price $250,000

Depreciation value

Delivered price (total cost)				250,000
Minus tire replacement cost				

Loc	Size	Qt	Amount
Front			
Rear			
Drive			

Total tires	-0-
Delivered price minus tire cost	250,000
Minus resale or trade in value	25,000
Net depreciation value	225,000

Ownership cost

Depreciation value

$$= \frac{\text{Net depreciation value (from above)}}{\text{Depreciation period in hours}}$$

$$= \frac{225,000}{10,000} = 22.50$$

Interest, insurance, taxes

Rate Int. 9%   Insc 4%   Taxes 5%

Estimated yearly use in hours 1,800

$$\frac{\text{Factor x delivered price}}{1000} = \frac{.058 \times 250,000}{1000}$$

Owning cost   14.50

Total ownership cost = depreciation cost + owning cost   37.00

Figure 15-K-23
Machine no. 701

Machine inventory

Mach. no. 701   **Type** Track excavator   **Brand** Ace - R200

Purchase date _____   **Purchase price** $250,000

Average hours per year use 1,800

HP 102   **Operating weight** 38,000 _____ / ton _____

Capacity full _____ CY   **Scraped** _____ CY

Rated load _____   **Rated RPM** _____

Weight distribution

Empty drive _____   **No. of drivers** _____   **Loaded drive** _____

Rear _____   **Rear** _____

Maximum height 9'9"   **Maximum reach** 29'   **Maximum depth** 20'

Dig unit width 36"

Gear / power / weight chart

Gear	Speed	Pounds of rim pull		Drawbar pull		RPM
		Rated	Maximum	Rated	Maximum	
1						
2						
3						
4						
5						
6						
7						
8						

Comments

Figure 15-K-24
Machine no. 701

Operating Costs No. 801

Fuel

Unit price	Used / hr		
.70	3.1 gal/hr		2.17

Lubricants / filters

Item	Unit price	Used / hr	
Engine	4.50 / gal	.05 gal / hr	0.23
Trans	6.00 / gal	.01 gal / hr	0.06
Finals			
Hyd	6.00 / gal	.07 gal / hr	0.42
Grease	1.25 / lb	.03 lb / hr	0.04
Filters	5 sets @ 22.50 / 1900		0.06
Other No. 101			
Total lubricants			2.98

Tires

$$\frac{\text{Replacement cost}}{\text{Estimated hours}} = \frac{1,000}{12,000}$$

0.08

Repairs

$$\frac{\text{Factor x del price - tires}}{1000} = \frac{.06 \times 69,000}{1000}$$

4.14

Other	
Total operation cost	7.20
Operator wages No. IV	10.00
Ownership cost	9.18
Total operation and ownership cost	26.38

Figure 15-K-25
Machine no. 801

Hourly Ownership Cost Estimate

Machine type _Boom truck   No. 801_

Purchase date _____ Purchase price _$70,000_

Depreciation value				
Delivered price (total cost)				70,000
Minus tire replacement cost				
Loc	Size	Qt	Amount	
Front				
Rear			500.00	
Drive			500.00	
Total tires				1000
Delivered price minus tire cost				69,000
Minus resale or trade in value				5,000
Net depreciation value				64,000

Ownership cost

Depreciation value

$$= \frac{\text{Net depreciation value (from above)}}{\text{Depreciation period in hours}}$$

$$= \frac{64,000}{12,000} = 5.33$$

Interest, insurance, taxes

Rate   Int. _9%_   Insc _4%_   Taxes _5%_

Estimated yearly use in hours _1,900_

$$\frac{\text{Factor} \times \text{delivered price}}{1000} = \frac{.055 \times 70,000}{1000}$$

Owning cost 3.85

Total ownership cost = depreciation cost + owning cost 9.18

Figure 15-K-26
Machine no. 801

| Machine inventory |

Mach. no. __801__   Type __Boom truck__   Brand __Right - 2000__

Purchase date _____   Purchase price __$70,000__

Average hours per year use __1,900__

HP __65__   Operating weight __9,000__ / ton _____

Capacity full _____ CY   Scraped _____ CY

Rated load _____   Rated RPM _____

Weight distribution

Empty drive _____   No. of drivers _____   Loaded drive _____

Rear _____   Rear _____

Maximum height _____   Maximum reach _____   Maximum depth _____

Dig unit width _____

| Gear / power / weight chart |

Gear	Speed	Pounds of rim pull		Drawbar pull		RPM
		Rated	Maximum	Rated	Maximum	
1						
2						
3						
4						
5						
6						
7						
8						

| Comments |

Figure 15-K-27
Machine no. 801

Operating Costs No. 901

Fuel

Unit price	Used/Hr		
.72	3.4 gal/hr		2.45

Lubricants / filters

Item	Unit price	Used / hr	
Engine	4.50 / gal	.05 gal / hr	0.23
Trans	6.00 / gal	.07 gal / hr	0.42
Finals			
Hyd	6.00 / gal	.06 gal / hr	0.36
Grease	1.25 / lb	.02 lb / hr	0.03
Filters	5 sets @ 25.00 / 8.00		0.16
Other Teeth	40 sets @ 3.50 / 8.00		0.18
Total lubricants			3.83

Tires

$$\frac{\text{Replacement cost}}{\text{Estimated hours}} = \frac{-0-}{-0-}$$

-0-

Repairs

$$\frac{\text{Factor} \times \text{del price} - \text{tires}}{1000} = \frac{.03 \times 95{,}000}{1000}$$

2.85

Other			
Total operation cost			6.68
Operator wages No. II			12.00
Ownership cost			20.13
Total operation and ownership cost			38.81

Figure 15-K-28
Machine no. 901

Hourly Ownership Cost Estimate

Machine type Trench compactor   No. 901
Purchase date _____   Purchase price $95,000

Depreciation value

Delivered price (total cost)				95,000
Minus tire replacement cost				

Loc	Size	Qt	Amount
Front			
Rear			
Drive			

Total tires	-0-
Delivered price minus tire cost	95,000
Minus resale or trade in value	10,000
Net depreciation value	85,000

Ownership cost

Depreciation value

$$= \frac{\text{Net depreciation value (from above)}}{\text{Depreciation period in hours}}$$

$$= \frac{85,000}{8,000} = 10.63$$

Interest, insurance, taxes

Rate Int. 9%   Insc 4%   Taxes 5%

Estimated yearly use in hours 800

$$\frac{\text{Factor} \times \text{Delivered price}}{1000} = \frac{.10 \times 95,000}{1000}$$

Owning cost = 9.50

Total ownership cost = depreciation cost + owning cost = 20.13

Figure 15-K-29
Machine no. 901

| Machine inventory |

Mach. no. __901__   Type __Trench compactor__   Brand __Ace - 102__

Purchase date _____   Purchase price __$95,000__

Average hours per year use __800__

HP __45__   Operating weight __7000__ / ton _____

Capacity full _____ CY   Scraped _____ CY

Rated load _____   Rated RPM _____

Weight distribution

Empty drive _____   No. of drivers _____   Loaded drive _____

Rear _____   Rear _____

Maximum height _____   Maximum reach _____   Maximum depth _____

Dig unit width __30"__

| Gear / power / weight chart |

Gear	Speed	Pounds of rim pull		Drawbar pull		RPM
		Rated	Maximum	Rated	Maximum	
1						
2						
3						
4						
5						
6						
7						
8						

| Comments |

Figure 15-K-30
Machine no. 901

Operating Costs No. 1001

Fuel   None

Unit price	Used / hr		

Lubricants / filters

Item	Unit price	Used / hr	
Engine			
Trans			
Finals			
Hyd			
Grease			
Filters			
Other _____			
Total lubricants			

Tires

$$\frac{\text{Replacement cost}}{\text{Estimated hours}} = \underline{\qquad}$$

Repairs

$$\frac{\text{Factor x del price - tires}}{1000} = \underline{\qquad}$$

Other		
Total operation cost		-0-
Operator wages (supervisor)		-0-
Ownership cost		2.53
Total operation and ownership cost		2.53

Figure 15-K-31
Machine no. 1001

Hourly Ownership Cost Estimate

Machine type Laser gun No. 1001

Purchase date _____ **Purchase price** $7,500.00

Depreciation value				
Delivered price (total cost)				7,500
Minus tire replacement cost				
Loc	Size	Qt	Amount	
Front				
Rear				
Drive				
Total tires				-0-
Delivered price minus tire cost				7,500
Minus resale or trade in value				-0-
Net depreciation value				7,500

Ownership cost

Depreciation value

$$= \frac{\text{Net depreciation value (from above)}}{\text{Depreciation period in hours}}$$

$$= \frac{7,500}{4,000}$$

= 1.88

Interest, insurance, taxes

Rate Int. _9%_ Insc _4%_ Taxes _5%_

Estimated yearly use in hours _1000_

$$\frac{\text{Factor} \times \text{delivered price}}{1000} = \frac{.10 \times 7,500}{1000}$$

Owning cost ..... .75

Total ownership cost = depreciation cost + owning cost 2.53

Figure 15-K-32
Machine no. 1001

| Machine inventory |

Mach. no. __1001__   Type __Laser gun__   Brand __Tong - 175__

Purchase date _____   Purchase price __7,500__

Average hours per year use __1000__

HP _____   Operating weight _____ / ton _____

Capacity full _____ CY   Scraped _____ CY

Rated load _____   Rated RPM _____

Weight distribution

Empty drive _____   No. of drivers _____   Loaded drive _____

Rear _____   Rear _____

Maximum height _____   Maximum reach _____   Maximum depth _____

Dig unit width _____

| Gear / power / weight chart |

Gear	Speed	Pounds of rim pull		Drawbar pull		RPM
		Rated	Maximum	Rated	Maximum	
1						
2						
3						
4						
5						
6						
7						
8						

| Comments |

Figure 15-K-33
Machine no. 1001

Operating Costs No. 1101

Fuel

Unit price	Used / hr		
			-0-

Lubricants / filters

Item	Unit price	Used / hr	
Engine			-0-
Trans			-0-
Finals			
Hyd			
Grease	1.25 / lb	0.05 lb/hr	0.06
Filters			-0-
Other _____			
Total lubricants			0.06

Tires

$$\frac{\text{Replacement cost}}{\text{Estimated hours}} = \frac{-0-}{-0-}$$ -0-

Repairs

$$\frac{\text{Factor x del price - tires}}{1000} = \frac{.01 \times 25{,}000}{1000}$$ 0.25

Other		
Total operation cost		0.31
Operator wages (supervisor)		-0-
Ownership cost		5.75
Total operation and ownership cost		6.06

Figure 15-K-34
Machine no. 1101

Hourly Ownership Cost Estimate

Machine type _Pull type sheepsfoot_ No. _1101_

Purchase date _____ Purchase price _$25,000_

Depreciation value

Delivered price (total cost)				25,000
Minus tire replacement cost				
Loc	Size	Qt	Amount	
Front				
Rear				
Drive				
Total tires				-0-
Delivered price minus tire cost				25,000
Minus resale or trade in value				-0-
Net depreciation value				25,000

Ownership cost

Depreciation value

$$= \frac{\text{Net depreciation value (from above)}}{\text{Depreciation period in hours}}$$

$$= \frac{25,000}{10,000} = 2.50$$

Interest, insurance, taxes

Rate Int. _9%_ Insc _4%_ Taxes _5%_

Estimated yearly use in hours _700_

$$\frac{\text{Factor x delivered price}}{1000} = \frac{.13 \times 25,000}{1000}$$

Owning cost 3.25

Total ownership cost = depreciation cost + owning cost 5.75

Figure 15-K-35
Machine no. 1101

| Machine inventory |

Mach. no. __1101__   Type __Pull sheepsfoot__   Brand __Ace HR-5__

Purchase date _____   Purchase price __$25,000__

Average hours per year use __700__

HP _____   Operating weight __22,000__ / ton _____

Capacity full _____ CY   Scraped _____ CY

Rated load _____   Rated RPM _____

Weight distribution

Empty drive _____   No. of drivers _____   Loaded drive _____

Rear _____   Rear _____

Maximum height _____   Maximum reach _____   Maximum depth _____

Dig unit width __12'__

| Gear / power / weight chart |

Gear	Speed	Pounds of rim pull		Drawbar pull		RPM
		Rated	Maximum	Rated	Maximum	
1						
2						
3						
4						
5						
6						
7						
8						

| Comments |

Figure 15-K-36
Machine no. 1101

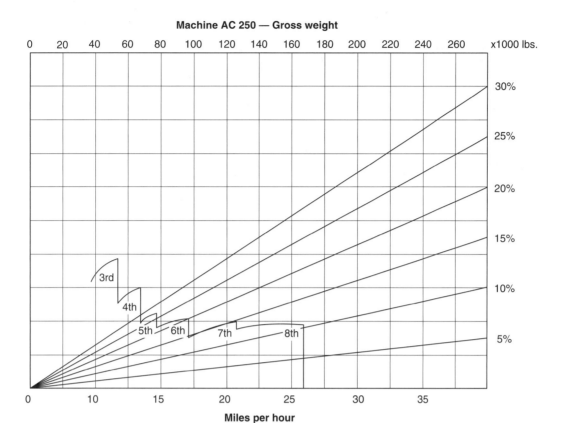

To use: Locate gross weight on top scale. Move down to intercept the effective grade line. Move to the left to intercept the retarder curve in gear range. Read down to intercept point on speed scale in mph.

Figure 15-K-37
Retarder chart

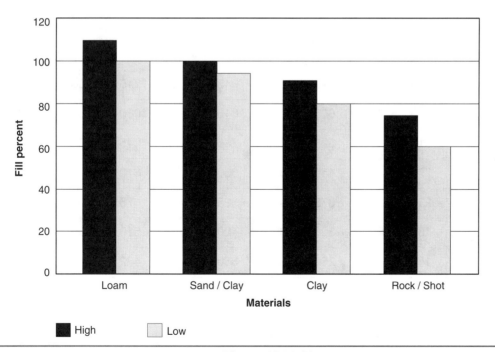

Figure 15-K-38
Bucket payload factor

Costs and Final Bid for the Sample Estimate **431**

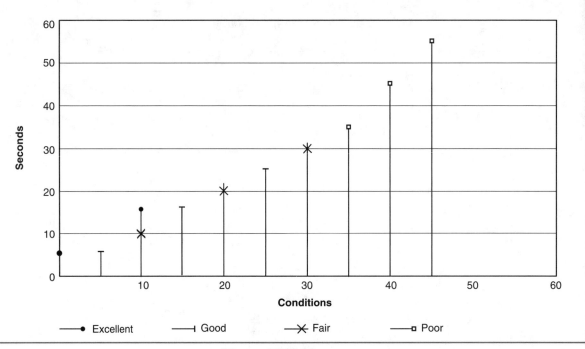

Figure 15-K-39
Backhoe cycle time

Estimated cycle times (seconds)	Estimated bucket payload									Cycles per hour
	1.0	1.5	2.0	2.5	3.0	3.5	4.0	4.5	5.0	
10										
11										
12	300									
13	270	404	540	675	810	945	1012	1215	1350	270
15	240	360	480	600	720	840	960	1080	1200	240
17	210	315	420	525	630	735	840	945	1050	210
20	180	270	350	450	540	630	720	810	900	180
24	150	225	300	375	450	525	600	675	750	150
30	120	180	240	300	360	420	480	510	600	120
35	102	154	205	256	308	360	410	462	513	102
40		135	180	225	270	315	360	405	450	90
45				200	240	280	320	360	400	78

Figure 15-K-40
Backhoe production per 60 minutes

Calculation Form

Line item __8" C.I.P.__   Segment _____

Mach. No.	O&O Cost / hr	Average Hours / Yr	O&O Cost / Year	% Overhead	Term Overhead	Overhead Cost / Hr	Total Hr / Cost
101	20.83	2200	45,826.00	5.0	5,147.50	2.34	23.17
301	53.71	1700	91,307.00	10.0	10,295.00	6.06	59.77
302	75.77	1400	106,078.00	11.8	12,148.10	8.68	84.23
401	43.12	1800	77,616.00	9.0	9,265.50	5.15	48.27
501	62.27	1450	90,291.00	10.0	10,295.00	7.10	69.37
502	62.84	1450	91,118.00	10.0	10,295.00	7.10	69.37
601	71.28	2000	142,560.00	16.0	16,472.00	8.24	79.52
701	91.00	1800	172,900.00	19.0	19,560.50	10.29	101.29
801	26.38	1900	50,122.00	5.5	5,662.25	2.98	29.36
901	38.81	800	31,048.00	3.0	3,088.50	3.86	42.67
1001	2.53	1000	2530.00	0.3	308.85	0.31	2.84
1101	6.06	700	4242.00	0.4	411.80	0.58	6.64
			905,638.00	100.00%	102,950.00		

page __1__ of __1__

Figure 15-L-1
Machine costs per hour

Blank Worksheets

■ ■ ■ ■ ■ ■ ■ ■ ■ ■

You can use these worksheets in estimating your jobs. Use your copy machine, or retype them into your computer, where you can customize them to fit your needs.

- Individual Grid Square Area and Volume Worksheets
- Grid Take-off – Existing Contour Only
- Grid Take-off – Proposed Contour Only
- Individual Grid Square Calculation Sheet
- Cut and Fill Prism Calculations Worksheet
- Quantities Take-off Sheet

Individual grid square area and volume worksheet
Grid square: _____   Area = l × w

Factors	Existing contour (symbol :)				(Proposed contour symbol:)			
	ul	ur	ll	lr	ul	ur	ll	lr
Out								
In								
Diff								
Dist								
Out±								
In±								
Point elevation								
Average elevation								

Fill volume (CY) = [(average proposed elevation − average existing elevation) × grid area] ÷ 27

Individual grid square area and volume worksheet
Grid square: _____   Area = l × w

Factors	Existing contour (symbol :)				(Proposed contour symbol:)			
	ul	ur	ll	lr	ul	ur	ll	lr
Out								
In								
Diff								
Dist								
Out±								
In±								
Point elevation								
Average elevation								

Fill volume (CY) = [(average proposed elevation − average existing elevation) × grid area] ÷ 27

Grid Take-off, Existing Contour Only

Sheet: _____ of _____

Prepared by (initials): _____   Date: _____
Approved by (initials): _____   Date: _____

Location	Low elevation	High elevation	Scale distance	Contour interval	Add elevation	Point elevation

Grid Take-off, Proposed Contour Only

Sheet: _____ of _____

Prepared by (initials): _____   Date: _____

Approved by (initials): _____   Date: _____

Location	Low elevation	High elevation	Scale distance	Contour interval	Add elevation	Point elevation

Note:

Individual Grid Square Calculation Sheet

Job number: _____ Project: _____ Prepared by (initials): _____ Date: _____

Sheet: _____ of _____ Approved by (initials): _____ Date: _____

Grid _____
Average depth:

Element	1	2	3	4
Proposed				
Existing				
Depth				

Grid _____
Average depth:

Element	1	2	3	4
Proposed				
Existing				
Depth				

Grid _____
Average depth:

Element	1	2	3	4
Proposed				
Existing				
Depth				

Grid _____
Average depth:

Element	1	2	3	4
Proposed				
Existing				
Depth				

Grid _____
Average depth:

Element	1	2	3	4
Proposed				
Existing				
Depth				

Cut and Fill Prism Calculations Worksheet

Project:_____ Date:_____

By:_____

All (cut or fill):_____ Checked by: _____

Grid	Corner	No.	Depth	Total depth (feet)	Notes
Totals					

Quantities Take-off Sheet

Project: _____ **Date:** _____

Quantities for: _____ **Sheet** ____ **of** ____

By: _____ **Checked:** _____ **Misc:** _____

Note:

Index

A

AASHTO 36
Access, equipment 198
Accessibility, job site 19
Accounting fees, overhead 234
Accuracy
 checking, effect on 78
 rounding, effect on 77-78
Adobe 33
Aerial maps .15
Altitude
 definition .216
 horsepower, effect on 216
 usable power, effect on 215
American Association of State
 Highway and Transportation
 Officials (AASHTO) 36
American Society of Testing
 Materials (ASTM) 36
American Soil Conservation
 Service (ASCS) 32
Angle of repose 156, 186, 189
Angle, reverse 187
Arc section, measuring by 50
Area
 circle 101, 150
 job site, formula for 137
 oblique triangle 105-108
 triangle 82
 using compensating lines . .104, 114-117
 using Trapezoidal Rule 119-120
 worksheet .86
Area available, stockpile 190-193
ASCS (American Soil Conservation
 Service) 32
Asphalt road, coefficient of traction . .215
ASTM (American Society of Testing
 Materials) .36
Atmospheric pressure 216
Available power, equipment 214
Average area, formula 55
Average depth, formula 96
Average elevation 89
Average end area, volume, formula . . .84
Average end method
 cut and fill areas 82
 trapezoidal prism 83
Average operating speed, compactors .228
Average slope line 158

B

Balance points
 engineer, to 196-197
 excavation, in 195-196
Bank cubic yards (BCY) 127
Bank material, defined 127
Bank run gravel 33
Barricades, traffic 23, 25
Baseline
 horizontal 207
 surveying .44
 vertical 208
Basement excavation
 calculating total volume 164
 equivalent area 158
 estimating 155-178
 finding real depth 168
 sample estimate 170-175
 slope angles 156
 wall dimensions 161
 work space allowance 157
 worksheet, volume calculations . . .166
Basement wall dimensions 160-161
BCY (bank cubic yards) 127
Bedding material, calculations 150
Bedding, trench 149
Bedrock 33
Beginning station 57
Bells, volume 178-179
Bench mark .45
Bench marks 45, 67-68
Bid price 237
Bid, sample 335-433
Bidding process 209
Blue Book values 231
Boost time, pusher units 226
Boring log 31-32
Borrow
 all about 181-193
 definition 181
 distance to site 19
 hauling 182
Borrow pit
 costs 181-182
 location .198
Boulders 33
Boundary lines 25
Braking force, effect of grade
 assistance on 213
Bridges, job site 19
Brush, job site 20
Buggy (polar planimeter) 47
Building and grounds, overhead 234
Bulldozer, production rates 227
Burning vegetation 25

C

Calculating
 cubic yard costs 7
 cut/fill, shortcut 96
 missing corner 99
 shortcuts .91
 using equivalent area 158
Calculator, use in estimating 12
Caliche 33
Carpenter's square, stockpile height,
 finding 187
Cast (soil characteristic) 34
CCY (compacted cubic yards) 128
Center of mass 196
 depth not uniform 203
 distance to edge 201
 example .208
 finding 200-201
 formulas 202, 205
 uniform depth 203
 vertical 203
Center to center dimensions, wall 161
Centerline, road slope 146-147
Channels, drainage 147
Checklist, site visit 17, 28-30
Circle
 area 101
 area formula 150
 center of mass 202
 haul distance for 201
Circumference, contour lines 100-101
Clay 33-34
Cobbles 33
Coefficient of traction
 example .216
 factors 215
Column headings, worksheet . . .86-89, 92
Compacted cubic yards (CCY) 128
Compacted material, defined 128
Compaction
 requirements 24
 soil 35
 Standard Proctor percent 130
 test diagrams 37
 testing 40
Compactor
 average operating speed 228
 production rates 227
Compensating lines 103
 formula to find area 114
 using to find volume 110-113
Computer, use in estimating 12
Concrete-lined ditch 147-148
Concrete road, coefficient of traction .215
Concrete, utility lines in 151
Condition of haul road
 effect on cycle time 218
 effect on rolling resistance 211
Conditions, surface 19
Cone, volume 186, 193
 formula .163
Constants
 brush volume, for 20
 planimeter 49
 scale 77
Construction material depth (TI) 142
Construction scheduling, site visit 18
Contour interval 64, 100
Contour lines
 characteristics 62-65
 circumference 100-101
 comparing 71
 elevation .66
 intermediate 65, 92
 measuring length 100
 reading 74
Contour maps, reading 61
Contour profile 69

Index **441**

Contractor's bidding process 209
Contractor's responsibilities,
  specified in plans 8
Contracts, overcut payment clause . . .153
Controlling traffic 23
Conversion chart, inches to
  decimal feet 11
Coordinate system, using to find
  volume 110-113
Corner elevations
  calculating .92
  interpolating 76-78
Corner volumes, excavating 163
Corners
  column headings 87
  grid square, identifying 86, 91
Cost information, collecting 15
Cost plus bids 237
Costs per cubic yard, formula 7
Costs, special, estimating for 7-10
Cross section method 46, 79
  estimating 99
  volume 84
Cross section
  payment based on 9
  sheets 43, 47
  worksheet 58
Cross slope 147
Crown, roadway 146
Cubic yard, cost per 209
Cubic yard estimates 6
Cubic yards per hour, production rate .219
Cut and fill
  areas 80
  combining 141
  cross section 43
  operations .42
  prism calculations worksheet 94-95
  shortcut worksheet 97
  under a structure 142
Cut calculations, worksheet 95
Cut depth, total, formula 140
Cut sections 42
Cut volume 89
Cycle time
  definition .217
  effect on cost 217
  equipment 218

D

Datum 61
Day operations, efficiency factors 220
Decimal feet, converting to 11
Decimal places 77
Density, soil 35
Depreciation, machine 230
Depth
  basement excavation 168
  calculating for corners 92-93
  calculation worksheet 93
Design, tire, effect on rolling
  resistance 210
Diameter of circle, formula 101
Difficulty, job .19
Dimensional systems 161
Dimensions, basement wall . . .160-161
Dirt road, coefficient of traction 215
Disposal
  spoil 182
  tires 233

Disposal site
  soil 24
  vegetation .25
Distance
  edge to center of mass 201
  effect on cycle time 218
Ditches
  drainage .147
  excavation 148
Downhill travel, total resistance 212
Drain slope 147
Drainage
  channels .147
  job site 39
  planning for 39
  problems 19-20
  slopes 143, 146-147
Drawbar pounds of pull 214
Dump sites 21
  distance to .19
  spoil 182

E

Earthmoving equipment *See Equipment*
Earthwork
  calculating net volumes 139-142
  estimates, types of 6-7
  estimating, skills needed 5
Easements, job site 25
Efficiency, effect on machine
  production 218
Efficiency factors, chart 220
Electrical lines, marking 22
Element, column heading 92
Elevation
  average 89
  between contour lines 66
  design 41
  final 41
  proposed 41
Elevation changes, cut/fill 142
Elevations
  project 69
  real 68
  sloping 142
Employee benefit costs 8
Encasement pipe, calculating volume .151
End area calculations
  arc section 50
  measuring strip 50
  planimeter .47
  stockpile .184
  volume 54, 186, 193
Ending station 57
Engineer, balance points, using 196
Engineer's scale 66-67
Engineers, soil 31
Engines, effect of altitude on 216
Equal depth contour method
  estimating 99
  worksheet 101
Equipment
  access 198
  life span 230
  load factors, calculating 131-132
  owning and operating
    costs 209, 230-237
  planning for 209
  production rates 218-230
  space for 19

  steep slopes, for 19
  trench 151
Equipment selection, importance of 7
Equivalent area, calculating
  excavation volumes 158
Estimate
  practice 239-333
  sample basement excavation . . .170-175
Estimates, keeping for future 14
Estimating
  basement excavation 155
  calculating cubic yard cost 7
  calculating volume 11
  corner volumes 163
  drainage ditch volumes 147
  excavation overcut volumes 152
  haul trips 130-131
  importance of accuracy 10
  machine production rates 219
  special costs 7-10
  tools and work area, organizing 12
  trench volumes 149
  underground structure excavation . . .155
  using shrink/swell factors 128-129
  using topo maps 99
  using your own plans 15-16
Estimating procedures 13-14
Estimating skills 5
Estimator, organizing the job 209
Excavation contractors, bidding
  process 209
Excavation equipment *See Equipment*
Excavation overcut 152
Excavation volume,
  sample estimate, basement 170-175
Excavator, balance points, using 196
Existing contour 86
Experience, importance of 18
Exterior dimensions, wall 161

F

Factors
  coefficient of traction 215
  efficiency .220
  equipment load 131-132
  hourly machine cost 231-232
  job efficiency 220
  load 131-132
  machine repair 234
  rolling resistance 211
  shrink/swell 127
  trench width 149
Field distance 144
Field visit 17
Fill
  imported 24
  providing from spoil 182
  trench 150
Fill and cut areas 80
Fill calculations, worksheet 94
Fill depth, total, formula 140
Fill sections 42
Fill volume, formula 90
Finish grade .196
Fixed costs (overhead) 234
Fixed time, definition 217
Flat-bottom ditches 147-148
Flooding, job site 25
Flow line elevation 150
Foliage, volume of 20

442 *Estimating Excavation*

Formulas
  area (from planimeter reading) 49
  area of circle 101, 150
  area of job site 137
  area of oblique triangle 104
  area, using compensating lines 114
  area, using the scale factor 54
  average area 55
  average depth 96
  average end method 84
  brush volume 20
  center of mass 202, 205
  cone, volume 163
  cost per cubic yard 7
  cubic yards, converting to 82
  cut and fill under a structure, total ..142
  cut depth, total 140
  depth, corner 94
  depreciation 231
  diameter of circle 101
  distance TOS to HS 159
  fill depth, total 140
  fill volume 90
  foliage volume 20
  grade resistance 213
  load factor 132
  machine production 220
  mound volume 109
  planimeter constant 49
  point elevation 89
  production rates 225
  pusher units 226
  Pythagorean theorem 177
  ramp volume 176
  resistance, total 212
  rimpull 216
  rolling resistance 211
  run, slope 159
  scale factor 54
  shaft volume 179
  shortcut, total depth 96
  shrink/swell factors 131
  spoil volume 183
  tangent of an angle 187
  topsoil volume in CY 137
  trench volume 130
  triangle, area 82
  triangle, oblique 176
  triangle, volume 99
  V-in, V-out 167
  volume 11
  volume in CY 75
  volume of equal depth contours 102
  volume using Trapezoidal Rule 117
  weight on drive wheels 216
Four-wheel tractor, weight on
  drive wheels 216
Friction, effect on rolling resistance ..210
Fuel costs 233

G

Gear ratio, effect on machine speed ..214
General quantities 7
Grade
  cycle time, effect on 218
  finish 196
Grade assistance, effect on braking
  power needed 213
Grade beams, volume 178-179
Grade line 41

Grade resistance
  definition 210, 212
  example 213, 223
  formula 213
  negative 213
Grader, motor, production rates 229
Graphic of contour 71
Gravel, definition 33
Gravel road, coefficient of traction ..215
Green heads 22-23
Grid overlay 73
Grid square corners
  calculating depth 92
  elevations 76-78
  identifying 86, 91
Grid system
  estimating with 72-73
  identifying 73-74
  scale 73
  volume calculations 75
Gross vehicle weight, percentage
  on drive wheels 216
Grubbing vegetation 20, 25
Gumbo 33

H

Half slope line (HS) 158
Hardpan 33
Haul distance 196
  asymmetrical borrow pit 206-208
  average 201, 208
  calculating 200-203, 205
  examples 199-200
  minimizing 200
  reducing 197-200
  symmetrical borrow pit 205-206
Haul road
  calculating productivity 222
  specifications 222
Haul road condition
  effect on cycle time 218
  importance of 218
Haul trips, estimating 130-131
Heads, markers 22
Highway conditions 24
Hill, finding volume 102
Holding tank excavation, estimating ..155
Horizontal baseline 207
Horizontal datum 61
Horizontal planes method, estimating ..99
Horizontal scale 44, 54
Horsepower
  altitude, effect on 216-217
  available power, effect on 214
Hourly operating cost 236
HS (half slope line) 158

I

Ice, coefficient of traction 215
Identifying the grid system 73-74
In elevation 87-88
Inches to decimal feet, conversion
  chart 11
Inflation, tire, effect on rolling
  resistance 210
Inside corner 87
  basement 162
Inside to inside dimensions, wall ...161
Insurance 231

Interest 231
Interim spoil 182, 184
Interim stockpiles, shrink and swell ..184
Interior dimensions, wall 161
Intermediate contour lines 65, 92
Interpolating elevations 76-78
Interpolation 76
  accuracy of 78
Irregular shapes, finding volume of ..103
ITT (interest, insurance, taxes) 231

J

Job difficulty 19
Job efficiency factors, chart 220
Job site
  accessibility 19
  analyzing 18
  conditions 18
  formula for area 137

K

Knox soil 33

L

Labor, local 23
Lake, finding volume of 119-125
LCY (loose cubic yards) 128
Legal fees, overhead 234
Legends, topo maps 72
Level travel, total resistance 212
Levels, surveying 45
Lift station excavation, estimating ..155
Lines
  contour 62-64
  zero 81
Load factors, formula for 132
Load time, pusher units 226
Loading the bid 237
Loam, definition 33-34
Local soil information, importance of ..33
Loess 33
Logs
  barricade 23
  boring 32
Loose cubic yards (LCY) 128
Loose material, defined 128
Lowboys, overhead 234
Lubricant costs 233
Lump sum estimates 6

M

Machine
  life span 230
  power 210
  speed, effect on operating cost ..214
  tire value 231
  weight, effect on machine speed ..214
Machine ownership and operating
  cost 230-237
  chart 236
  summary 235
Machine production 218-230
  formula 220
Maintenance, effect on rolling
  resistance 210

Management, overhead 234
Map scale 100
Map symbols .72
Maps, locating and using 15
Maps, survey .43
Markers, survey type 22
Mass, center of *See Center of mass*
Material
   borrow pit 182
   effect on machine production 218
   job site storage 22
   selling 184
   weights, chart 132
Measuring methods
   arc method 50, 52-53
   measuring strip 50
   planimeter 47-49
Measuring tools
   engineer's scale 66
   rubber band 67
   strip 50-51
Middle section, stockpile 184
Midpoint
   horizontal .207
   vertical 208
Modified Proctor Test 36
Moisture, soil 36-38
   optimum 26
   problems 38
Monotypic soils 132
Monuments 67-68
Motor grader 218
   production rates 229
Mound, finding volume 103-109
Muck 33
Mud 33

N

National Geodetic Vertical Datum 62
Naturally-aspirated engines, effect
   of altitude on 216
Net cut/fill depths 140
Net earthwork volumes,
   calculating 139-142
Night operations, efficiency factors . .220

O

Oblique triangle, finding area of .105-108
Obstructions, job site 19
Operating costs
   fuel and lubricants 233
   repairs 233
   tires 233
Operating gear, effect on available
   power 214
Organic soils .34
Out elevation 87-88
Outside corner 87
   basement .162
Outside to outside dimensions, wall . .161
Overcut
   payment clause 153
   trenching .152
Overfilling 10
Overhead 234-237
   calculating 236
   cost per hour 236
Ownership costs 230-233
   depreciation 230
   insurance .231
   interest 231
   overhead .234
Owning and operating costs,
   equipment 209, 230-233

P

Pay yards 133
Payment
   overcut, for 153
   services, for 9
Payroll, overhead 234
Pea gravel 33
Peat 33
Pebbles 33
Percolation test 38-39
Performance records, equipment,
   importance of 210
Permanent bench marks (BM) 68
Permits 8, 25
Personnel, planning for 209
Phone lines, marking 22
Piers, volume 178-179
Pipe wall thickness, importance in
   estimating 151
Plan and profile area take-off 41-60
Plan and profile sheets 43, 45-46
   road project 204
Plan difference 142
Plan distance 144
Plan reading 8
Plan sheets 43
Plan view, sample basement
   excavation 171
Planes
   existing elevation 79
   proposed elevation 79
   trapezoidal shape 83
Planimeter 47-49
   anchoring 48
   constant 49
   reading 49
Planimetric maps 61
Planning slopes 143
Planning team, contractors 209
Plans, reviewing 17
Plasticity, soil 35
Plumb bob, stockpile height, finding . .187
Pneumatic roller, average operating
   speed 228
Point elevation, formula 89
Point of optimum moisture 36
Polar planimeter 47
Pond, finding volume 100
Pounds of pull, traction 214-215
Pounds of push, traction 215
Power, available 214
Power, machine, definition 210
Price per cubic yard 237
Prism
   stockpile middle section 184
   trapezoidal .83
   truncated 79
   volume 192
Proctor tests .35
Production calculations, material . . .226
Production rate
   calculating 219
   formula 225
   reducing .219
Production rates, machine 218-221
   bulldozer .227
   compactor 227
   formulas .225
   motor grader 229
   pusher units 226
Productivity calculations
   cycle time 225
   gear 224
   haul roads 221
   operating speed 224
   production rates 225
   resistance .222
   travel times 225
Productivity rates, equipment 209
Profiles, finish 196-197
Profit 237
Project elevations 69
Project size 22
Properties of soils 31-40
Proposed contour 86
Public records, using 15
Pull, pounds of 214
   effect on machine speed 214
   traction 215
Purchase price, machine 231
Push, pounds of, traction 215
Pusher units
   formula 226
   productivity rates 226
Pythagorean theorem 177

Q

Quadrangle (quad) sheets 62-63
Quantities
   general 7
   special 7
Quantities take-off, sample
   basement excavation 172, 174
Quantities take-off sheet 60
   sample 166
Quantities, tracking overcut 152
Quantity, calculating quantity 91
Quantity estimates 6

R

Ramps
   combination 175, 178
   estimating 175-178
   inside 175
   outside 175-178
Real depth, basement excavation 169
Real elevations 68
Record keeping, importance of 14-15
Recorder of Deeds, checking with 25
Rectangle, center of mass 202
Red heads 22-23
Relief maps 61
Relief markings 62
Repair cost, machine 233-234
Repose, angle of 186, 189
Resistance data, machine 214
Resistance, grade, definition 210
Resistance, rolling, definition 210
Resistance, total, formulas 212
Return time, pusher units 226
Reverse angle method, stockpile
   height 187-190
Ribbon (soil characteristic) 34

Rimpull
  chart 214
  definition .214
  formula 216
Rise 143
River, job site 25
Road condition, effect on rolling
  resistance .211
Road surface
  coefficients of traction 215
  effect on rolling resistance 211
Roads, slope 146-147
Rock
  undercutting for 9
  weathered .33
Rocks, job site 19
Rolling resistance
  definition .210
  estimating 210
  example .223
  factors for wheeled machines 211
  formulas .211
Rounding, effect on accuracy 77-78
Rubber band measuring tool 67
Run 143
Run/rise ratio, slope 143

S

Safety
  public 24
  slopes/trenches 146-147
Safety equipment 23
Sample
  bid 335-433
  estimate, basement excavation .170-175
  take-off 239-333
Samples, soil 35
Sand, definition 33-34
Sand road, coefficient of traction 215
Sandy loam .34
Scale
  choosing 43-44
  grid 73
  horizontal .44
  using to interpolate elevations 76
  vertical 44
Scale factor .54
Scheduling, equipment 209
Scraper, rolling resistance 211
Security, job site 24
Self-propelled compactors, average
  operating speed 228
Semicircle
  center of mass 202
  haul distance for 201
Sewage discharge 38
Sewer lines, job site 22
Shaft, volume 179
Shale 33
Sheepsfoot roller, average operating
  speed 228
Shoring, slopes/trenches 146
Shortcut
  calculating quantity 91
  formula for total depth 96
Shrink/swell factors 127, 128
  conversion chart 129
  formula for 131
  using materials weights 131
Shrinkage, spoil 183

Silt 33-34
Site problems, anticipating 18
Site visit 17
  checklist 17, 28-30
  sample 24
Size, site 22
Skills needed, estimating 5
Slope
  calculating degree of 143
  calculating volume of topsoil . .144-146
  effect on grade resistance 212
  run/rise ratio 143
  safety 146-147
  total run formula 159
Slope angles, selecting 156
Slope lines .142
  drainage .146
  estimating length, chart 145
Slope volume, outside basement wall .156
Slopes, roadway 146-147
Snow, coefficient of traction 215
Soil
  characteristics 34
  classifications 32-34
  compaction 35-36
  hauling, cost of 197-200
  loading, cost of 197
  moisture 36-39
  monotypic 132
  movement 39
  properties of 31-40
  stability .35
  states 127
  testing 31-32
  type 182
  types, maximum safe slope 147
  unstable .21
  weight charts, obtaining 131
Soils engineer 21, 40
Special quantities 7
Specifications
  sample basement excavation 170
  topsoil quantities 137
Specifications, reading 8
Speed, machine, effect on
  operating cost 214
Spoil
  all about 181-193
  definition 181
  disposal 182
  dump site 182
  interim 182, 184
  shrink and swell 183
Stability, soil 35
Standard Proctor, compaction percent .130
Standard Proctor Test 35
Stations, surveying 44
  alternate names 204
  beginning 57
  ending 57
Steep slopes, equipment for 19
Stockpile
  selling 184
  set area, volume 190-193
  unknown height, volume 186-190
  volume 184-186
Stockpile locations, topsoil 136
Storage, job site 22
Storm lines, job site 22
Stream beds, job site 21, 25
Strip, measuring 50-51
Subcontour lines 74

Subsurface conditions 21
Superelevation 147
Superintendents, overhead 234
Suppliers, local 23
Surface conditions 19
Surface structures, cut/fill under 142
Surface, road, effect on rolling
  resistance 211
Survey costs 8-9
Survey maps 43
Survey ties .22
  utilities 25
Swell, spoil 183

T

Take-off
  arc section 50
  measuring strip 50
  plan and profile 41-60
  planimeter 47-49
  sample 239-333
  topo maps, from 71-102
  worksheets 85
Tangent of an angle
  formula .187
  table 189
Tape measure, stockpile height,
  finding .187
Template 71
Temporary bench marks (TBM) 68
Test borings 24-26
Test hole, topsoil 136
Testing
  compaction 40
  Proctor 36
  scheduling around 40
  shrink/swell factors 128-130
  soil 31-32
Ties, survey 22
Till 33
Time, effect on machine production . .218
Tire design, effect on rolling
  resistance 210
Tire inflation, effect on rolling
  resistance 210
Tires
  coefficient of traction 215
  hourly cost 233
  replacement costs 233
  value 231
Toe of slope 142
Top of slab 148
Top of slope 142
Topographical (topo) maps 61
  estimating quantities, using 99
  shortcuts, calculating 91
  subcontour lines 74
  symbols 62, 72
  volume, using to find 110
Topsoil 135-142
  disposal 136
  layers 135
  quantities, calculating 137
  replaced, volume of 138-139
  stockpiling 136
  storage on site 22
  strip depth (TO) 142
  stripped, volume of 137-138
  value of 182
  volume, slope 144-146

Index 445

Total corner depth, formula 94
Total depth, shortcut formula 96
Total resistance
   example 223
   formulas 212
Total volume of cut/fill, formula 96
Track equipment, efficiency factors ..220
   rolling resistance factor 211
   slopes, on 19
   weight on drive wheels 216
   working speed 19
Tracks, coefficient of traction 215
Traction
   effect on usable power 215
   coefficients of, factors 215
Tractor
   overhead cost 234
   rolling resistance 211
   weight on drive wheels 216
Traffic conditions, job site 19
Traffic control 19, 23
Transfer time, pusher units 226
Trapezoidal prism 83
Trapezoidal Rule
   using to find area 119-120
   using to find volume 117-121
Trees, job site 20
Trench boxes 146
Trenches
   calculating fill 150
   concrete lined 151
   estimating overcut 152
   excavation equipment 151
   formula for volume 130
   slope safety 146-147
   utility 149
   width factors 149
Triangle
   area 82
   center of mass 202
   finding volume 98-99
Truck, rolling resistance 211
Truncated prism 79
Two-wheel tractor, weight on drive
   wheels 216
Tying down utility lines 22-23

U

Undercutting 9-10
Underground structure excavation,
   estimating 155
Unstable slopes, equipment for 19
Unstable soil 21
Uphill travel, total resistance 212
U.S. Coast and Geodetic Survey 68
U.S. Geodetic Survey 67-68
U.S. Geological Survey (USGS) 62
   maps 15

Usable power 215
Utilities, overhead 234
Utility easements 25
Utility lines 25
   excavating for 150
   job site 22
   locating 8
   marking 22
   set in concrete 151
Utility trenches 149

V

V-in/V-out formulas 167
V-out calculations, sample 173
Vandalism, job site 24
Variable time, definition 218
Vee ditches 147-148
Vegetation, job site 19-20
   burning 25
Vernier, planimeter 49
Vertical center of mass 203
Vertical datum 61
Vertical scale 44, 54
Vertical wall excavations 165
Visit, site 17
Void ratio 38
Volume
   average area 55-56
   average end area 84
   average end method 82, 83
   bells 178-179
   cone 163, 186, 193
   coordinate system, using 110-113
   cross section method 79, 84
   cut and fill areas 82, 96
   end area 193
   end area calculations 54
   end areas, combined 184
   equal depth contours, formula 102
   equivalent area, calculating by 160
   exterior basement excavation 160
   formula 11
   formula using Trapezoidal Rule 117
   grade beams 178-179
   hill 102
   interpolation 76-78
   irregular areas 103
   missing corner, calculating 99
   mound 103-109
   mound, using average depth 109
   mound, using compensating
     lines 105-113
   piers 178-179
   pond 100
   prism 184, 192
   ramp 175-178
   replaced topsoil 138-139
   sample basement excavation ...170-175

   shaft 179
   sloping wall basement
     excavation 165-168
   small lake 119-125
   spoil 183-184
   stockpile 184-186
   stockpile of set area 190-193
   stockpile of unknown height ...186-190
   stripped topsoil 137-138
   topo maps, using 110
   topsoil, slope 144-146
   trapezoidal prism 83
   trees, job site 20
   trench bedding 150
   trench formula 130
   triangle 98-99
   vertical wall basement excavations .165
   worksheet 86

W

Wall dimensions, basement 160-161
Water
   drainage 19-20
   job site 25
   problems 38
Water lines, marking 22
Water table, job site 21, 38
Weathered rock 33
Weight, machine, effect on machine
   speed 214
Weight of drive wheels, formula 216
Weight on wheels, calculating 212
Wheel equipment
   efficiency factors 220
   rolling resistance factor 211
Wheel scraper
   grade resistance 213
   usable power 215
Work space allowance, basement
   excavation 157
Workers, local 23
Worksheet
   column headings 86-89, 92
   cut/fill prism calculations 94-95
   depth calculation 93
   equal depth contour volume 101
   individual grid square area 86
   take-off 85
   shortcut for cut/fill 97
   volume 86

X, Y, Z

Yards, pay 133
Yellow heads 22-23
Zero line 80, 91-92

Practical References for Builders

Excavation & Grading Handbook Revised

The foreman's, superintendent's and operator's guide to highway, subdivision and pipeline jobs: how to read plans and survey stake markings, set grade, excavate, compact, pave and lay pipe on nearly any job. Includes hundreds of informative, on-the-job photos and diagrams that even experienced pros will find invaluable. This new edition has been completely revised to be current with state-of-the-art equipment usage and the most efficient excavating and grading techniques. You'll learn how to read topo maps, use a laser level, set crows feet, cut drainage channels, lay or remove asphaltic concrete, and use GPS and sonar for absolute precision. For those in training, each chapter has a set of self-test questions, and a Study Center CD-ROM included has all 250 questions in a simple interactive format to make learning easy and fun. **512 pages, 8½ x 11, $42.00**

CD Estimator

If your computer has *Windows*™ and a CD-ROM drive, CD Estimator puts at your fingertips over 135,000 construction costs for new construction, remodeling, renovation & insurance repair, home improvement, framing & finish carpentry, electrical, concrete & masonry, painting, and plumbing & HVAC. Monthly cost updates are available at no charge on the Internet. You'll also have the *National Estimator* program — a stand-alone estimating program for *Windows*™ that *Remodeling* magazine called a "computer wiz," and *Job Cost Wizard*, a program that lets you export your estimates to QuickBooks Pro for actual job costing. A 60-minute interactive video teaches you how to use this CD-ROM to estimate construction costs. And to top it off, to help you create professional-looking estimates, the disk includes over 40 construction estimating and bidding forms in a format that's perfect for nearly any *Windows*™ word processing or spreadsheet program. **CD Estimator is $78.50**

Markup & Profit: A Contractor's Guide

In order to succeed in a construction business, you have to be able to price your jobs to cover all labor, material and overhead expenses, *and* make a decent profit. The problem is knowing what markup to use. You don't want to lose jobs because you charged too much, and you don't want to work for free because you charged too little. If you know how to calculate markup, you can apply it to your job costs to find the right sales price for your work. This book gives you tried and tested formulas, with step-by-step instructions and easy-to-follow examples, so you can easily figure the markup that's right for *your* business. Includes a CD-ROM with forms and checklists for your use. **320 pages, 8½ x 11, $32.50**

Construction Estimating Reference Data

Provides the 300 most useful manhour tables for practically every item of construction. Labor requirements are listed for sitework, concrete work, masonry, steel, carpentry, thermal and moisture protection, doors and windows, finishes, mechanical and electrical. Each section details the work being estimated and gives appropriate crew size and equipment needed. Includes a CD-ROM with an electronic version of the book with *National Estimator*, a stand-alone *Windows*™ estimating program, plus an interactive multimedia video that shows how to use the disk to compile construction cost estimates. **432 pages, 11 x 8½, $39.50**

National Electrical Estimator

This year's prices for installation of all common electrical work: conduit, wire, boxes, fixtures, switches, outlets, loadcenters, panelboards, raceway, duct, signal systems, and more. Provides material costs, manhours per unit, and total installed cost. Explains what you should know to estimate each part of an electrical system. Includes a CD-ROM with an electronic version of the book with *National Estimator*, a stand-alone *Windows*™ estimating program, plus an interactive multimedia video that shows how to use the disk to compile construction cost estimates. **552 pages, 8½ x 11, $57.75. Revised annually**

Getting Financing & Developing Land

Developing land is a major leap for most builders – yet that's where the big money is made. This book gives you the practical knowledge you need to make that leap. Learn how to prepare a market study, select a building site, obtain financing, guide your plans through approval, and then control your building costs so you can ensure yourself a good profit. Includes a CD-ROM with forms, checklists, and a sample business plan you can customize and use to help you sell your idea to lenders and investors. **232 pages, 8½ x 11, $39.00**

Pipe & Excavation Contracting

Shows how to read plans and compute quantities for both trench and surface excavation, figure crew and equipment productivity rates, estimate unit costs, bid the work, and get the bonds you need. Explains what equipment will deliver maximum productivity for a job, how to lay all types of water and sewer pipe, and how to switch your business to excavation work when you don't have pipe contracts. Covers asphalt and rock removal, working on steep slopes or in high groundwater, and how to avoid the pitfalls that can wipe out your profits on any job. **400 pages, 5½ x 8½, $29.00**

Basic Engineering for Builders

This book is for you if you've ever been stumped by an engineering problem on the job, yet wanted to avoid the expense of hiring a qualified engineer. Here you'll find engineering principles explained in non-technical language and practical methods for applying them on the job. With the help of this book you'll be able to understand engineering functions in the plans and how to meet the requirements, how to get permits issued without the help of an engineer, and anticipate requirements for concrete, steel, wood and masonry. See why you sometimes have to hire an engineer and what you can undertake yourself: surveying, concrete, lumber loads and stresses, steel, masonry, plumbing, and HVAC systems. This book is designed to help you, the builder, save money by understanding engineering principles that you can incorporate into the jobs you bid. **400 pages, 8½ x 11, $36.50**

National Construction Estimator

Current building costs for residential, commercial, and industrial construction. Estimated prices for every common building material. Provides manhours, recommended crew, and gives the labor cost for installation. Includes a CD-ROM with an electronic version of the book with *National Estimator*, a stand-alone *Windows*™ estimating program, plus an interactive multimedia video that shows how to use the disk to compile construction cost estimates. **672 pages, 8½ x 11, $57.50. Revised annually**

National Earthwork & Heavy Equipment Estimator

Complete labor, material, and manhour prices for estimating most earthmoving projects. You'll find how to read site plans, what's necessary in site preparation, how to determine site cut and fill quantities, how to determine soil swell and compaction, how to work with topsoil, slabs and paving, road work, curbs and gutters, and trench excavation. Includes factors affecting equipment production and how to include them in your estimate. Contains production rates for excavators, trucks, dozers, scrapers, compactors, graders and pavers, as well as blasting, ripping rock, and soil stabilization. Includes a FREE CD-ROM with an electronic version of the book, a stand-alone estimating program and a multimedia tutorial. Monthly price updates on the Web are free and automatic all during 2008. Also includes *Job Cost Wizard* that lets you turn your estimates into invoices and export them into *Quickbooks Pro*. **340 pages, 8½ x 11, $57.00. Revised annually**

National Building Cost Manual

Square-foot costs for residential, commercial, industrial, and farm buildings. Quickly work up a reliable budget estimate based on actual materials and design features, area, shape, wall height, number of floors, and support requirements. Includes all the important variables that can make any building unique from a cost standpoint. **248 pages, 8½ x 11, $33.00. Revised annually**

The Contractor's Legal Kit

Stop "eating" the costs of bad designs, hidden conditions, and job surprises. Set ground rules that assign those costs to the rightful party ahead of time. And it's all in plain English, not "legalese." For less than the cost of an hour with a lawyer you'll learn the exclusions to put in your agreements, why your insurance company may pay for your legal defense, how to avoid liability for injuries to your sub and his employees or damages they cause, how to collect on lawsuits you win, and much more. It also includes a FREE computer disk with contracts and forms you can customize for your own use. **352 pages, 8½ x 11, $69.95**

Land Development, Tenth Edition

The industry's bible. Nine chapters cover everything you need to know about land development from initial market studies to site selection and analysis. New and innovative design ideas for streets, houses, and neighborhoods are included. Whether you're developing a whole neighborhood or just one site, you shouldn't be without this essential reference.
360 pages, 6 x 9, $55.00

Construction Forms & Contracts

125 forms you can copy and use — or load into your computer (from the FREE disk enclosed). Then you can customize the forms to fit your company, fill them out, and print. Loads into *Word* for *Windows*™, *Lotus 1-2-3*, *WordPerfect*, *Works*, or *Excel* programs. You'll find forms covering accounting, estimating, fieldwork, contracts, and general office. Each form comes with complete instructions on when to use it and how to fill it out. These forms were designed, tested and used by contractors, and will help keep your business organized, profitable and out of legal, accounting and collection troubles. Includes a CD-ROM for *Windows*™ and Mac™.
400 pages, 8½ x 11, $41.75

Greenbook Standard Specifications for Public Works Construction

Since 1967, twelve previous editions of the popular "Greenbook" have been used as the official specification, bidding and contract document for many cities, counties and public agencies throughout the West. New federal regulations mandate that all public construction use metric documentation. This complete reference, which meets this new requirement, provides uniform standards of quality and sound construction practice easily understood and used by engineers, public works officials, and contractors across the U.S. Includes hundreds of charts and tables.
480 pages, 8½ x 11, $69.95

National Concrete & Masonry Estimator

Since you don't get every concrete or masonry job you bid, why generate a detailed list of materials for each one? The data in this book will allow you to get a quick and accurate bid, and allow you to do a detailed material takeoff, only for the jobs on which you're the successful bidder. Includes assembly prices for bricks, and labor and material prices for brick bonds, brick specialties, concrete blocks, CMU, concrete footings and foundations, concrete on grade, concrete specialties, concrete beams and columns, beams for elevated slabs, elevated slab costs, and more. Includes a CD-ROM with an electronic version of the book with *National Estimator*, a stand-alone *Windows*™ estimating program, plus an interactive multimedia video that shows how to use the disk to compile construction cost estimates. **672 pages, 8½ x 11, $59.00. Revised annually**

Download all of Craftsman's most popular costbooks for one low price with the Craftsman Site License
http://www.craftsmansitelicense.com

 Craftsman Book Company
6058 Corte del Cedro
P.O. Box 6500
Carlsbad, CA 92018

☎ **24 hour order line**
1-800-829-8123
Fax (760) 438-0398

_____
Name
_____
Company
_____
Address
_____
City/State/Zip
○ This is a residence
Total enclosed_____(In California add 7.25% tax)
We pay shipping when your check covers your order in full.

In A Hurry?
We accept phone orders charged to your
○ Visa, ○ MasterCard, ○ Discover or ○ American Express
Card#_____
Exp. date_____Initials_____

Tax Deductible: Treasury regulations make these references tax deductible when used in your work. Save the canceled check or charge card statement as your receipt.

Order online www.craftsman-book.com
Free on the Internet! Download any of Craftsman's estimating databases for a 30-day free trial!
www.craftsman-book.com/downloads

10-Day Money Back Guarantee

○ 36.50 Basic Engineering for Builders
○ 78.50 CD Estimator
○ 39.50 Construction Estimating Reference Data with FREE *National Estimator* on a CD-ROM
○ 41.75 Construction Forms & Contracts with a CD-ROM for *Windows*™ and *Mac*™
○ 69.95 Contractor's Legal Kit
○ 42.00 Excavation & Grading Handbook Revised
○ 39.00 Getting Financing & Developing Land
○ 69.95 Greenbook Standard Specifications for Public Works Construction
○ 55.00 Land Development, Tenth Edition
○ 32.50 Markup & Profit: A Contractor's Guide
○ 33.00 National Building Cost Manual
○ 59.00 National Concrete & Masonry Estimator with FREE *National Estimator* on a CD-ROM
○ 57.50 National Construction Estimator with FREE *National Estimator* on a CD-ROM
○ 57.00 National Earthwork & Heavy Equipment Estimator with FREE *National Estimator* on a CD-ROM
○ 57.75 National Electrical Estimator with FREE *National Estimator* on a CD-ROM
○ 29.00 Pipe & Excavation Contracting
○ 39.50 Estimating Excavation
○ FREE   Full Color Catalog

Prices subject to change without notice

Craftsman Book Company
6058 Corte del Cedro
P.O. Box 6500
Carlsbad, CA 92018

☎ 24 hour order line
1-800-829-8123
Fax (760) 438-0398

Name _____
Company _____
Address _____
City/State/Zip _____
○ This is a residence
Total enclosed_____(In California add 7.25% tax)

We pay shipping when your check covers your order in full.

In A Hurry?
We accept phone orders charged to your
○ Visa, ○ MasterCard, ○ Discover or ○ American Express

Card#_____

Exp. date_____Initials_____
Tax Deductible: Treasury regulations make these references tax deductible when used in your work. Save the canceled check or charge card statement as your receipt.

Order online www.craftsman-book.com
Free on the Internet! Download any of Craftsman's estimating databases for a 30-day free trial! www.craftsman-book.com/downloads

10-Day Money Back Guarantee

- ○ 36.50 Basic Engineering for Builders
- ○ 78.50 CD Estimator
- ○ 39.50 Construction Estimating Reference Data with FREE *National Estimator* on a CD-ROM
- ○ 41.75 Construction Forms & Contracts with a CD-ROM for *Windows*™ and *Mac*™
- ○ 69.95 Contractor's Legal Kit
- ○ 42.00 Excavation & Grading Handbook Revised
- ○ 39.00 Getting Financing & Developing Land
- ○ 69.95 Greenbook Standard Specifications for Public Works Construction
- ○ 55.00 Land Development, Tenth Edition
- ○ 32.50 Markup & Profit: A Contractor's Guide
- ○ 33.00 National Building Cost Manual
- ○ 59.00 National Concrete & Masonry Estimator with FREE *National Estimator* on a CD-ROM
- ○ 57.50 National Construction Estimator with FREE *National Estimator* on a CD-ROM
- ○ 57.00 National Earthwork & Heavy Equipment Estimator with FREE *National Estimator* on a CD-ROM
- ○ 57.75 National Electrical Estimator with FREE *National Estimator* on a CD-ROM
- ○ 29.00 Pipe & Excavation Contracting
- ○ 39.50 Estimating Excavation
- ○ FREE Full Color Catalog

Prices subject to change without notice

Craftsman Book Company
6058 Corte del Cedro
P.O. Box 6500
Carlsbad, CA 92018

☎ 24 hour order line
1-800-829-8123
Fax (760) 438-0398

Name _____
Company _____
Address _____
City/State/Zip _____
○ This is a residence
Total enclosed_____(In California add 7.25% tax)

We pay shipping when your check covers your order in full.

In A Hurry?
We accept phone orders charged to your
○ Visa, ○ MasterCard, ○ Discover or ○ American Express

Card#_____

Exp. date_____Initials_____
Tax Deductible: Treasury regulations make these references tax deductible when used in your work. Save the canceled check or charge card statement as your receipt.

Order online www.craftsman-book.com
Free on the Internet! Download any of Craftsman's estimating databases for a 30-day free trial! www.craftsman-book.com/downloads

10-Day Money Back Guarantee

- ○ 36.50 Basic Engineering for Builders
- ○ 78.50 CD Estimator
- ○ 39.50 Construction Estimating Reference Data with FREE *National Estimator* on a CD-ROM
- ○ 41.75 Construction Forms & Contracts with a CD-ROM for *Windows*™ and *Mac*™
- ○ 69.95 Contractor's Legal Kit
- ○ 42.00 Excavation & Grading Handbook Revised
- ○ 39.00 Getting Financing & Developing Land
- ○ 69.95 Greenbook Standard Specifications for Public Works Construction
- ○ 55.00 Land Development, Tenth Edition
- ○ 32.50 Markup & Profit: A Contractor's Guide
- ○ 33.00 National Building Cost Manual
- ○ 59.00 National Concrete & Masonry Estimator with FREE *National Estimator* on a CD-ROM
- ○ 57.50 National Construction Estimator with FREE *National Estimator* on a CD-ROM
- ○ 57.00 National Earthwork & Heavy Equipment Estimator with FREE *National Estimator* on a CD-ROM
- ○ 57.75 National Electrical Estimator with FREE *National Estimator* on a CD-ROM
- ○ 29.00 Pipe & Excavation Contracting
- ○ 39.50 Estimating Excavation
- ○ FREE Full Color Catalog

Prices subject to change without notice

Mail This Card Today
For a Free Full Color Catalog

Over 100 books, annual cost guides and estimating software packages at your fingertips with information that can save you time and money. Here you'll find information on carpentry, contracting, estimating, remodeling, electrical work, and plumbing.

All items come with an unconditional 10-day money-back guarantee. If they don't save you money, mail them back for a full refund.

Name _____
Company _____
Address _____
City/State/Zip _____

Craftsman Book Company / 6058 Corte del Cedro / P.O. Box 6500 / Carlsbad, CA 92018

BUSINESS REPLY MAIL
FIRST CLASS MAIL PERMIT NO. 271 CARLSBAD, CA

POSTAGE WILL BE PAID BY ADDRESSEE

 Craftsman Book Company
6058 Corte del Cedro
P.O. Box 6500
Carlsbad, CA 92018-9974

NO POSTAGE
NECESSARY
IF MAILED
IN THE
UNITED STATES

BUSINESS REPLY MAIL
FIRST CLASS MAIL PERMIT NO. 271 CARLSBAD, CA

POSTAGE WILL BE PAID BY ADDRESSEE

 Craftsman Book Company
6058 Corte del Cedro
P.O. Box 6500
Carlsbad, CA 92018-9974

NO POSTAGE
NECESSARY
IF MAILED
IN THE
UNITED STATES

BUSINESS REPLY MAIL
FIRST CLASS MAIL PERMIT NO. 271 CARLSBAD, CA

POSTAGE WILL BE PAID BY ADDRESSEE

 Craftsman Book Company
6058 Corte del Cedro
P.O. Box 6500
Carlsbad, CA 92018-9974